An Illustrated Guide to Electrical Safety

6th Edition

Michael Kovacic
John Grzywacz

Editors

American Society of Safety Engineers Des Plaines, Illinois

Disclaimer

While the publisher and editors have used their best efforts in preparing this book, they make no representations or warranties with respect to the accuracy or completeness of the contents, and specifically disclaim any implied warranties of fitness for a particular purpose. The information herein is provided with the understanding that the editors are not hereby engaged in rendering professional or legal services. Moreover, the mention of any product herein does not constitute an endorsement or recommendation by the American Society of Safety Engineers, and was done solely at the discretion of the Editors. Note that any excerpt from a National Fire Protection Association Standard reprinted herein is not the complete and official position of the NFPA on the referenced subject(s), which is represented only by the cited Standard in its entirety.

Library of Congress Cataloging-in-Publication Data

An illustrated guide to electrical safety / Michael Kovacic and John Grzywacz, editors. -- 6th ed.

p. cm.

Includes bibliographical references and index.

ISBN 978-1-885581-59-4 (alk. paper)

1. Electrical engineering--Safety measures. 2. Industrial safety. 3. Electrical engineering--Safety measures--Standards--United States. 4. Electric apparatus and appliances--Safety measures. I. Kovacic, Michael. II. Grzywacz, John. III. American Society of Safety Engineers.

TK152.I493 2011

621.3028'9--dc23

2011032649

Managing Editor: Michael F. Burditt, ASSE
Cover Design: REEDdesign studio
Cover Photos: Michael Burditt

Printed in the United States of America

17 16 15 14 13 12 11 7 6 5 4 3 2 1

Contents

List of Figures

Contents

List of Figures

List of Tables

Foreword

In 1983, ASSE recognized the value of the original OSHA Guide publication and reprinted it with certain additions in the form of footnotes and an appendix. These additions were identified by the prefix "ASSE." They were provided for explanatory purposes only and were not intended to imply any alteration of the OSHA requirements. This first edition consisted only of Part I (Division [a]) titled, DESIGN SAFETY STANDARDS FOR ELECTRICAL INSTALLATIONS.

Part II (Division [b]) of the electrical standard, SAFETY RELATED WORK PRACTICES, was made available by OSHA in 1990. It soon became apparent that a Second Edition of the Guide was needed to include Part II and to also update the changes made over the decade by OSHA in Part I. The additions and comments for Part I of the first edition were provided by William S. Watkins, P.E., who assisted in the preparation of the original Guide. The additions and comments for Part II and the update of Part I in the second edition were provided by E.K. Greenwald, Ph.D., P.E. Both of these gentlemen had worked as consulting engineers and had designed electrical installations for industrial, commercial and institutional facilities for many years.

As pointed out in the OSHA Preface, which follows, an understanding of the basic principles of electricity would be helpful in using this Guide. Therefore, a few brief explanations have been added where the need seemed to be especially critical. In previous editions, the ASSE Appendix included an article explaining "grounding," which had originally appeared in *Professional Safety* magazine. Because the regulations on grounding for electrical safety are the most misunderstood and most violated of all OSHA regulations, in this edition the grounding article is included in the main text, where it is better integrated with the other commentary on the regulations.

The third edition was edited by Bert M. Cohn, P.E., a fire protection and safety engineer in private practice. Minor revisions to OSHA Subpart S were incorporated, and recognition was given to new material in Subpart R, Special Industries, affecting electrical installations and work practices. Substantial updating of commentary in this book occurred as a result of changes in the 1996 edition of the National Electrical Code®.

Only the material herein copied directly from CFR, Subpart S, Divisions (a) and (b) is official; and only the authority having jurisdiction can make judgments on the application of any of these requirements. The editors make no claim of any official standing with OSHA. Comments added by the original editors of this Guide, as well as the material added by ASSE, represent the opinion of the various editors. It is hoped that these comments will be helpful, but neither the editors nor ASSE can be held responsible for attempts to use these materials as if they were the complete, official electrical safety requirements.

Having said this, we would still like to point out that the present policy of OSHA activity is based on the "common sense" approach. Therefore, we suggest that the material we are providing can be helpful in avoiding OSHA penalties if it provides better understanding of what electrical hazards are and how they can best be minimized. Moreover, as the title of this Guide implies, its aim is to improve electrical safety, not merely to avoid citations.

Finally, it must be stated that electricity is a complicated subject. There is no substitute for qualified persons, applying recognized good practice, in preventing electrical accidents.

Throughout this book the abbreviation NEC® is used. It refers to the National Electrical Code®, NFPA 70, published by National Fire Protection Association, Quincy, MA.

The reader will note that frequent reference is made to the National Electrical Code® (NEC®) in this Guide, even though the NEC® is not a part of the OSHA regulations. Subpart S of Part 1910 of the Code of Federal Regulations was revised, effective April 16, 1981, in an effort to shorten and simplify the regulations, and the NEC® was not specifically included. The intent of the revision was to extract only the most directly applicable requirements of the NEC®, and to restate them in a performance-oriented form. Therefore, it is still appropriate to be aware of the relationship of Subpart S to the NEC® in order to know how the performance requirements can best be met in the workplace. Subpart S was based on the 1978 Edition of the NEC®, which is now out of print. Most authorities are now enforcing later editions, and OSHA recognizes the latest edition where the resulting installation would not be less safe.

There are significant differences, with respect to employee safety, between the various editions of the NEC®. The reader should be aware that there are a few electrical requirements in various sections of the OSHA standards that are not repeated in Subpart S. For example, spray booths have some additional requirements stated in CFR 1910.107.

There are some explanations in the Guidelines as published by OSHA that are not entirely correct, and many illustrations (included to explain some particular point) do not show the only, or even the best, installation methods. However, ASSE decided to add notes and corrections rather than making any substantive changes in the original copy in order to avoid the position of disputing OSHA statements. To do otherwise would open up a long and inconclusive process of examining every detail for consensus among many different authorities. Certainly, some electrical experts will not agree with all of the comments we have added, just as we don't agree with all of the material in the original version. The reader can obtain the greatest benefit from this document by remembering that it is not intended to be used as a court of last resort on technical details, but rather as a means of improving understanding of some of the commonsense requirements for electrical safety.

The Fourth and Fifth Editions were edited by Thomas M. Kovacic, an electrical safety specialist with TMK and Associates, Inc. Beginning with the Fourth Edition, it was decided to make additional corrections to some of the original OSHA comments and to improve on some of the graphics that were part of the original OSHA publication. These changes do not reflect a conflict with the original OSHA comment, but rather they bring the comment into closer harmony with OSHA's current position. Since the original publication dates back to 1980, the technology and graphic resources were not readily available to OSHA to produce high quality graphics with their budget. Also beginning with the Fourth Edition, the actual OSHA text is highlighted with a gray background

The Fifth Edition continued with this approach. Many new improvements were made to the comments compiled over the years to make them more precise and easier to read. Numerous printing errors were corrected. A significant number of graphics were redone to improve on the information presented as well as graphic quality.

The current Sixth Edition approaches a new milestone with both the passing of previous editor Thomas M. Kovacic and OSHA publishing new electrical standards in 2007 with a revision to Sections 1910.302-.308 of Subpart S.

OSHA had been preparing for some time to update the decades old electrical installation standards contained in Subpart S and in 2004 published a proposal in the federal register. In addition to the public comments received from the notice of proposed rulemaking, OSHA relied heavily on the installation specifications of Part I of NFPA 70E, Standard for Electrical Safety for Employee Workplaces, 2000 Edition. The installation specifications contained in NFPA 70E were requirements ported over from the NEC®, which had been identified as specifically applying to the safety of personnel (as opposed to the protection of property or equipment). Since OSHA jurisdiction and mandates only apply to people safety, OSHA had requested that NFPA identify and publish these identified electrical requirements in NFPA 70E for OSHA to use in updating its electrical standards. Regrettably, and presumably since the OSHA final rule has been published, NFPA will no longer publish a chapter on Installation Safety Requirements in NFPA 70E, and the future edition will contain only Electrical Safety Related Work Practice requirements.

The final rule was published in Federal Register #72:7135-7221, entitled "Electrical Standard; Final Rule" on February 14th of 2007 and became effective (enforceable by OSHA) August 13, 2007. Many remain, even now, unaware that the new rule even exists. It is our hope that this book can provide knowledge and guidance on implementation of OSHA regulations.

Preface

This Illustrated Electrical Safety Standards guide was originally developed by the Occupational Safety and Health Administration (OSHA) to serve as a supplement to OSHA's Electrical Safety Standards, 29 CFR 1910, Subpart S, Electrical. It is intended to serve as a guide to the Electrical Safety Standards to address the needs of the layman, both employer and employee alike. This safety standards guide provides additional explanation and clarification for use by individuals who have little or limited training or familiarity with the field of electricity. However, no attempt has been made to explain the basic principles of electricity, and an understanding of these principles would be useful. Some of the more technical provisions are explained to a level of detail appropriate to achieve an appreciation of the hazards involved and an understanding of the correct safeguards or precautions that should be employed. The Illustrated Guide follows the format of 29 CFR 1910, Subpart S as it would appear in the *Federal Register* as revised February 14, 2007. The number of sections in the Standards have been singled out for explanation. Explanations and illustrations follow these sections.

This document can be used for many purposes: training, education and information. It can be used by a variety of people in a variety of positions: the layman, worker, employer, compliance safety and health officer, union official, educator, and others. The intent, no matter how it is used or who uses it, is to provide a better understanding of the OSHA safety requirements in hopes of reducing the prospects of electrical injury.

Acknowledgments

This Guide was originally prepared in 1980 by the Office of Electrical and Electronic Engineering Safety Standards of the U.S. Occupational Safety and Health Administration (OSHA) and its Office of Training and Education. The base document was prepared by JRB Associates Inc., McLean, Virginia, under a contract with OSHA.

Editing and updates of this Sixth Edition have been done by Michael Kovacic and John Grzywacz to include the 2007 Subpart S (1910.302-.308) updates to the OSHA electrical standards and important changes found in the NFPA 70E, Standard for Electrical Safety in the Workplace. Mr. Kovacic is an electrical safety specialist with TMK and Associates, Inc., Chardon, Ohio, an occupational safety firm, and a full-time occupational safety instructor specializing in electrical safety and lockout/tagout. Mr. Grzywacz worked for OSHA for 20 years providing investigative guidance to OSHA compliance inspectors and safety professionals.

The Society would also like to gratefully acknowledge L. Rene' Graves, CSP, Texas Instruments, Inc., and John J. Kolak, CSP, Praxis Corporation, for their technical reviews of the draft manuscript.

Subpart S
Electrical - General

Authority: Authority: Secs. 4, 6, 8, Occupational Safety and Health Act of 1970 (29 U.S.C. 653, 655, 657); Secretary of Labor's Order No. 8-76 (41 FR 25059), 1-90 (55 FR 9033), 5-2002 (67 FR 65008), 5-2007 (72 FR 31160), as applicable; 29 CFR part 1911.

Source: 46 FR 4056, Jan. 16, 1981, unless otherwise noted.

Part 1
Design Safety Standards for Electrical Systems

§ 1910.301 Introduction

[46 FR 4056, Jan 16, 1982; 46 FR 40185, Aug. 7, 1981]

This subpart addresses electrical safety requirements that are necessary for the practical safeguarding of employees in their workplaces and is divided into four major divisions as follows:

(a) **Design safety standards for electrical systems.** These regulations are contained in §§ 1910.302 through 1910.330. Sections 1910.302 through 1910.308 contain design safety standards for electric utilization systems. Included in this category are all electric equipment and installations used to provide electric power and light for employee workplaces. Sections 1910.309 through 1910.330 are reserved for possible future design safety standards for other electrical systems.

(b) **Safety-related work practices.** These regulations will be contained in §§ 1910.331 through 1910.360.

(c) **Safety-related maintenance requirements.** These regulations will be contained in §§ 1910.361 through 1910.380.

(d) **Safety requirements for special equipment.** These regulations will be contained in §§ 1910.381 through 1910.398.

(e) **Definitions.** Definitions applicable to each division are contained in § 1910.399.

§ 1910.302 Electric Utilization Systems

[46 FR 4056, Jan. 16, 1981; 46 FR 40185, Aug. 7, 1981; 72 FR 7190, Feb. 14, 2007]

(a)	**Scope.**
(a)(1)	**Covered**. The provisions of §§ 1910.302 through 1910.308 of this subpart cover electrical installations and utilization equipment installed or used within or on buildings, structures, and other premises including:
(a)(1)(i)	Yards,
(a)(1)(ii)	Carnivals,
(a)(1)(iii)	Parking and other lots,
(a)(1)(iv)	Mobile homes,
(a)(1)(v)	Recreational vehicles,
(a)(1)(vi)	Industrial substations,
(a)(1)(vii)	Conductors that connect the installations to a supply of electricity, and
(a)(1)(viii)	Other outside conductors on the premises.

Covered

The electrical installations covered by this OSHA Standard include yards, carnivals, parking and other lots, mobile homes, recreational vehicles, utility substations, such as the one in Figure 1, conductors which connect these installations to their source of electricity and other exterior conductors used on these premises.

Requirements for these types of installations are found throughout 1910.302-.308. The source of electrical supply is very often the serving electric utility company, and coverage by this subpart would begin at the service (as defined in 1910.399). However, where the source of supply is other than the serving utility company, such as with generators, battery systems, photovoltaic systems and fuel cells, as some examples, Subpart S and National Electrical Code® (NEC®) requirements would most likely apply.

(a)(2)	**Not covered.** The provisions of §§ 1910.302 through 1910.308 of this subpart do not cover:
(a)(2)(i)	Installations in ships, watercraft, railway rolling stock, aircraft, or automotive vehicles other than mobile homes and recreational vehicles.
(a)(2)(ii)	Installations underground in mines.
(a)(2)(iii)	Installations of railways for generation, transformation, transmission, or distribution of power used exclusively for operation of rolling stock or installations used exclusively for signaling and communication purposes.
(a)(2)(iv)	Installations of communication equipment under the exclusive control of communication utilities, located outdoors or in building spaces used exclusively for such installations; or
(a)(2)(v)	Installations under the exclusive control of electric utilities for the purpose of communication or metering; or for the generation, control, transformation, transmission, and distribution or electric energy located in buildings exclusively by utilities for such purposes or located outdoors on property owned or leased by the utility or on public highways, streets, roads, etc., or outdoors by established rights on private property.

Not Covered

The "exempted" types of installations in both the previous and current subparts are identical to those "exempted" by the National Electrical Code® and NFPA 70E. Not covered are electrical systems on ships, watercraft, railway rolling stock, aircraft, autos, and underground mining installations. OSHA does not have jurisdiction over mines in general, regardless of whether the mining activity takes place above ground or underground. Under the Mine Safety and Health Act (MSH Act) (30 U.S.C. 801 et seq.), the Mine Safety and Health Administration (MSHA) regulates safety and health in mines. Further information concerning the Interagency Agreement between MSHA and OSHA can be found on the OSHA website at:
http://www.osha.gov/pls/oshaweb/owadisp.show_document?p_id=1639&p_table=DIRECTIVES.

The agreement is designated CPL 02-00-042, titled "Interagency Agreement Between the Mine Safety and Health Administration and the OSHA-USDOL", and was implemented 03/14/1980. Also, in general, the electrical equipment used by railroads for control of their rolling stock and signaling and communications equipment is not covered.

Although Subpart S does not cover installations in ships, it does cover installations used on ships if the installation is shore-based (that is, not part of the vessel's original, internal electrical system) (see § 1918.1(b)(3)). For example, Section 1910.303(g)(2) (guarding live parts) applies to the shore-based wiring of the shipyard and to any wiring taken onto the ship when it is supplied by shore-based wiring. It does not apply to the ship's permanent wiring and the new final rule would apply in this case. Electrical installations in marine terminals are also covered under Subpart S (see § 1917.1(a)(2)(iv)). In addition, 1910.5 applies to shipyard employment and Part 1915 contains some requirements related to electrical safety. These provisions continue to apply in lieu of any corresponding requirements in Subpart S of Part 1910. Conversely, where there is no specific electrical installation requirement for shipyard employment in Part 1915, Subpart S of Part 1910 applies.

Similarly, the equipment used by electric utilities to generate, transmit and distribute electrical power is also exempt from these standards. Installations under the exclusive control of communications utilities are covered by the OSHA Telecommunication Standard 29 CFR 1910.268 and installations for the purpose of communication or metering or the generation, control, transformation, transmission, and distribution or electric energy are covered by the OSHA Standard 29 CFR 1910.269 Power Generation Transmission and Distribution and the National Electrical Safety Code (ANSI/IEEE C2).

Figure 1 – An example of a utility substation covered under 1910.269.

1910.302

Title: Electric Utilization Sytems

Subpart: S
SubPart Title: Electrical - Design Safety Standards for Electrical Systems

Aside from the obvious fact that all the industries excluded are covered by additional standards, codes and regulations, OSHA provided some additional insight into the reasoning behind excluding these systems:

"These exempted installations present special design considerations that are not adequately addressed in Subpart S. For example, electric power transmission and distribution installations are typically installed where unqualified persons will not have access to them, and the only employees working on them are highly trained and skilled. Additionally, public safety considerations demand that these installations be capable of quick repair when weather or equipment failure disrupts electrical service.

Installations in ships, watercraft, railway rolling stock, aircraft, or automotive vehicles (other than mobile homes and recreational vehicles) are designed to be transportable. These transportability considerations make many of the design requirements in Subpart S irrelevant or infeasible. For example, attaching the grounded circuit conductor and the equipment grounding conductor to a permanent grounding electrode on a transportable wiring system is generally not feasible. Thus, some of the provisions in final § 1910.304(g), which contains requirements for grounding electrical systems, are inappropriate for the wiring of ships, watercraft, railway rolling stock, aircraft, or automotive vehicles."

(b) **Extent of application.**

(b)(1) Requirements applicable to all installations. The following requirements apply to all electrical installations and utilization equipment, regardless of when they were designed or installed.

Sections:

1910.303(b) .... Examination, installation, and use of equipment.
1910.303(c).............................. Electrical connections - Splices
1910.303(d) .... Arcing parts
1910.303(e) .... Marking
1910.303(f) except (f)(4) and
(f)(5) .... Disconnecting means and circuits
1910.303(g)(2).......................... 600 volts or less - Guarding of live parts
1910.304(a)(3).......................... Use of grounding terminals and devices
1910.304(f)(1)(i), (f)(1)(iv), and
(f)(1)(v)..................................... Overcurrent protection – 600 volts, nominal or less
1910.304(g)(1)(ii), 1910.304(g)(1)(iii), 1910.304(g)(1)(iv), and
1910.304(g)(1)(v)...................... Grounding – Systems to be grounded
1910.304(g)(4).......................... Grounding - Grounding connections.
1910.304(9)(5).......................... Grounding - Grounding path.
1910.304(g)(6)(iv)(A) through 1910.304(g)(6)(iv)(D), and
1910.304(g)(6)(vi)...................... Grounding - Supports, enclosures, & equipment to be grounded
1910.304(g)(7).......................... Grounding - Nonelectrical equipment
1910.304(g)(8)(i)....................... Grounding - Methods of grounding fixed equipment
1910.305(g)(1).......................... Flexible cords and cables - Use of flexible cords and cables
1910.305(g)(2)(ii)and
1910.305(g)(2)(iii)...................... Flexible cords and cables - Identificaion, splices, & terminations
1910.307, except as specified in
1910.307(b) .... Hazardous (classified) locations.

1910.302

Title: Electric Utilization Sytems

Subpart: S

SubPart Title: Electrical - Design Safety Standards for Electrical Systems

OSHA's mission was to obtain a safe workplace for employees; in this case, safe from electrical hazards. There are many buildings which were constructed prior to the establishment and/or enforcement of these electrical requirements. Therefore, it was necessary for OSHA to review certain requirements to determine the degree of hazard each represented and make compliance necessary, regardless of when the facility was built or the equipment installed. The results are the requirements listed above.

The new electrical rule has added "extent of application" sections with additional requirements linked to installations made after specific years. This was needed in order for OSHA to satisfy the necessary regulatory economic feasibility analysis which was required in order to promulgate the new final rule.

The revised installation specifications sections of Subpart S (1910.301-308) are based on NFPA 70E – Standard for Electrical safety Requirements for Employee Workplaces, 2000 Edition. However, it is recommended that the latest edition of NFPA 70E (2009 as of this printing) and the latest National Electrical Code® (2011, as of this printing) be used when correcting any non-compliant items. Normally, the only issue would be cost. Although the National Electrical Code® is upgraded every three years, there have been little or no changes in these areas that would have a significant impact on cost.

(b)(2) **Requirements applicable to installations made after March 15, 1972.** Every electrical installation and all utilization equipment installed or overhauled after March 15, 1972, shall comply with the provisions of §§ 1910.302 through 1910.308, except as noted in paragraphs (b)(3) and (b)(4) of this section.

OSHA introduced the term "overhaul" to include the types of activities that would trigger compliance with the otherwise grandfathered provisions of Subpart S for older installations. This was done in an effort to simplify all the elements of "major replacement, modification, or rehabilitation" into a single term without making any substantive change to the way in which Subpart S applies. To further clarify, OSHA has added the definition of overhaul in definitions section 1910.399. **Overhaul** *means to perform a major replacement, modification, repair, or rehabilitation similar to that involved when a new building or facility is built, a new wing is added, or an entire floor is renovated.*

(b)(3) **Requirements applicable only to installations made after April 16, 1981.** The following requirements apply only to electrical installations and utilization equipment installed after April 16, 1981:

Sections:

1910.303(h)(4)	Over 600 volts, nominal - Entrance and access to workspace
1910.304(f)(1)(vii) and 1910.304(f)(1)(viii)	Overcurrent protection - 600 volts, nominal, or less
1910.304(g)(9)(i)	Grounding - Grounding of systems and circuits of 1000 volts and over (high voltage)
1910.305(j)(6)(ii)(D)	Equipment for general use - Capacitors
1910.306(c)(9)	Elevators, dumbwaiters, escalators, moving walks, wheelchair lifts, and stairway chair lifts - Interconnection between multicar controllers
1910.306(i)	Electrically driven or controlled irrigation machines
1910.306(j)(5)	Swimming pools, fountains, and similar installations - Fountains
1910.308(a)(1)(ii)	Systems over 600 volts, nominal - Aboveground wiring methods
1910.308(c)(2)	Class 1, Class 2 and Class 3 remote control, signaling, and power-limited circuits - Marking
1910.308(d)	Fire alarm systems

1910.302(b)(3), containing requirements which only apply to installations made after April 16, 1981, has not been changed from the previous edition. No provisions under this paragraph have been added to or removed from the list.

When promulgated, OSHA delayed the enforcement of these requirements until April 16, 1981, since that issue of Subpart S was to be published before the 1981 National Electrical Code® would have been legally adopted as the electrical standard for the majority of states. Based upon comments from various state governments, OSHA determined that the majority had adopted the latest National Electrical Code® by the end of the first calendar quarter.

So as not to leave the impression that OSHA is the only agency that requires compliance with these standards, it is important to keep in mind that nearly all 50 states adopt the National Electrical Code® as the minimum standard. Additionally, the majority of the states adopt the National Fire Code as the minimum standard. The National Fire Code® (NFC®) includes NFPA 70E - *Standard for Electrical Safety in the Workplace*, which is extremely important to achieving an effective electrical safety program. Therefore, everyone is legally obligated to comply with these requirements even if OSHA did not exist.

(b)(4) **Requirements applicable only to installations made after August 13, 2007.** The following requirements apply only to electrical installations and utilization equipment installed after August 13, 2007:

1910.303(f)(4) .... Disconnecting means and circuits - Capable of accepting a lock

1910.303(f)(5) .... Disconnecting means and circuits - Marking for series combination ratings

1910.303(g)(1)(iv) and

1910.303(g)(1)(vii) .... 600 Volts, nominal, or less - Space about electric equipment

1910.303(h)(5)(vi) .... Over 600 volts, nominal - Working space and guarding

1910.304(b)(1) .... Branch circuits - Identification of multiwire branch circuits

1910.304(b)(3)(i) .... Branch circuits - Ground-fault circuit interrupter protection for personnel

1910.304(f)(2)(i)(A), 1910.304(f)(2)(i)(B), (but not the introductory text to 1910.304(f)(2)(i)), and

1910.304(f)(2)(iv)(A) .... Overcurrent protection - Feeders and branch circuits over 600 volts, nominal

1910.305(c)(3)(ii) .... Switches - Connection of switches

1910.305(c)(5) .... Switches - Grounding

1910.306(a)(1)(ii) .... Electric signs and outline lighting - Disconnecting means

1910.306(c)(4) .... Elevators, dumbwaiters, escalators, moving walks, wheelchair lifts, and stairway chair lifts - Operation

1910.306(c)(5) .... Elevators, dumbwaiters, escalators, moving walks, wheelchair lifts, and stairway chair lifts - Location

1910.306(c)(6) .... Elevators, dumbwaiters, escalators, moving walks, wheelchair lifts, and stairway chair lifts - Identification and signs

1910.306(c)(7) .... Elevators, dumbwaiters, escalators, moving walks, wheelchair lifts, and stairway chair lifts - Single-car and multicar installations

1910.306(j)(1)(iii) .... Swimming pools, fountains, and similar installations-Receptacles

1910.306(k) .... Carnivals, circuses, fairs, and similar events

1910.308(a)(5)(v) and

1910.308(a)(5)(vi)(B) .... Systems over 600 volts, nominal - Interrupting and isolating devices

1910.308(a)(7)(vi) .... Systems over 600 volts, nominal - Tunnel installations

1910.308(b)(3)	Emergency power systems - Signs
1910.308(c)(3)	Class 1, Class 2, and Class 3 remote control, signaling, and power-limited circuits - Separation from conductors of other circuits
1910.308(f)	Solar photovoltaic systems

These provisions are based on requirements that have been added to the NEC® since the 1978 edition. OSHA believes that an increase in employee protection will result from compliance with them in new installations. According to OSHA research, employers would incur minimal costs to achieve this increase in new installations. In local jurisdictions requiring compliance with the NEC®, there should be no additional costs involved, because the installations would already conform to the new OSHA requirements. The Agency further believes that even in other jurisdictions, the vast majority of installations already comply with the latest edition of the NEC®, because compliance with the latest Code® is standard industry practice. OSHA, however, does not believe that it is reasonably necessary and appropriate to require existing installations to conform to these provisions, particularly given the cost and difficulty associated with retrofitting older installations.

(c) **Applicability of requirements for disconnecting means.** The requirement in §1910.147(c)(2)(iii) that energy isolating devices be capable of accepting a lockout device whenever replacement or major repair, renovation or modification of a machine or equipment is performed, and whenever new machines or equipment are installed after January 2, 1990, applies in addition to any requirements in § 1910.303 through § 1910.308 that disconnecting means be capable of being locked in the open position under certain conditions.

§ 1910.303 General

[46 FR 4056, Jan. 16, 1981; 46 FR 40185, Aug. 7, 1981; 72 FR 7191, Feb. 14, 2007, 73 FR 64205, Oct. 29, 2008]

(a) Approval. The conductors and equipment required or permitted by this subpart shall be acceptable only if approved, as defined in § 1910.399.

OSHA requires all conductors and equipment to have third party certification by an independent testing laboratory. Such testing laboratories are referred to as Nationally Recognized Testing Laboratories (NRTL) by OSHA. Some of OSHA's recognized NRTLs include Canadian Standards Association (CSA), also known as CSA International, FM Approvals LLC (FM), formerly Factory Mutual Research Corporation, Intertek Testing Services NA, Inc. (ITSNA), formerly ETL, MET Laboratories, Inc. (MET), and Underwriters Laboratories Inc. (UL). A complete list of OSHA's NRTLs be found on the OSHA website (http://www.osha.gov/dts/otpca/nrtl/) under Office of Technical Programs and Coordination Activities (OTPCA), Nationally Recognized Testing Laboratories (NRTLs). Figure 2 provides examples of the listing symbols used by the currently recognized NRTLs, as well as some that are no longer recognized by OSHA. It is important to recognize that these NRTLs, and OSHA's recognition of them, are subject to change.

Figure 2 – Symbols of common Nationally Recognized Testing Laboratories (NRTLs).

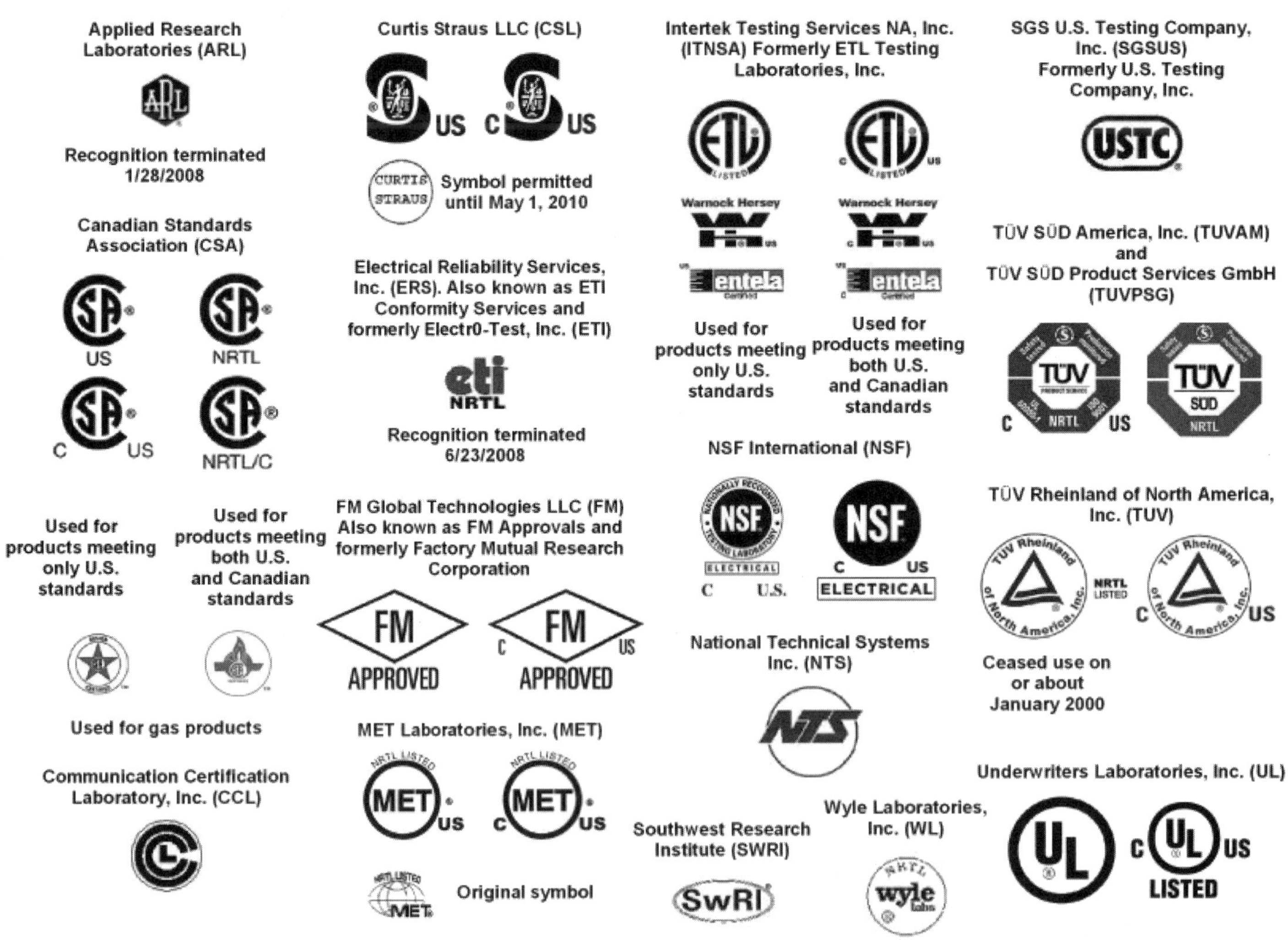

An exception to this general rule would be "custom-made" equipment that is not available "off the shelf" and has no manufacturing safety specification or UL standard that applies to this type of equipment. Typically such custom equipment is manufactured using individually listed components. It is important that such individually listed components be applied in accordance with their intended use.

Occasionally product manufacturers will mistakenly indicate that their product is "OSHA Approved." OSHA does not "approve" any product. OSHA does not have the authority, or the resources, to perform such a service. Likewise, NRTLs do not "approve" any product. The NRTL simply performs tests to verify that the equipment conforms to safety specification standards. The various terms or words associated with this process are:

Acceptable. An installation or equipment is acceptable to the Assistant Secretary of Labor, and approved within the meaning of this Subpart S:

(i) If it is accepted, or certified, or listed, or labeled, or otherwise determined to be safe by a nationally recognized testing laboratory, such as, but not limited to, Underwriters' Laboratories, Inc. and Factory Mutual Engineering Corp.; or

(ii) With respect to an installation or equipment of a kind which no nationally recognized testing laboratory accepts, certifies, lists, labels, or determines to be safe, if it is inspected or tested by another Federal agency, or by a State, municipal, or other local authority responsible for enforcing occupational safety provisions of the National Electrical Code®, and found in compliance with the provisions of the National Electrical Code® as applied in this Subpart; or

(iii) With respect to custom-made equipment or related installations which are designed, fabricated for, and intended for use by a particular customer, if it is determined to be safe for its intended use by its manufacturer on the basis of test data which the employer keeps and makes available for inspection to the assistant Secretary and his authorized representatives. Refer to 1910.7 for definition of nationally recognized testing laboratory.

Accepted. An installation is "accepted" if it has been inspected and found by a nationally recognized testing laboratory to conform to specified plans or to procedures of applicable codes.

Approved. Acceptable to the authority enforcing this Subpart. The authority enforcing this Subpart is the Assistant Secretary of Labor for Occupational Safety and Health. The definition of "acceptable" indicates what is acceptable to the Assistant Secretary of Labor, and therefore approved within the meaning of this Subpart.

Approved for the purpose. Approved for a specific purpose, environment, or application described in a particular standard requirement.

Suitability of equipment or materials for a specific purpose, environment or application may be determined by a nationally recognized testing laboratory, inspection agency or other organization concerned with product evaluation as part of its listing and labeling program. (See "Labeled" or "Listed".)

Certified. Equipment is "certified" if it

(a) has been tested and found by a nationally recognized testing laboratory to meet nationally recognized standards or to be safe for use in a specified manner, or

(b) is of a kind whose production is periodically inspected by a nationally recognized testing laboratory, and

(c) it bears a label, tag or other record of certification.

Listed. Equipment is "listed" if it is of a kind mentioned in a list which, (a) is published by a nationally recognized laboratory which makes periodic inspection of the production of such equipment, and (b) states such equipment meets nationally recognized standards or has been tested and found safe for use in a specified manner.

Labeled. Equipment is "labeled" if there is attached to it a label, symbol, or other identifying mark of a nationally recognized testing laboratory which, (a) makes periodic inspections of the production of such equipment, and (b) whose labeling indicates compliance with nationally recognized standards or tests to determine safe use in a specified manner.

(b) **Examination, installation, and use of equipment.**

(b)(1) **Examination.** Electrical equipment shall be free from recognized hazards that are likely to cause death or serious physical harm to employees. Safety of equipment shall be determined using the following considerations:

(b)(1)(i) Suitability for installation and use in conformity with the provisions of this subpart;

Note to paragraph (b)(1)(i) of this section: Suitability of equipment for an identified purpose may be evidenced by listing or labeling for that identified purpose.

(b)(1)(ii) Mechanical strength and durability, including, for parts designed to enclose and protect other equipment, the adequacy of the protection thus provided;

(b)(1)(iii) Wire-bending and connection space;

(b)(1)(iv) Electrical insulation;

(b)(1)(v) Heating effects under all conditions of use;

(b)(1)(vi) Arcing effects;

(b)(1)(vii) Classification by type, size, voltage, current capacity, and specific use; and

(b)(1)(viii) Other factors that contribute to the practical safeguarding of persons using or likely to come in contact with the equipment.

1910.303 (a) & (b) illustrate the "performance approach" which OSHA used for Subpart S. The ultimate in a performance standard would simply state "make it safe," but this Subpart goes a little further in itemizing most of the hazards to be considered. The first sentence of (b)(1) is especially basic, and OSHA can use this as a "general duty" clause to cite hazards not specifically covered by this Subpart. It is a reasonable argument that a situation prohibited by the NEC® is a recognized hazard; therefore, it is prudent to comply with the National Electrical Code® even though it is not specifically included in OSHA's regulations. For example, Subpart S does not provide any specifics for wire bending and connection space within electrical equipment. The NEC® does, however, provide this information in the sections appropriate to the type of cabinet or box (i.e. Articles 312, 314, 404, 408 and 430). Since the NEC® is enforced by OSHA through its "general duty clause", application of the NEC® sections still remain a requirement, even when not specifically covered by Subpart S.

The wording in 1910.303(b)(1) is nearly identical to the wording in Section 5(a)(1) of the OSHA Act. In other words, it repeats OSHA's general duty clause "Free from recognized hazards that are causing or are likely to cause death or serious physical harm to employees". Subpart S in 1910 and Subpart K in 1926 are the only two subparts that do this. This provides additional support for the OSHA Compliance Officer to "step out" of the OSHA regulations and reference other consensus standards when a specific finding has not been addressed by the OSHA regulations.

(b)(2) **Installation and use.** Listed or labeled equipment shall be used or installed in accordance with any instructions included in the listing or labeling.

Approximately 500 citations are written by Federal OSHA each year for noncompliance with 1910.303(b)(2), enough to make this a top 25 general industry citation. This requirement has been cited as "serious". Frequently it is not as easy as one may assume to find all the pertinent instructions on the use and application of equipment. In addition, in many cases, the equipment is so common that a person may not feel it is necessary to read the instructions. However, reading and complying with the instructions could make the difference between a compliant and non-compliant installation. A non-compliant application could result in a serious hazard. More often than not, it usually requires the simple reading and following the labels on the product, such as "use indoors in dry locations".

(b)(3) **Insulation integrity.** Completed wiring installations shall be free from short circuits and from grounds other than those required or permitted by this subpart.

Insulation is the material which prevents the transfer of electrical current from a conductor to other conductive objects (also known as a dielectric). One of the most common causes of problems in both high and low voltage systems is a breakdown (or failure) of conductor insulation. Technical specifications on conductor insulation, based short circuit currents and thermal-withstand abilities can be found in the Insulated Cable Engineers Association's document ICEA P-32-382.

(b)(4) **Interrupting rating.** Equipment intended to interrupt current at fault levels shall have an interrupting rating sufficient for the nominal circuit voltage and the current that is available at the line terminals of the equipment.

Equipment intended to interrupt current at other than fault levels shall have an interrupting rating at nominal circuit voltage sufficient for the current that must be interrupted.

The interrupting rating of an overcurrent device (fuse or circuit breaker) is the highest current, at its rated voltage, that the device is intended to interrupt under fault conditions. An example is shown in Figure 3. While the term is not defined in the subpart, it is defined by the NEC® and NFPA 70E. The rating of the device is determined by standard test conditions. Overcurrent devices used in electrical systems without the proper interrupting rating can fail in a catastrophic manner under fault conditions. The importance of using properly rated overcurrent devices is further emphasized by the danger of arc-flash, discussed further under Safety-Related Work Practices.

Figure 3 – Example of the interrupting rating marked on an overcurrent device. Photo provided courtesy of Cooper Bussmann.

(b)(5) Circuit impedance and other characteristics. The overcurrent protective devices, the total impedance, the component short-circuit current ratings, and other characteristics of the circuit to be protected shall be selected and coordinated to permit the circuit protective devices used to clear a fault to do so without the occurrence of extensive damage to the electrical components of the circuit. This fault shall be assumed to be either between two or more of the circuit conductors, or between any circuit conductor and the grounding conductor or enclosing metal raceway.

Short-circuit current ratings are marked on electrical equipment as part of the listing and labeling process. As it is difficult to know what exactly constitutes "extensive damage", the National Electrical Code® now includes a further clarification statement: *"Listed products applied in accordance with their listing shall be considered to meet the requirements of this section."* Therefore, provided that the equipment in question is applied in accordance with the manufacturer and NRTL instructions and ratings, and is used in an electrical system properly, with appropriate overcurrent protection, listed equipment is generally assumed to meet this requirement by the Authority Having Jurisdiction (AHJ).

Figure 4 – NEMA Enclosure Specifications for Indoor Nonhazardous Locations.

[From NEMA 250-2003]

Comparison of Specific Applications of Enclosures for Indoor Nonhazardous Locations

Provides a Degree of Protection Against the Following Conditions	Type of Enclosure									
	1 *	2 *	4	4X	5	6	6P	12	12K	13
Access to hazardous parts	X	X	X	X	X	X	X	X	X	X
Ingress of solid foreign objects (falling dirt)	X	X	X	X	X	X	X	X	X	X
Ingress of water (Dripping and light splashing)	...	X	X	X	X	X	X	X	X	X
Ingress of solid foreign objects (Circulating dust, lint, fibers, and flyings **)	...	...	X	X	...	X	X	X	X	X
Ingress of solid foreign objects (Settling airborne dust, lint, fibers, and flyings **)	...	...	X	X	X	X	X	X	X	X
Ingress of water (Hosedown and splashing water)	...	...	X	X	...	X	X	...	...	...
Oil and coolant seepage	...	...	...	..	...	...	...	X	X	X
Oil or coolant spraying and splashing	...	...	...	...	...	...	...	...	...	X
Corrosive agents	...	...	...	X	...	...	X	...	...	...
Ingress of water (Occasional temporary submersion)	...	...	...	...	...	X	X	...	...	...
Ingress of water (Occasional prolonged submersion)	...	...	...	...	...	...	X	...	...	...

* These enclosures may be ventilated.

** These fibers and flyings are nonhazardous materials and are not considered Class III type ignitable fibers or combustible flyings. For Class III type ignitable fibers or combustible flyings see the National Electrical Code, Article 500.

(b)(6) **Deteriorating agents.** Unless identified for use in the operating environment, no conductors or equipment shall be located in damp or wet locations; where exposed to gases, fumes, vapors, liquids, or other agents that have a deteriorating effect on the conductors or equipment; or where exposed to excessive temperatures.

Figure 5 – NEMA Enclosure Specifications for Outdoor Nonhazardous Locations.

[From NEMA 250-2003]

Comparison of Specific Applications of Enclosures for Outdoor Nonhazardous Locations

Provides a Degree of Protection Against the Following Conditions	Type of Enclosure									
	3	3X	3R*	3RX*	3S	3SX	4	4X	6	6P
Access to hazardous parts	X	X	X	X	X	X	X	X	X	X
Ingress of water (Rain, snow, and sleet **)	X	X	X	X	X	X	X	X	X	X
Sleet ***	...	...	...	...	X	X	...	...	...	...
Ingress of solid foreign objects (Windblown dust, lint, fibers, and flyings)	X	X	...	...	X	X	X	X	X	X
Ingress of water (Hosedown)	...	...	...	...	...	...	X	X	X	X
Corrosive agents	...	X	...	X	...	X	...	X	...	X
Ingress of water (Occasional temporary submersion)	...	...	...	...	...	...	...	...	X	X
Ingress of water (Occasional prolonged submersion)	...	...	...	...	...	...	...	...	...	X

* These enclosures may be ventilated.
** External operating mechanisms are not required to be operable when the enclosure is ice covered.
*** External operating mechanisms are operable when the enclosure is ice covered.

This section was long overdue for OSHA and welcomed by its inspectors, as electrical equipment and hardware is frequently found in damp or wet or other locations where the equipment is not intended for the location. This section was previously only found in Subpart K, for construction, but now applies to general industry, as well. All listed electrical equipment will have a NEMA (National Electrical Manufacturers Association) that determines the environments in which it can be used. The specifics of each enclosure type are described by NEMA (http://www.nema.org/). Figures 4 and 5 list comparisons types of enclosures for indoor and outdoor locations respctively. The deteriorating equipment photographed in Figure 6 was not designed for the outdoor environment to which it was subjected.

(b)(7) **Mechanical execution of work.** Electric equipment shall be installed in a neat and workmanlike manner.

The requirement for installing electrical equipment in a neat and workmanlike manner first appeared in the 1897 NEC®. Proper installations indicate that a professional with pride in their work did the job. While the decision of whether an installation meets the requirement is left to the AHJs experience and judgement, the accepted industry standard which describes such methods is ANSI/NECA 1-2006, entitled "Standard Practice for Good Workmanship in Electrical Contracting" (See Figure 7).

Figure 6 – Example of equipment affected by deteriorating agents without the proper NEMA rating for the environment.

(b)(7)(i) Unused openings in boxes, raceways, auxiliary gutters, cabinets, equipment cases, or housings shall be effectively closed to afford protection substantially equivalent to the wall of the equipment.

This reference closely mirrors 1910.305(b)(1)(i) which required unused openings in cabinets, boxes, and fittings to be effectively closed. Openings, such as knockouts and unused circuit breaker openings must be closed to prevent contact with live parts, to contain arcs and sparks during faults, and for similar circumstances. Unused openings may be closed through the used of devices such as knockout plugs and circuit breaker filler plates, available from a wide range of manufacturers. The NEC® also specifies that this requirement is for those openings "other than those intended for the operation of equipment, those intended for mounting purposes, or those permitted as part of the design for listed equipment."

(b)(7)(ii) Conductors shall be racked to provide ready and safe access in underground and subsurface enclosures that persons enter for installation and maintenance.

(b)(7)(iii) Internal parts of electrical equipment, including busbars, wiring terminals, insulators, and other surfaces, may not be damaged or contaminated by foreign materials such as paint, plaster, cleaners, abrasives, or corrosive residues.

(7)(iv) There shall be no damaged parts that may adversely affect safe operation or mechanical strength of the equipment, such as parts that are broken, bent, cut, or deteriorated by corrosion, chemical action, or overheating. The white residue in Figure 10 was caused by exposure to a corrosive chemical.

Figure 7 – ANSI/NECA 1-2010, Standard Practice for Good Workmanship in Electrical Contracting.

Figure **8 – Example of one type of knockout plug.**

Figure 9 – Unused opening in a disconnect switch.

Contaminates within equipment are frequently indicative of improper enclosures or missing covers, knockouts, hub plugs, etc. (see examples, Figures 8 and 9). The NEMA enclosure guide can be consulted for enclosure ratings. This is also an indicator of poor maintenance and/or preventative maintenance. NFPA 70B – Recommended Practice for Electrical Equipment Maintenance, can be consulted for maintenance issues. Note the residue at the terminations and the collection of corroded material within the enclosure pictured in Figure 10.

Equipment with damaged parts such as the switch shown in Figure 11 would also be in violation of the approval requirements of 1910.303(a) since the damaged equipment would in all likelihood fail to meet the NRTL safety testing requirements required of undamaged equipment. The figure here illustrates damage that would adversely affect the safe operation of the equipment, as the handle could not be properly operated.

(b)(8) **Mounting and cooling of equipment.**

(b)(8)(i) Electric equipment shall be firmly secured to the surface on which it is mounted.

Note to paragraph (b)(8)(i) of this section: Wooden plugs driven into holes in masonry, concrete, plaster, or similar materials are not considered secure means of fastening electric equipment.

Electrical equipment is provided with instructions and means for proper mounting by the manufacturer. It is important that equipment is securely mounted to provide for safe operation. Figures 12 and 13 show improperly secured equipment.

Figure 10 – Example of corrosive chemical residue **causing damage to internal parts of electrical equipment.**

Figure 11 – Note the damaged handle on this disconnect switch.

Figure 12 – Example of electrical equipment not firmly secured.

Figure 13 – Example of electrical equipment not firmly secured.

(b)(8)(ii) Electric equipment that depends on the natural circulation of air and convection principles for cooling of exposed surfaces shall be installed so that room airflow over such surfaces is not prevented by walls or by adjacent installed equipment. For equipment designed for floor mounting, clearance between top surfaces and adjacent surfaces shall be provided to dissipate rising warm air.

(b)(8)(iii) Electric equipment provided with ventilating openings shall be installed so that walls or other obstructions do not prevent the free circulation of air through the equipment.

Lack of proper ventilation can lead to serious overheating of electrical equipment. Overheating effectively cuts down on the life of electrical equipment and often leads to catastrophic failure of the equipement. Even when ventilation openings are free, covering top and adjacent surfaces can prevent the proper convection of air and heat dissipation. An actual example is shown in Figure 14. Alternatively, ventilation openings can also provide paths for flame, gasses and plasma during an arc flash incident, and should be considered as part of a proper hazard analysis under Safe Work Practices discussed later.

Lack of maintenance can also lead to ventilation issues when ventilation openings do not allow for air circulatiom. Figure 15 shows one example. i

Figure 14 – Even with ventilation openings clear, equipment may not be able to dissipate heat properly.

Figure 15 – Example of lack of maintenance leading to inadequate ventilation of electrical equipment.

(c) Electrical connections.

(c)(1) **General.** Because of different characteristics of dissimilar metals: **(c) Electrical connections.**

(c)(1)(i) Devices such as pressure terminal or pressure splicing connectors and soldering lugs shall be identified for the material of the conductor and shall be properly installed and used;

(c)(1)(ii) Conductors of dissimilar metals may not be intermixed in a terminal or splicing connector where physical contact occurs between dissimilar conductors (such as copper and aluminum, copper and copper-clad aluminum, or aluminum and copper-clad aluminum) unless the device is identified for the purpose and conditions of use; and

(c)(1)(iii) Materials such as solder, fluxes, inhibitors, and compounds, where employed, shall be suitable for the use and shall be of a type that will not adversely affect the conductors, installation, or equipment.

Terminations and wire connectors are tested for proper application under UL 486A and 486B, entitled “Wire Connectors.” These standards provide detailed information on tightening torques based types of connectors and wire size/type. Additionally, devices and connectors are marked for the type of wire they were listed for use with. Devices designed to intermix copper and aluminum will be marked “CU/ALR” by the manufacturer, to indicate that the product has been designed and tested for heat-cycling requirements.

Industry reports, trade magazines and field observations show that electrical connection failure is a significant cause of “electrical” fires. One NFPA report shows that of the 1.8 million fires in 1999, 26% could be attributed to “electrical causes”. Unless specifically labeled for use with more than one conductor, circuit breakers and other terminals may not have have multiple wire connections (Figure 16).

1910.303

Title: General Requirements

Subpart: S

SubPart Title: Electrical - Design Safety Standards for Electrical Systems

Industry reports, trade magazines and field observations show that electrical connection failure is a significant cause of "electrical" fires. One NFPA report shows that of the 1.8 million fires in 1999, 26% could be attributed to "electrical causes". Unless specifically labeled for use with more than one conductor, circuit breakers and other terminals may not have have multiple wire connections (Figure 16).

(c)(2)	**Terminals**.
(c)(2)(i)	Connection of conductors to terminal parts shall ensure a good connection without damaging the conductors and shall be made by means of pressure connectors (including set-screw type), solder lugs, or splices to flexible leads. However, No. 10 or smaller conductors may be connected by means of wire binding screws or studs and nuts having upturned lugs or equivalent.
(c)(2)(ii)	Terminals for more than one conductor and terminals used to connect aluminum shall be so identified.

Figure 16 – None of the circled connections are identified for multiple wire termination.

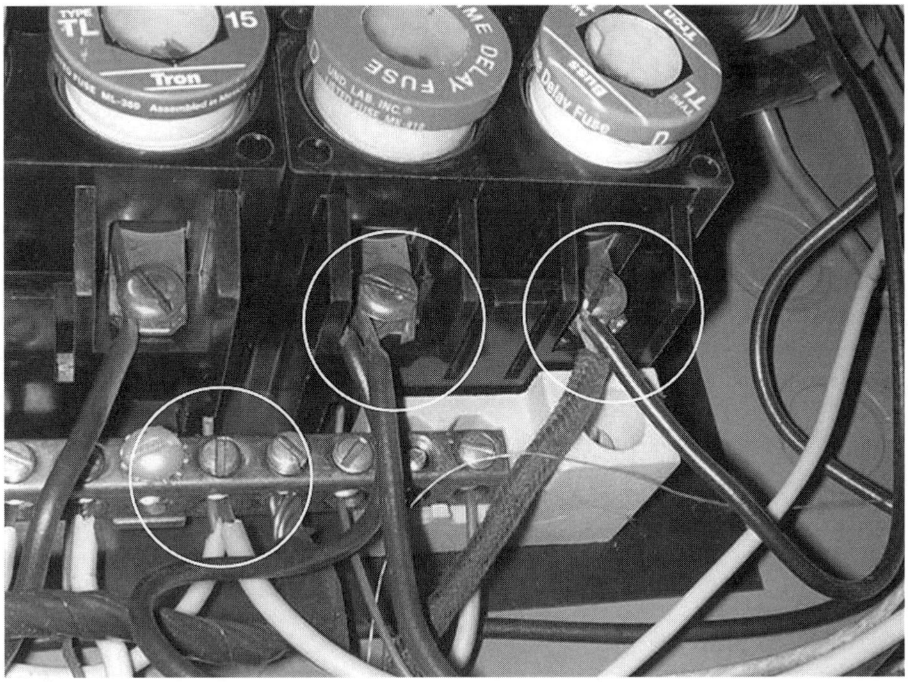

(c)(3)	**Splices.**
(c)(3)(i)	Conductors shall be spliced or joined with splicing devices identified for the use or by brazing, welding, or soldering with a fusible metal or alloy. Soldered splices shall first be so spliced or joined as to be mechanically and electrically secure without solder and then soldered. All splices and joints and the free ends of conductors shall be covered with an insulation equivalent to that of the conductors or with an insulating device identified for the purpose.
(c)(3)(ii)	Wire connectors or splicing means installed on conductors for direct burial shall be listed for such use.

Figure 17 – Sample mechanical splicing devices.

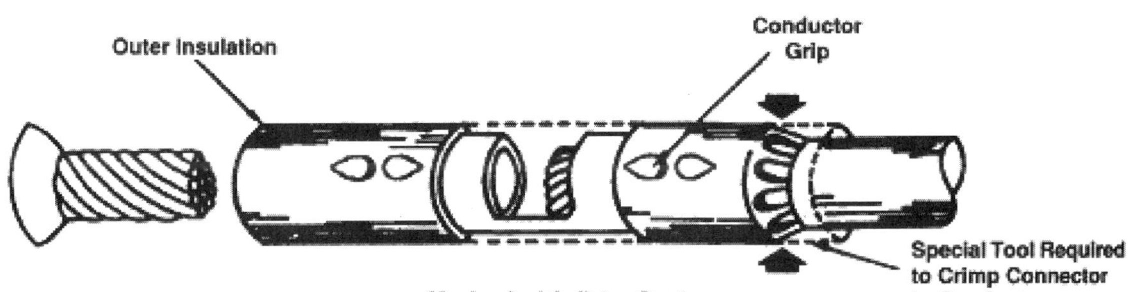

Mechanical Splicing Device

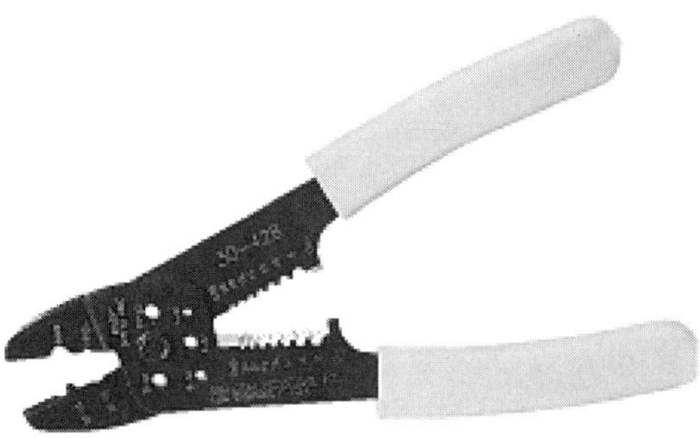

Splicing device with crimping tool

Split Bolt Tap Joint Connector

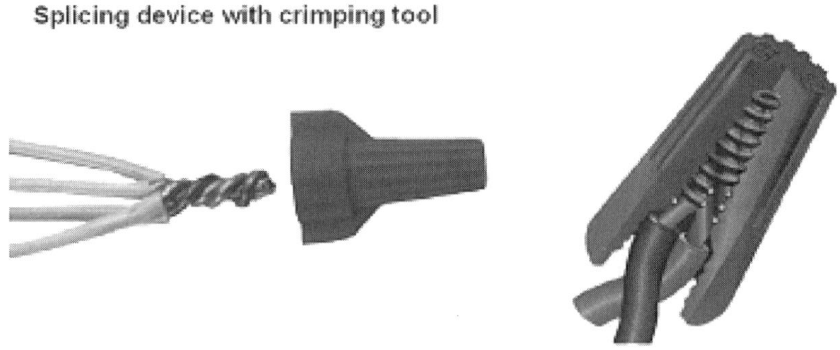

Solderless screw-on connectors (commonly called wire nuts) can be used instead of soldering joints. These connectors are generally made of plastic, and are threaded inside to screw onto bare wires, forming a splice

When electrical conductors are joined together by splicing, the connection must be strong and safe. It must also provide a path for the continuous flow of electricity. These requirements can be met by using a mechanical splicing device or by welding, brazing, or soldering the connection with a fusible metal or alloy. Figures 17-19 show some but not all of the materials and methods used in splicing. In general, these diagrams are likely to be useful only to individuals actually making an installation.

Mechanical splicing devices range from common wire nuts to those that employ a compression-type fitting to hold the conductors securely and to provide a suitable, safe splice. Figure 17 shows some of the more common mechanical splicing devices.

Figure 18 – Conductors mechanically joined and soldered.

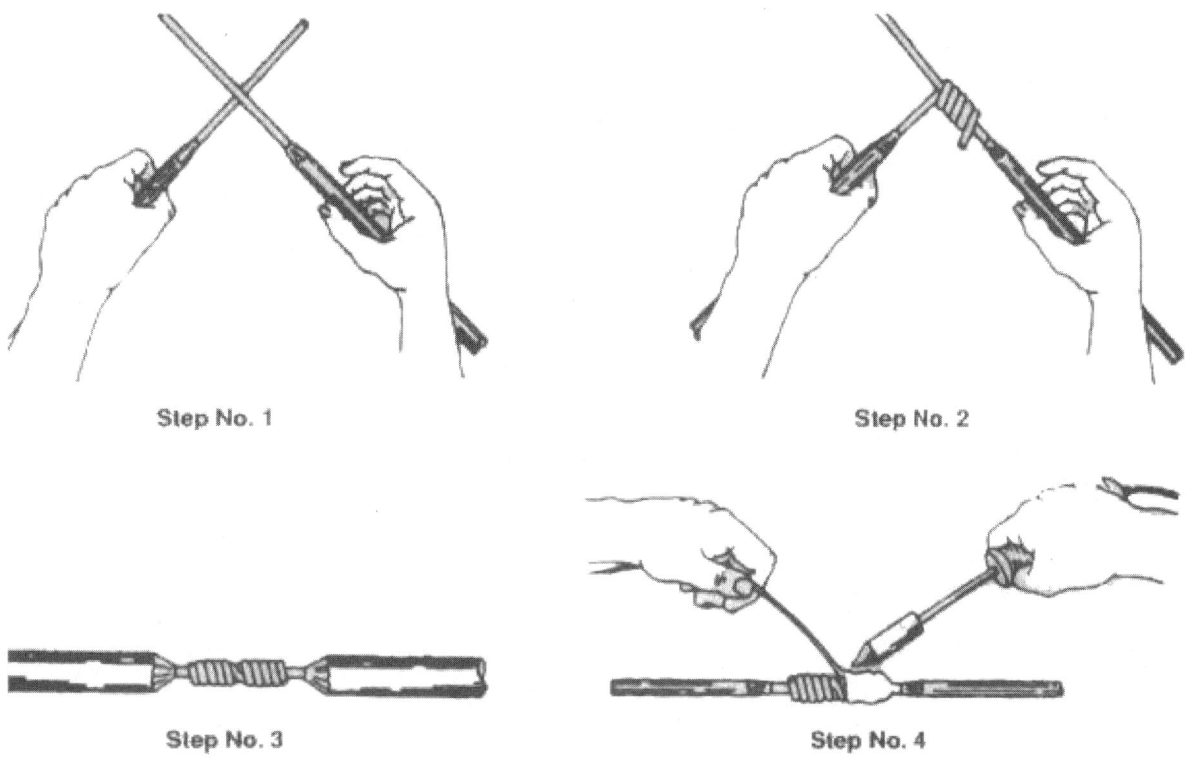

A safe splice can also be made by welding or brazing the ends of the conductors together, forming a permanent bond between the metals. Soldered splices, however, are not secure connections unless they are first properly joined mechanically before soldiering to achieve the same strength as that of the conductors being joined. This connection must also provide a path for the continuous flow of electricity. The solder can then be applied. One proper splicing method is illustrated in Figure 18.

To prevent electric shock or fire, all splices and free ends of conductors must be covered with an insulation that will withstand the same environmental conditions and voltages to which the original conductors are subjected. Suitable insulation for low voltage conductors (less than 600 volts) includes tape (rubber latex), end caps that are approved for the purpose, and shrink sleeves. Shrink sleeves provide a tight-fitting, uniform insulation when heated. Figure 19 illustrates the proper application of electrical tape.

There are additional requirements for the splicing of flexible cord found in Section 1910.305(g)(2)(ii) and Section 400.9 in the 2011 Edition of the National Electrical Code®.

Figure 19 – Insulating a splice with electrical tape.

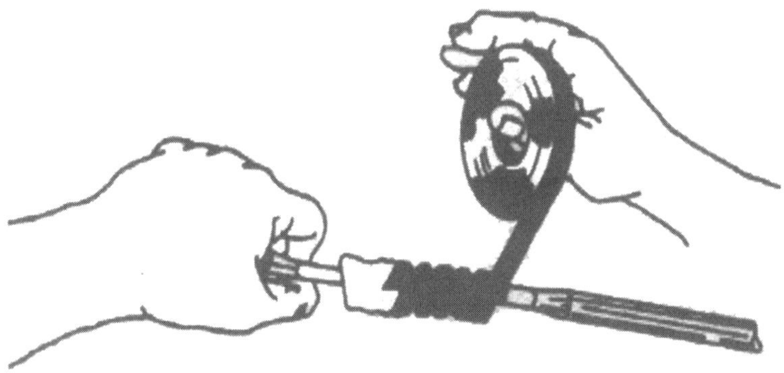

(d)	**Arcing parts.** Parts of electric equipment which in ordinary operation produce arcs, sparks, flames or molten metal shall be enclosed or separated and isolated from all combustible material.
(e)	**Marking.**
(e)(1)	**Identification of manufacturer and ratings.** Electric equipment may not be used unless the following markings have been placed on the equipment:
(e)(1)(i)	The manufacturer's name, trademark, or other descriptive marking by which the organization responsible for the product may be identified; and
(e)(1)(ii)	Other markings giving voltage, current, wattage, or other ratings as necessary.
(e)(2)	**Durability.** The marking shall be of sufficient durability to withstand the environment involved.

An important requirement in this marking is the manufacturer's name, trademark, or other descriptive marking by which the organization responsible for the product may be identified in the event of a product recall. It would not be unusual for qualified electricians or engineers to deal with technical issues even though lacking nameplate information.

Occasionally it becomes necessary for product manufacturers to issue recall or modification notices regarding a safety issue with their products. If such a notice was issued by a product manufacturer, it would be important for an employer to know whether such products were being used at their facility.

Figure 20 – Generator nameplate

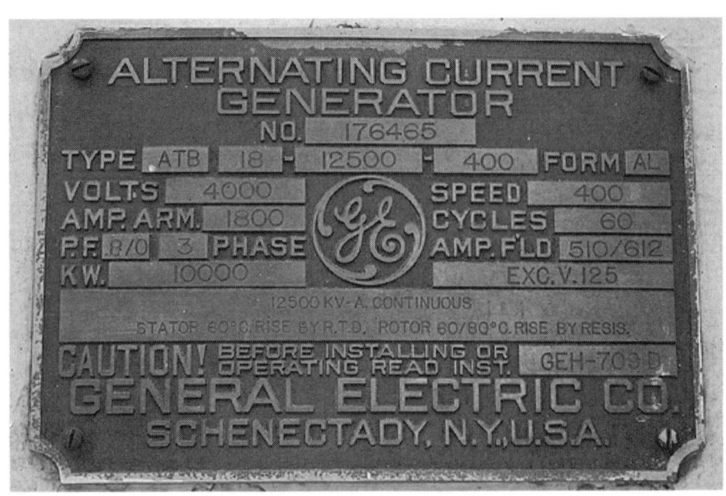

1910.303

Title: General Requirements

Subpart: S

SubPart Title: Electrical - Design Safety Standards for Electrical Systems

(f)	**Disconnecting means and circuits.**
(f)(1)	**Motors and Appliances.** Each disconnecting means required by this subpart for motors and appliances shall be legibly marked to indicate its purpose, unless located and arranged so the purpose is evident.
(f)(2)	**Services, feeders, and branch circuits.** Each service, feeder, and branch circuit, at its disconnecting means or overcurrent device, shall be legibly marked to indicate its purpose, unless located and arranged so the purpose is evident.
(f)(3)	**Durability of markings.** The markings required by paragraphs (f)(1) and (f)(2) of this section shall be of sufficient durability to withstand the environment involved.
(f)(4)	**Capable of accepting a lock.** Disconnecting means required by this subpart shall be capable of being locked in the open position.
(f)(5)	**Marking for series combination ratings.**
(f)(5)(i)	Where circuit breakers or fuses are applied in compliance with the series combination ratings marked on the equipment by the manufacturer, the equipment enclosures shall be legibly marked in the field to indicate that the equipment has been applied with a series combination rating.
(f)(5)(ii)	The marking required by paragraph (f)(5)(i) of this section shall be readily visible and shall state "Caution - Series Combination System Rated__Amperes. Identified Replacement Component Required."

Figure 21 – Each disconnect and circuit requires identification. The sequence of switches could be the opposite of that shown; the exact position does not always identify the purpose of each one.

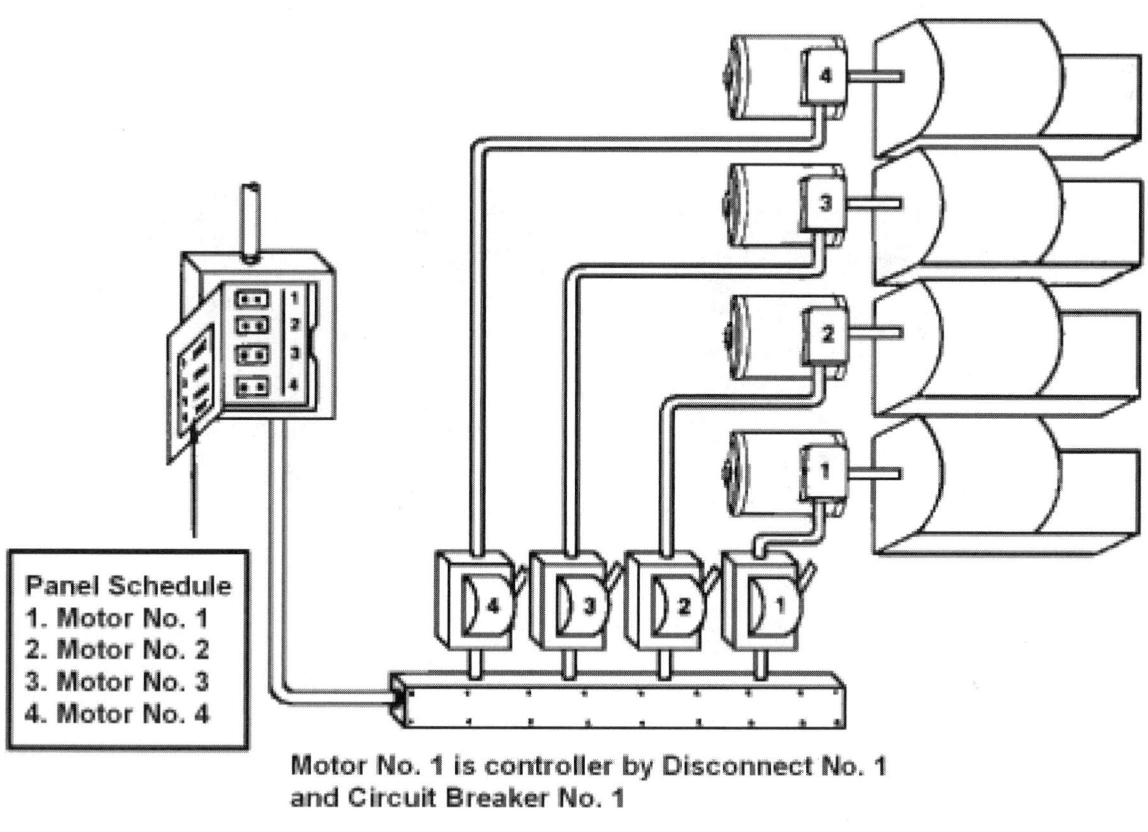

Identification of Disconnecting Means

Accurate identification of disconnects is critical for an effective lockout/tagout program as required by 1910.333 and 1910.147.

A disconnecting means is a switch that is used to disconnect the conductors of a circuit from the source of electricity. Disconnect switches are important because they enable a circuit to be opened, stopping the flow of electricity, and thus effectively protect workers and equipment. Each disconnect switch required for a service, feeder, or branch circuit must be clearly labeled to indicate the circuit's function and the label or marking should be located at the point where the circuit originates. For example, on a panel that controls several motors or on a motor control center, each disconnect must be clearly marked to indicate the motor to which each circuit is connected. In Figures 21 and 22, the number 2 circuit breaker in the panel box supplies power only to disconnect number 2, which in turn controls the power to motor number 2. This power to motor number 2 can be shut off by the number 2 circuit breaker or the number 2 disconnect.

In Figures 21 and 22, the motors may not be within sight of the disconnects or may be arranged in such a way that the purpose is not evident and identification would be required. In most instances more than just a number would be required such as "Shredder Motor #1", "Decoiling Motor #2." Identification should be specific rather than general. A branch circuit serving receptacles in Office 302 should be labeled as such, not simply labeled "receptacles." If the purpose of the circuit is obvious, identification of the disconnect is not required. (See Figure 23). All labels and markings must be durable enough to withstand weather, sunlight, chemicals, heat, corrosion, or any other environment to which they may be exposed, during the expected lifetime of the equipment. Care should be taken not to duplicate names of equipment within the facility.

Figure 22 – Actual picture of disconnect switches illustrated in Figure 21.

Figure 23 – Disconnect switch located on machine; load is obvious and marking is not required.

Working Clearances

(g) **600 Volts, nominal, or less.** This paragraph applies to electric equipment operating at 600 volts, nominal, or less to ground.

(g)(1) **Space about electric equipment.** Sufficient access and working space shall be provided and maintained about all electric equipment to permit ready and safe operation and maintenance of such equipment.

(g)(1)(i) Working space for equipment likely to require examination, adjustment, servicing, or maintenance while energized shall comply with the following dimensions, except as required or permitted elsewhere in this subpart:

Section 1910.303(g)(1)(i) applies to the persons whose duties require them to work on the electrical equipment and coincides with NEC® Section 110.26. The presumption is that this is a qualified person (for example, an electrician) who has been trained to perform the work safely. The intent of the section is to guarantee adequate space to ensure worker safety while any of the listed tasks are required. This space is required to be maintained regardless of whether work is occurring or not.

It should additionally be noted that this section is where Subpart S first separates voltage ranges. This section applies to systems and equipment operating at *600 volts, nominal, or less*, while 1910.303(h) applies to systems and equipment operating at *over 600 volts, nominal.* While there are various terms used to describe different ranges of voltages throughout the industry (low voltage, high voltage, medium voltage, extra-low voltage, etc.), Subpart S and the NEC® simply specify voltage ranges in these and similar sections, and do not make use of specific terms such as "Low Voltage" or "High Voltage". Table S-1 establishes the minimum clear distances that are required around electrical equipment rated at 600 volts or less.

(g)(1)(i)(A) The depth of the working space in the direction of access to live parts may not be less than indicated in Table S-1. Distances shall be measured from the live parts if they are exposed or from the enclosure front or opening if they are enclosed;

(g)(1)(i)(B) The width of working space in front of the electric equipment shall be the width of the equipment or 762 mm (30 in.), whichever is greater. In all cases, the working space shall permit at least a 90-degree opening of equipment doors or hinged panels; and

In some cases, the door of the enclosure may be larger than minimum required working space, measured from the front of the enclosure. Therefore, in order to allow a 90-degree opening of the door, the working space would be required to be increased to at least the size of the enclosure door.

(g)(1)(i)(C) The work space shall be clear and extend from the grade, floor, or platform to the height required by paragraph (g)(1)(vi) of this section. However, other equipment associated with the electrical installation and located above or below the electric equipment may extend not more than 153 mm (6 in.) beyond the front of the electric equipment.

(g)(1)(ii) Working space required by this standard may not be used for storage. When normally enclosed live parts are exposed for inspection or servicing, the working space, if in a passageway or general open space, shall be suitably guarded.

(g)(1)(iii) At least one entrance of sufficient area shall be provided to give access to the working space about electric equipment.

(g)(1)(iv) For equipment rated 1200 amperes or more and over 1.83 m (6.0 ft) wide, containing overcurrent devices, switching devices, or control devices, there shall be one entrance not less than 610 mm (24 in.) wide and 1.98 m (6.5 ft) high at each end of the working space, except that:

(g)(1)(iv)(A) Where the location permits a continuous and unobstructed way of exit travel, one means of exit is permitted; or

(g)(1)(iv)(B) Where the working space required by paragraph (g)(1)(i) of this section is doubled, only one entrance to the working space is required; however, the entrance shall be located so that the edge of the entrance nearest the equipment is the minimum clear distance given in Table S–1 away from such equipment.

These requirements are intended to provide appropriate access to the electrical equipment. However, the primary intent is to allow qualified workers the opportunity to escape the area in the event of an arc flash event (see Figure 24).

(g)(1)(v) Illumination shall be provided for all working spaces about service equipment, switchboards, panelboards, and motor control centers installed indoors. Additional lighting fixtures are not required where the working space is illuminated by an adjacent light source. In electric equipment rooms, the illumination may not be controlled by automatic means only.

(g)(1)(vi) The minimum headroom of working spaces about service equipment, switchboards, panelboards, or motor control centers shall be as follows:

(g)(1)(vi)(A) For installations built before August 13, 2007, 1.91 m (6.25 ft); and

(g)(1)(vi)(B) For installations built on or after August 13, 2007, 1.98 m (6.5 ft), except that where the electrical equipment exceeds 1.98 m (6.5 ft) in height, the minimum headroom may not be less than the height of the equipment.

TABLE S-1
MINIMUM DEPTH OF CLEAR WORKING SPACE AT ELECTRIC EQUIPMENT, 600 V OR LESS

Nominal Voltage to Ground	Minimum clear distance for condition[2,3]					
	Condition A		Condition B		Condition C	
	m	ft	m	ft	m	ft
0 - 150	[1]0.9	[1]3.0	[1]0.9	[1]3.0	0.9	3.0
151 - 600	[1]0.9	[1]3.0	1.0	3.5	1.2	4.0

Notes to Table S-1:

1. Minimum clear distances may be 0.7 m (2.5 ft) for installations built before April 16, 1981.
2. Conditions A, B, and C are as follows:
   - Condition A - Exposed live parts on one side and no live or grounded parts on the other side of the working space, or exposed live parts on both sides effectively guarded by suitable wood or other insulating material. Insulated wire or insulated busbars operating at not over 300 volts are not considered live parts.
   - Condition B - Exposed live parts on one side and grounded parts on the other side.
   - Condition C - Exposed live parts on both sides of the workspace (not guarded as provided in Condition A) with the operator between.
3. Working space is not required in back of assemblies such as dead-front switchboards or motor control centers where there are no renewable or adjustable parts (such as fuses or switches) on the back and where all connections are accessible from locations other than the back. Where rear access is required to work on deenergized parts on the back of enclosed equipment, a minimum working space of 762 mm (30 in.) horizontally shall be provided.

Figure 24 – Requirements for entrance and egress from working space.

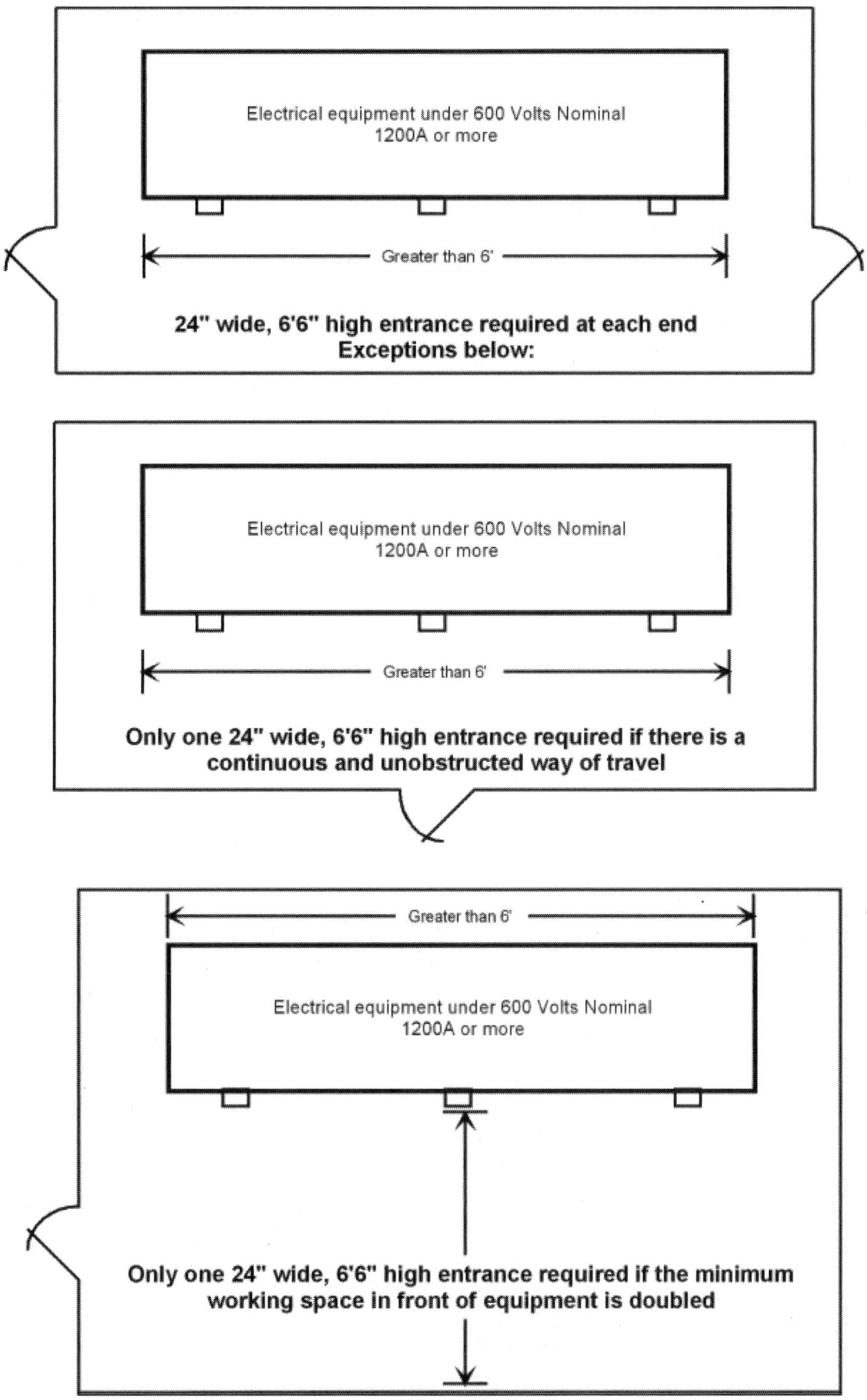

The first requirements for working clearances appeared in the 1937 National Electrical Code®, long before the introduction of the OSHA regulations. While OSHA appears to have significantly changed Section 1910.303(g), none of the changes or additions made to the section are outside the requirements of the NEC® and NFPA 70E. OSHA simply brought the federal regulations into harmony with existing NFPA requirements. Minimum clear distances are the shortest distances which must be provided to afford a worker enough room to work safely near energized electrical equipment. Footnotes 1, 2 and 3 to Table S-1 provide additional detail about the working clearance requirements. Specifically, these distances are established according to the type of electrical hazard presented by the arrangement of equipment, as described in conditions (A), (B), and (C) of Footnote 2. Condition (A) describes a location where there are live electrical parts on one side of the worker and no live or grounded parts on the other side (see Figure 25). If, for example, a worker is servicing a panelboard that has exposed live busbars inside and there is a wall constructed of nonconductive materi al such as wood or sheetrock behind him, a minimum safe working clearance of 3 feet, measured from the live parts to the wall,

Figure 25 – Condition (A)—Live parts on one side and no live or grounded parts on the other side.

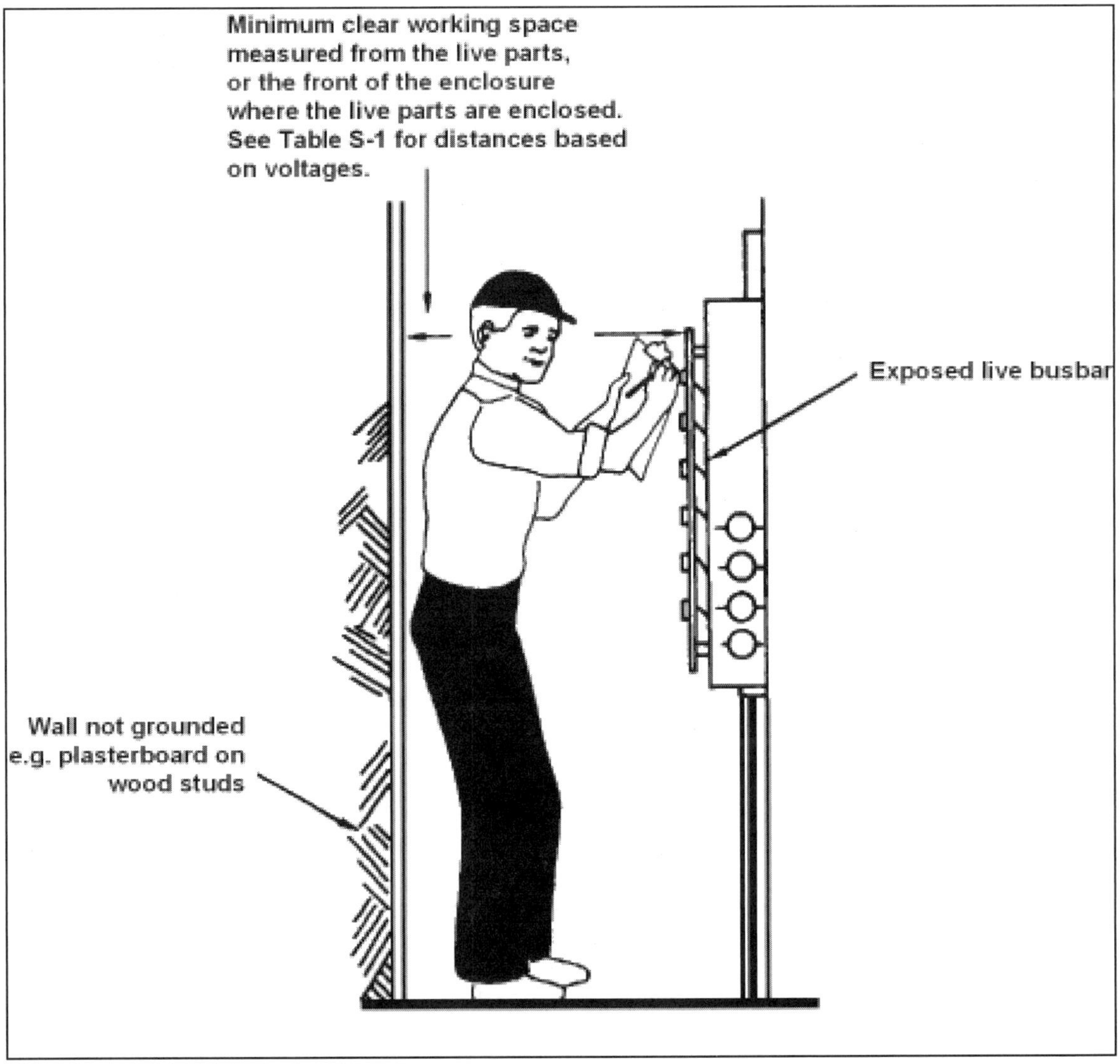

is required. Additionally, if there are effectively guarded live electrical parts on both sides of the worker, then condition (A) for safe clearances also applies. Live parts that are effectively guarded include insulated conductors (operating at 300 volts or less) and exposed switchgear guarded with an insulating barrier.

However, electrical equipment installed before April 16, 1981, is permitted to have a minimum clear working distance of 2 feet 6 inches (or 2½ feet) under condition (A) for both voltage ranges and condition (B) (Figure 26) for the lower voltage range. This clearance distance was originally established by the National Electrical Code® and was eventually increased from 2'6" in the 1975 NEC® to 3' in the 1978 NEC®. Since the NEC® is not retroactive OSHA still allowed the reduced clearance exemption in the current new rule. Note that in the 2002 and later editions of the National Electrical Code®, measurements are expressed in Metric (SI) format (e.g. meters) with the United States Inch-Pound unit

Figure 26 – Condition (B)—Exposed live parts on one side and grounded parts on the other side.

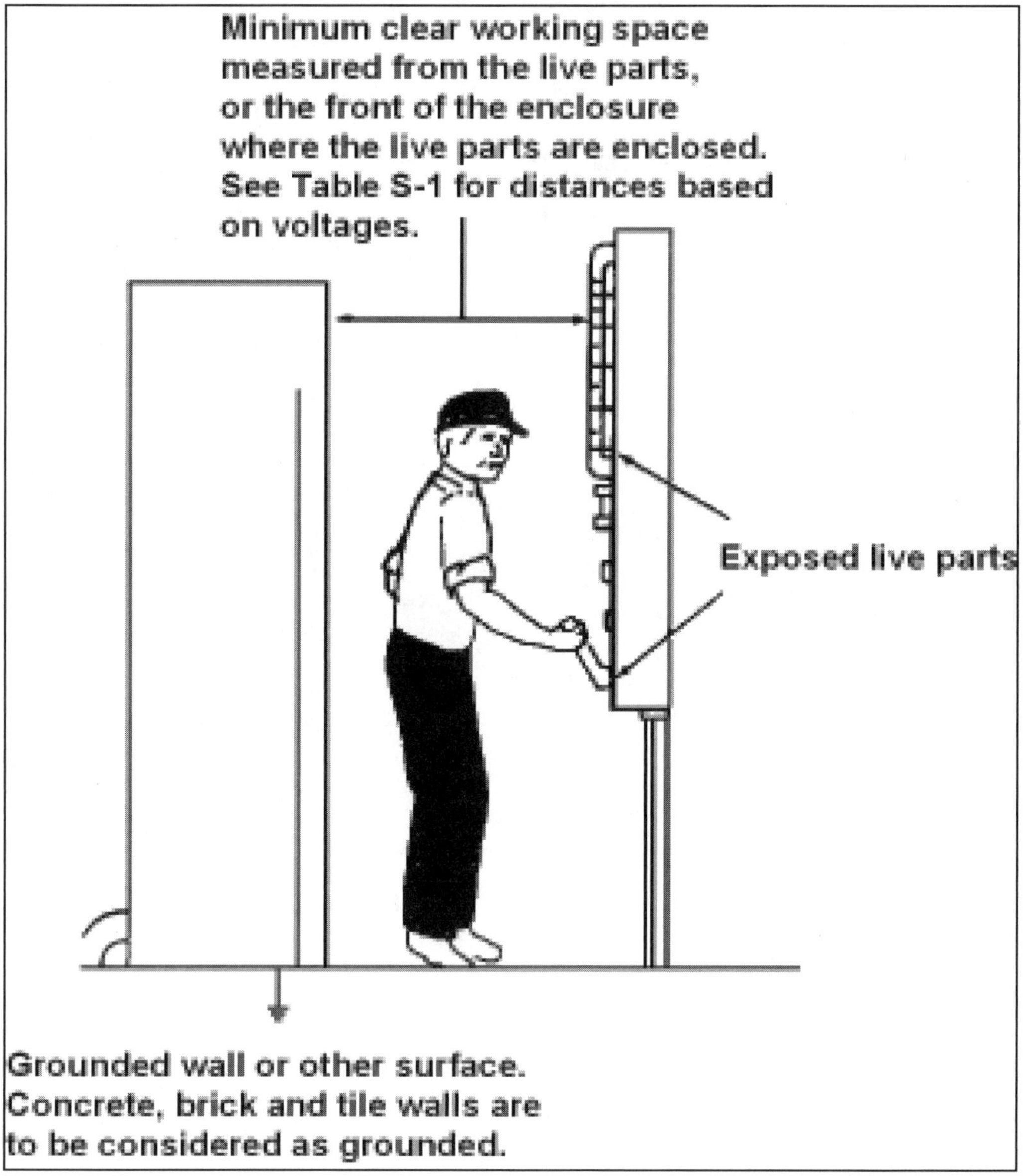

number (e.g. feet) in parenthesis. It is the intent of NFPA to bring the National Electrical Code® into global harmonization. OSHA adopted this implementation of the metric system with the revision. This occurred for a number of reasons. NFPA shows both metric al such as wood or sheetrock behind him, a minimum safe working clearance of 3 feet, measured from the live parts to the wall, is required. Additionally, if there are effectively guarded live electrical parts on both sides of the worker, then condition (A) for safe clearances also applies. Live parts that are effectively guarded include insulated conductors (operating at 300 volts or less) and exposed switchgear guarded with an insulating barrier.

However, electrical equipment installed before April 16, 1981, is permitted to have a minimum clear working distance of 2 feet 6 inches (or 2½ feet) under condition (A) for both voltage ranges and condition (B) (Figure 26) for the lower voltage range. This clearance distance was originally established by the National Electrical Code® and was eventually increased from 2'6" in the 1975 NEC® to 3' in the 1978 NEC®. Since the NEC® is not retroactive OSHA still allowed the reduced clearance exemption in the current new rule. Note that in the 2002 and later editions of the National Electrical Code®, measurements are expressed in Metric (SI) format (e.g. meters) with the United States Inch-Pound unit number (e.g. feet) in parenthesis. It is the intent of NFPA to bring the National Electrical Code® into global harmonization. OSHA adopted this implementation of the metric system with the revision. This occurred for a number of reasons. NFPA shows both metric and US measurements, a number of foreign countries have, or are in the process of implementing regulations based on the OSHA standard, and Executive Order 12770, signed by President George Bush on July 25, 1991 ordered all federal agencies to convert to metric per the Metric Conversion Act of 1975.

Distance in Tables S-1 and S-2 is measured from the live parts involved if exposed, or from the enclosure front or opening if the live parts are enclosed. Voltage specified is "voltage to ground". See definition in Section 1910.399. For example, if live switchgear operating at 150 volts or less to ground are in front of the worker, and a grounded equipment enclosure is behind, the minimum clear working distance must be 3 feet (2½ feet for equipment installed prior to April 16, 1981.) For equipment operating at 151 to 600 volts under the same conditions, the working distance must be 3 feet 6 inches (or 3½ ft.). These distances allow enough room to work, reducing the possibility of simultaneous contact with the live parts and conductive surfaces that are grounded. Concrete, brick, and tile walls are considered to be grounded. Although the statement was in NFPA 70E, OSHA inadvertently omitted the notation that "concrete, brick and tile are considered grounded" under Table S-1 but did include it under Table S-2 condition B. Since the physical electrical properties of concrete, brick and tile have not changed they should always be considered grounded or conductive.

Condition (C), illustrated in Figure 27, describes a situation where there are exposed live parts on both sides of the workspace. For example, if two switchboards are installed opposite each other, the minimum clear workspace between them is required to be 3 feet for equipment operating at 150 volts or less to ground. For equipment operating at 151 to 600 volts to ground, the minimum clear work space is 4 feet. For equipment operating at over 600 volts see Table S-2.

Working clearances have always been important. However, these requirements take on new emphasis with the better understanding of safety-related work practices. This is especially true with regard to the improved PPE requirements in the 2000 Edition of NFPA 70E. A proper hazard analysis will take into consideration the effect that working space, egress to and from the work space, and proper illumination will have on the overall safety of the worker.

Figure 27 – Condition (C)—Exposed live unguarded parts on both sides.

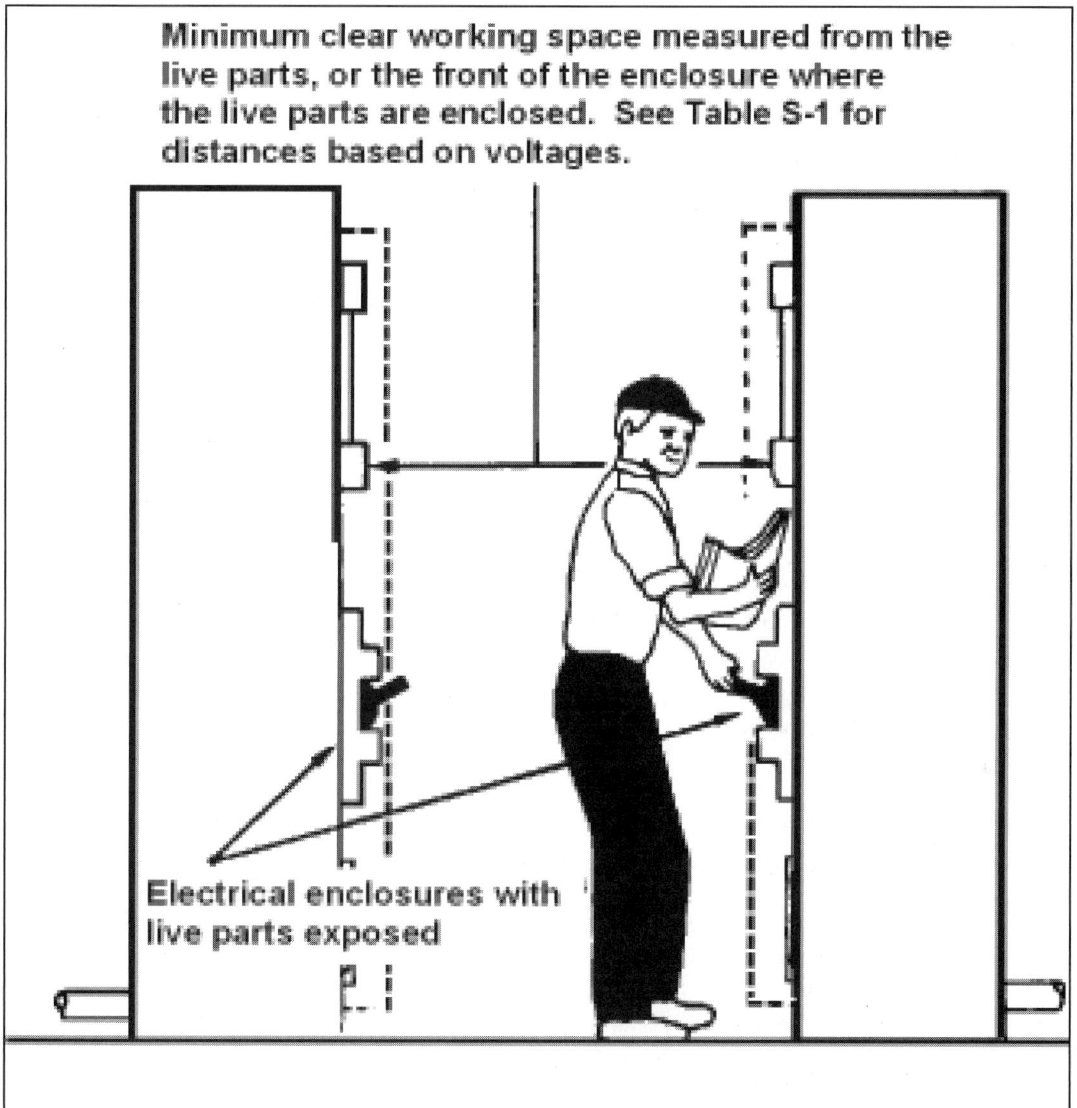

(g)(1)(vii) Switchboards, panelboards, and distribution boards installed for the control of light and power circuits, and motor control centers shall be located in dedicated spaces and protected from damage.

(g)(1)(vii)(A) For indoor installation, the dedicated space shall comply with the following:

(g)(1)(vii)(A)(1) The space equal to the width and depth of the equipment and extending from the floor to a height of 1.83 m (6.0 ft) above the equipment or to the structural ceiling, whichever is lower, shall be dedicated to the electrical installation. Unless isolated from equipment by height or physical enclosures or covers that will afford adequate mechanical protection from vehicular traffic or accidental contact by unauthorized personnel or that complies with paragraph (g)(1)(vii)(A)(2) of this section, piping, ducts, or equipment foreign to the electrical installation may not be located in this area;

(g)(1)(vii)(A)(2) The space equal to the width and depth of the equipment shall be kept clear of foreign systems unless protection is provided to avoid damage from condensation, leaks, or breaks in such foreign systems. This area shall extend from the top of the electric equipment to the structural ceiling;

(g)(1)(vii)(A)(3) Sprinkler protection is permitted for the dedicated space where the piping complies with this section; and

(g)(1)(vii)(A)(4) Control equipment that by its very nature or because of other requirements in this subpart must be adjacent to or within sight of its operating machinery is permitted in the dedicated space.

Note to paragraph (g)(1)(vii)(A) of this section: A dropped, suspended, or similar ceiling that does not add strength to the building structure is not considered a structural ceiling.

(g)(1)(vii)(B) Outdoor electric equipment shall be installed in suitable enclosures and shall be protected from accidental contact by unauthorized personnel, or by vehicular traffic, or by accidental spillage or leakage from piping systems. No architectural appurtenance or other equipment may be located in the working space required by paragraph (g)(1)(i) of this section.

These required spaces are known as *dedicated equipment spaces.* They exist for the physical protection of electrical equipment, to provide future access and to prevent the entrance of non-electrical equipment.

(g)(2) **Guarding of live parts.**

(g)(2)(i) Except as elsewhere required or permitted by this subpart, live parts of electric equipment operating at 50 volts or more shall be guarded against accidental contact by use of approved cabinets or other forms of approved enclosures, or by any of the following means:

(g)(2)(i)(A) By location in a room, vault, or similar enclosure that is accessible only to qualified persons;

(g)(2)(i)(B) By suitable permanent, substantial partitions or screens so arranged so that only qualified persons will have access to the space within reach of the live parts. Any openings in such partitions or screens shall be so sized and located that persons are not likely to come into accidental contact with the live parts or to bring conducting objects into contact with them;

(g)(2)(i)(C) By placement on a suitable balcony, gallery, or platform so elevated and otherwise located as to prevent access by unqualified persons; or

(g)(2)(i)(D) By elevation of 2.44 m (8.0 ft) or more above the floor or other working surface.

In contrast to 1910.303(g)(1), note that 1910.303(g)(2) above is intended to protect **any person** (not necessarily a–qualified person) who might be in the vicinity of electrical equipment. This can only be done by a proper enclosure (or other specified means) to ensure that only a qualified person has access to the live parts.

(g)(2)(ii) In locations where electric equipment is likely to be exposed to physical damage, enclosures or guards shall be so arranged and of such strength as to prevent such damage.

(g)(2)(iii) Entrances to rooms and other guarded locations containing exposed live parts shall be marked with conspicuous warning signs forbidding unqualified persons to enter.

Signage, such as that pictured in Figure 28, is required where unqualified persons may be exposed to electrical hazards. Depending upon your geographical location, consideration should also be given to providing the warnings in another language besides English.

Figure 28 – Example of warning sign for unqualified persons.

(h)	**Over 600 volts, nominal.**
(h)(1)	**General.** Conductors and equipment used on circuits exceeding 600 volts, nominal, shall comply with all applicable provisions of the paragraphs (a) through (g) of this section and with the following provisions which supplement or modify the preceding requirements. However, paragraphs (h)(2), (h)(3), and (h)(4) of this section do not apply to the equipment on the supply side of the service point.
(h)(2)	**Enclosure for electrical installations.**
(h)(2)(i)	Electrical installations in a vault, room, or closet or in an area surrounded by a wall, screen, or fence, access to which is controlled by lock and key or other approved means, are considered to be accessible to qualified persons only. The type of enclosure used in a given case shall be designed and constructed according to the hazards associated with the installation.
(h)(2)(ii)	For installations other than equipment described in paragraph (h)(2)(v) of this section, a wall, screen, or fence shall be used to enclose an outdoor electrical installation to deter access by persons who are not qualified. A fence may not be less than 2.13 m (7.0 ft) in height or a combination of 1.80 m (6.0 ft) or more of fence fabric and a 305-mm (1-ft) or more extension utilizing three or more strands of barbed wire or equivalent.
(h)(2)(iii)	The following requirements apply to indoor installations that are accessible to other than qualified persons:

(h)(2)(iii)(A)	The installations shall be made with metal-enclosed equipment or shall be enclosed in a vault or in an area to which access is controlled by a lock;
(h)(2)(iii)(B)	Metal-enclosed switchgear, unit substations, transformers, pull boxes, connection boxes, and other similar associated equipment shall be marked with appropriate caution signs; and
(h)(2)(iii)(C)	Openings in ventilated dry-type transformers and similar openings in other equipment shall be designed so that foreign objects inserted through these openings will be deflected from energized parts.
(h)(2)(iv)	Outdoor electrical installations having exposed live parts shall be accessible to qualified persons only.
(h)(2)(v)	The following requirements apply to outdoor enclosed equipment accessible to unqualified employees:
(h)(2)(v)(A)	Ventilating or similar openings in equipment shall be so designed that foreign objects inserted through these openings will be deflected from energized parts;
(h)(2)(v)(B)	Where exposed to physical damage from vehicular traffic, suitable guards shall be provided;
(h)(2)(v)(C)	Nonmetallic or metal-enclosed equipment located outdoors and accessible to the general public shall be designed so that exposed nuts or bolts cannot be readily removed, permitting access to live parts;
(h)(2)(v)(D)	Where nonmetallic or metal-enclosed equipment is accessible to the general public and the bottom of the enclosure is less than 2.44 m (8.0 ft) above the floor or grade level, the enclosure door or hinged cover shall be kept locked; and
(h)(2)(v)(E)	Except for underground box covers that weigh over 45.4 kg (100 lb), doors and covers of enclosures used solely as pull boxes, splice boxes, or junction boxes shall be locked, bolted, or screwed on.

All of these requirements carry two intents; protect personnel from the hazards of exposure to high voltage electricity, and to protect the equipment from damage by unqualified personnel, mobile equipment, etc. The photo in Figure 29 shows one example of preventing unauthorized entrance through the use of fencing and a barrier.

Figure 29 – Entrance to this outdoor substation is prevented by a locked fence with barbed-wire and protected from mobile equipment by a suitable barrier.

1910.303

Title: General Requirements

Subpart: S

SubPart Title: Electrical - Design Safety Standards for Electrical Systems

Entrances and Access to Workspace

(h)(3)	Workspace about equipment. Sufficient space shall be provided and maintained about electric equipment to permit ready and safe operation and maintenance of such equipment. Where energized parts are exposed, the minimum clear workspace may not be less than 1.98 m (6.5 ft) high (measured vertically from the floor or platform), or less than 914 mm (3.0 ft) wide (measured parallel to the equipment). The depth shall be as required in paragraph (h)(5)(i) of this section. In all cases, the workspace shall be adequate to permit at least a 90-degree opening of doors or hinged panels.
(h)(4)	Entrance and access to workspace. (See §1910.302(b)(3)).
(h)(4)(i)	At least one entrance not less than 610 mm (24 in.) wide and 1.98 m (6.5 ft) high shall be provided to give access to the working space about electric equipment.
(h)(4)(i)(A)	On switchboard and control panels exceeding 1.83 m (6.0 ft) in width, there shall be one entrance at each end of such boards unless the location of the switchboards and control panels permits a continuous and unobstructed way of exit travel, or unless the work space required in paragraph (h)(5)(i) of this section is doubled. **(h)(4)(i)(B)** Where one entrance to the working space is permitted under the conditions described in paragraph (h)(4)(i)(A) of this section, the entrance shall be located so that the edge of the entrance nearest the switchboards and control panels is at least the minimum clear distance given in Table S–2 away from such equipment.
(h)(4)(i)(C)	Where bare energized parts at any voltage or insulated energized parts above 600 volts, nominal, to ground are located adjacent to such entrance, they shall be suitably guarded.
(h)(4)(ii)	Permanent ladders or stairways shall be provided to give safe access to the working space around electric equipment installed on platforms, balconies, mezzanine floors, or in attic or roof rooms or spaces.

Entrances and Access to Workspaces

In any workspace where there is electric equipment operating at over 600 volts, there must be at least one entrance/exit large enough to allow the worker to safely enter the work area and to easily escape from it in the event of an emergency. This entrance must be at least 24 inches wide (610 mm) and 6 feet, 6 inches (2.0 m) high. If a switchboard or control panel is more than 6 feet wide (1.8 m), there must be one entrance at each end of the board. Both of these entrances should also be at least 24 inches wide and 6 feet, 6 inches high. Keep in mind that 1910 and the NEC® are minimum standards. Therefore, larger areas around electrical equipment may be needed. Figure 30 on the next page illustrates the requirements for entrance and egress from workspace where the electrical equipment is greater than 6 feet wide.

For new installations, where the entrance to a room containing equipment operating at over 600 volts has a personnel door, the door(s) shall open in the direction of egress and be equipped with panic bars, pressure plates, or other devices that are normally latched but open under simple pressure (NEC® Sections 110.26(C)(3) and 110.33(A)(3)).

Any exposed energized electric parts operating at any voltage and located near the entrances must be guarded to prevent accidental contact by the worker. Any insulated energized parts operating at more than 600 volts and located next to the doors must also be guarded.

Figure 30 – Requirements for entrance and egress from workspaces.

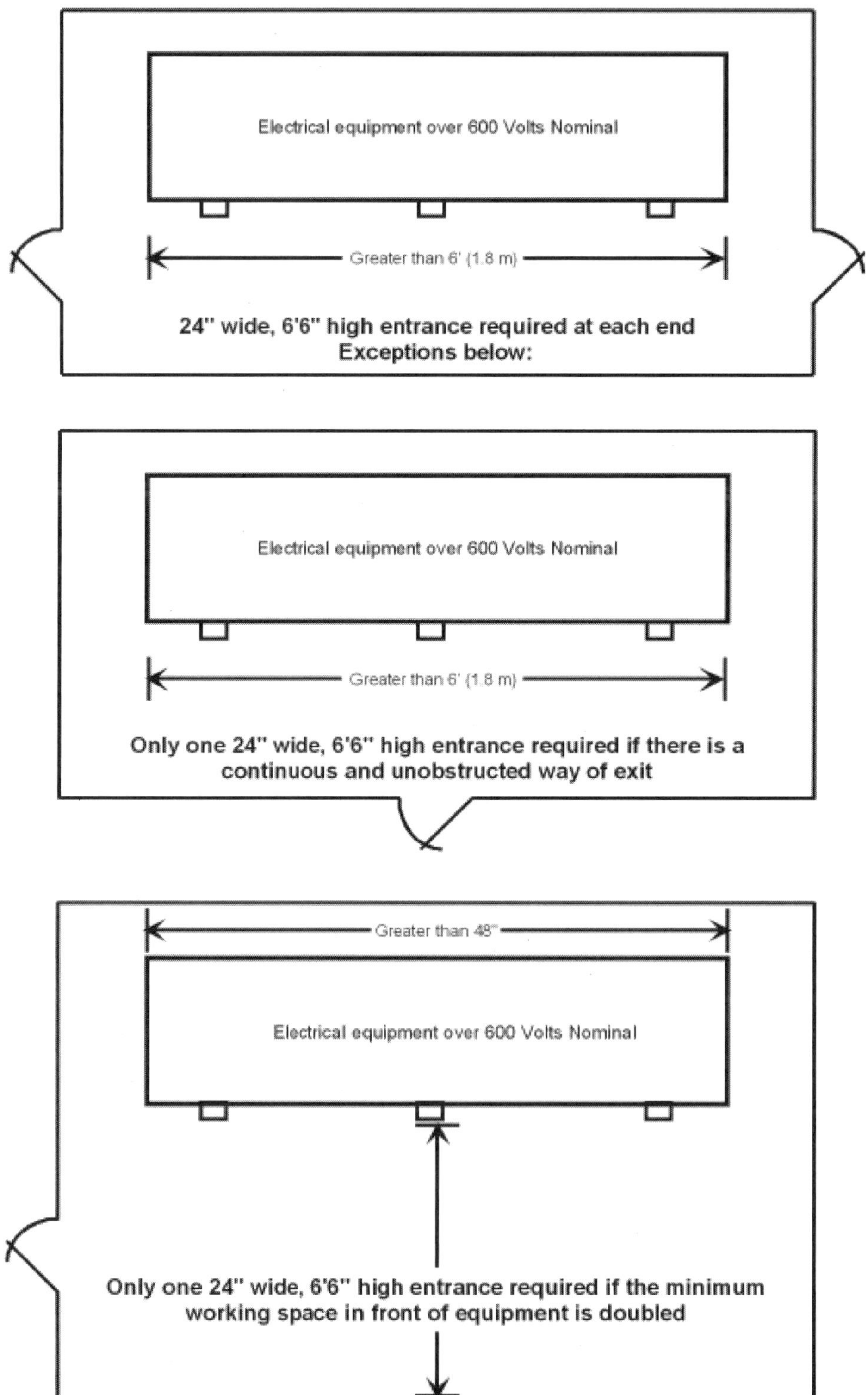

(h)(5) **Working space and guarding.**

(h)(5)(i) Except as elsewhere required or permitted in this subpart, the minimum clear working space in the direction of access to live parts of electric equipment may not be less than specified in Table S-2. Distances shall be measured from the live parts, if they are exposed, or from the enclosure front or opening, if they are enclosed.

(h)(5)(ii) If switches, cutouts, or other equipment operating at 600 volts, nominal, or less, are installed in a room or enclosure where there are exposed live parts or exposed wiring operating at over 600 volts, nominal, the high-voltage equipment shall be effectively separated from the space occupied by the low-voltage equipment by a suitable partition, fence, or screen. However, switches or other equipment operating at 600 volts, nominal, or less, and serving only equipment within the high-voltage vault, room, or enclosure may be installed in the high-voltage enclosure, room, or vault if accessible to qualified persons only.

(h)(5)(iii) The following requirements apply to the entrances to all buildings, rooms, or enclosures containing exposed live parts or exposed conductors operating at over 600 volts, nominal:

(h)(5)(iii)(A) The entrances shall be kept locked unless they are under the observation of a qualified person at all times; and

(h)(5)(iii)(B) Permanent and conspicuous warning signs shall be provided, reading substantially as follows:

"DANGER - HIGH VOLTAGE - KEEP OUT."

(h)(5)(iv) Illumination shall be provided for all working spaces about electric equipment.

(h)(5)(iv)(A) The lighting outlets shall be arranged so that persons changing lamps or making repairs on the lighting system will not be endangered by live parts or other equipment.

(h)(5)(iv)(B) The points of control shall be located so that persons are prevented from contacting any live part or moving part of the equipment while turning on the lights.

(h)(5)(v) Unguarded live parts above working space shall be maintained at elevations not less than specified in Table S-3.

(h)(5)(vi) Pipes or ducts that are foreign to the electrical installation and that require periodic maintenance or whose malfunction would endanger the operation of the electrical system may not be located in the vicinity of service equipment, metal-enclosed power switchgear, or industrial control assemblies. Protection shall be provided where necessary to avoid damage from condensation leaks and breaks in such foreign systems.

Note to paragraph (h)(5)(vi) of this section: Piping and other facilities are not considered foreign if provided for fire protection of the electrical installation.

Figure 31 – Working spaces behind electrical equipment.

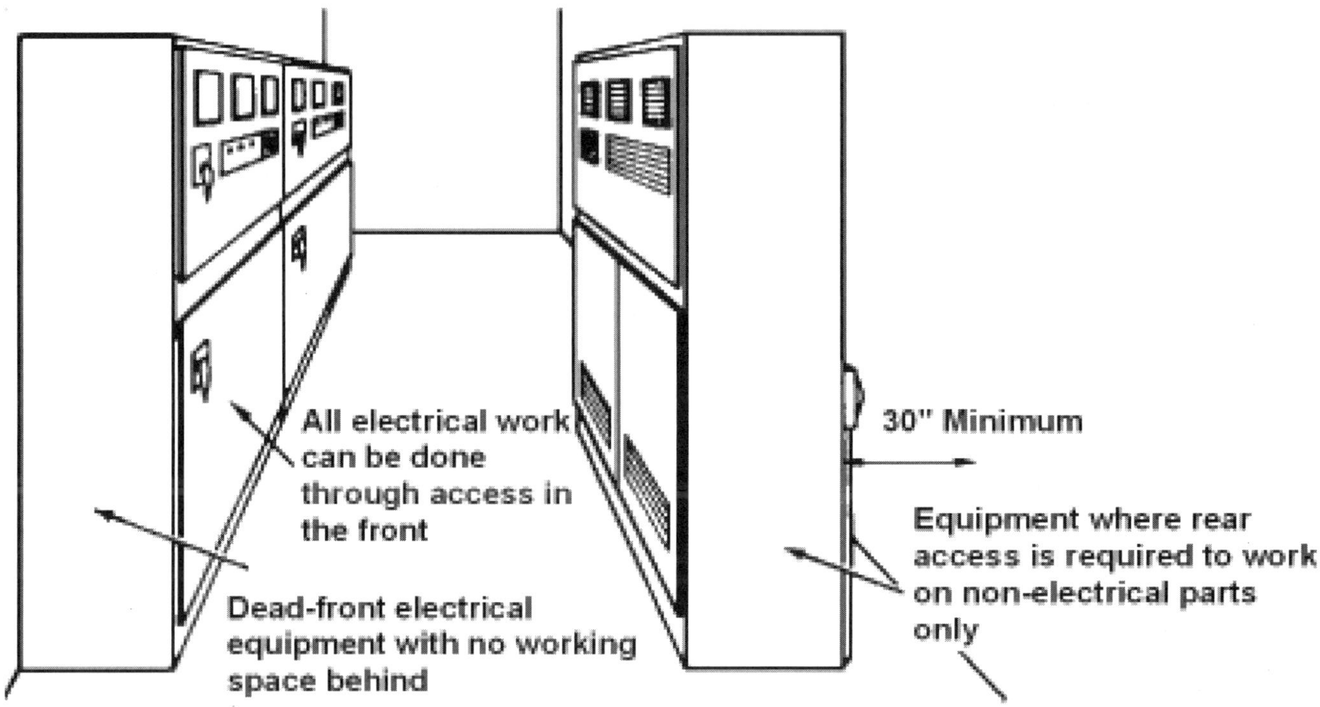

Working Space About Electrical Equipment over 600 volts.

Table S-2 shows the minimum clear working distances required in front of electric equipment that is rated over 600V and is conditional, just as for under 600V. However, working space is not required behind electric equipment such as dead-front switchboards or control panels where parts that may need to be replaced or adjusted, and all connections, can be reached from locations other than the back. If it is necessary for workers to go behind enclosed equipment to work on de-energized parts, the depth of the working space must be no less than 30 inches so that the worker has enough room to move around and to escape in an emergency situation. Figure 31 depicts both situations.

Table S-2
Minimum Depth of Clear Working Space at Electric Equipment, Over 600 V

Nominal Voltage to Ground	Minimum clear distance for condition[2 3]					
	Condition A		Condition B		Condition C	
	m	ft	m	ft	m	ft
601 - 2,500 V	0.9	3.0	1.2	4.0	1.5	5.0
2,501 - 9,000 V	1.2	4.0	1.5	5.0	1.8	6.0
9,001V - 25 kV	1.5	5.0	1.8	6.0	2.8	9.0
Over 25 - 75kV[1]	1.8	6.0	2.5	8.0	3.0	10.0
Above 75kV[1]	2.5	8.0	3.0	10.0	3.7	12.0

Notes to Table S-2:

1. Minimum depth of clear working space in front of electric equipment with a nominal voltage to ground above 25,000 volts may be the same as that for 25,000 volts under Conditions A, B, and C for installations built before April 16, 1981.
2. Conditions A, B, and C are as follows:
   - Condition A - Exposed live parts on one side and no live or grounded parts on the other side of the working space, or exposed live parts on both sides effectively guarded by suitable wood or other insulating material. Insulated wire or insulated busbars operating at not over 300 volts are not considered live parts.
   - Condition B - Exposed live parts on one side and grounded parts on the other side. Concrete, brick, and tile walls are considered as grounded surfaces.
   - Condition C - Exposed live parts on both sides of the work space (not guarded as provided in Condition A) with the operator between.
3. Working space is not required in back of equipment such as dead-front switchboards or control assemblies that has no renewable or adjustable parts (such as fuses or switches) on the back and where all connections are accessible from locations other than the back. Where rear access is required to work on the deenergized parts on the back of enclosed equipment, a minimum working space 762 mm (30 in.) horizontally shall be provided.

Conditions A, B, and C of Table S-2 are described in Figures 25, 26, and 27. They are identical to those pictured for low voltage systems. Only the distance varies according to the different system voltages.

Table S-3
Elevation of Unguarded Live Parts Above Working Space

Nominal Voltage Between Phases	Elevation	
	m	ft
601 - 7,500 V	[1]2.8	[1]9.0
7,501 V - 35 kV	2.8	9.0
Over 35 kV	2.8 + 9.5 mm/kV over 35 kV	9.0 + 0.37 in./kV over 35 kV

[1]The minimum elevation may be 2.6 m (8.5 ft) for installations built before August 13, 2007. The minimum elevation may be 2.4 m (8.0 ft) for installations build before April 16, 1981, if the nominal voltage between phases is in the range of 601 - 6600 volts.

§ 1910.304 Wiring Design and Protection

[46 FR 4056, Jan. 16, 1981; 46 FR 40185, Aug. 7, 1981, as amended at 55 FR 32015, Aug. 6, 1990; 72 FR 7195, Feb. 14, 2007; 73 FR 64205, Oct. 29, 2008]

Before using the material which follows we suggest that readers make certain they understand the differences among the three conductors present in almost every branch circuit. These conductors are the ungrounded conductor, the grounded conductor and the grounding conductor.

The ungrounded conductor is sometimes called the phase conductor or the unidentified conductor. It is also referred to by some as the "hot" conductor. The ungrounded conductor connects the overcurrent device or switch-leg to the load and makes up approximately one-half of the normal branch circuit current pathway. Color for the ungrounded conductor is not specified by the NEC® or by OSHA.

The grounded conductor (frequently called the "neutral" by the qualified person) forms the other half of the circuit and is intentionally grounded at the service entrance or source of a separately-derived system. When the circuit is operating normally, there is current flowing in the grounded conductor. The NEC® requires the grounded conductor to have an outer color of white or gray. The words "natural gray" were deleted from the 2002 Edition of the National Electrical Code® and replaced with only "gray." It is important to note that the color gray may have been used for other purposes in the past. Therefore qualified persons must determine whether this is the case at their individual facility and make a decision whether or not gray may be used for the grounded conductor.

These two conductors, the ungrounded and grounded, form the basic loop of the branch circuit.

By contrast, the equipment grounding conductor is not part of the current pathway during normal circuit operation. The equipment grounding conductor, frequently called the grounding conductor by the qualified person, is used to connect all non-current-carrying metal parts of equipment and enclosures to the system ground or grounding electrode at the service entrance. Its primary purpose is to prevent electrocution to the user of the equipment. It accomplishes this task by providing an effective low impedance return pathway to the power source for any fault current in order to operate the protective device (circuit breaker or fuse), instantaneously shutting off or deenergizing the circuit. This equipment grounding conductor can be a green, green with a yellow stripe, or bare wire or even the conduit or raceway housing the circuit.

(a)	**Use and identification of grounded and grounding conductors.**
(a)(1)	**Identification of conductors.**
(a)(1)(i)	A conductor used as a grounded conductor shall be identifiable and distinguishable from all other conductors.
(a)(1)(ii)	A conductor used as an equipment grounding conductor shall be identifiable and distinguishable from all other conductors.

Identification of Conductors.

The grounded conductor is a normally current carrying circuit conductor from the supply source that is also connected to earth through the system ground. The equipment grounding conductor is not an energized conductor under normal conditions. Instead, it acts as a safeguard against electrocution in the event of insulation failure or faults in the other circuit conductors. The equipment grounding conductor is energized only if there is a leak or fault in the normal current path, and it directs this current back to the source. Directing the fault current back to the source enables protective devices, such as circuit breakers or fuses, to operate, thus preventing fires and reducing the hazard of electrical shocks or electrocution.

The grounded and equipment grounding conductors of an electrical circuit must be marked or color-coded in a way that allows qualified persons to identify them and tell them apart from each other and from the other conductors in the circuit.

Figure 32 shows a cutaway illustration of a distribution panelboard. One means by which each conductor's use is identified and made distinguishable from the other circuit conductors is the use of color-coding. The basic rule in Section 200.6 of the National Electrical Code® requires that an insulated grounded conductor shall be identified by a continuous white or gray outer finish or by three continuous white stripes on other than green insulation.

1910.304

Title: Wiring design and protection.

Subpart: S

SubPart Title: Electrical - Design Safety Standards for Electrical Systems

This basic rule applies to wire sizes No. 6 or smaller. For wire sizes larger than No. 6, distinctive white or gray marking encircling the conductor at the terminations is also permissible. NEC® Section 250.119 specifies how the equipment grounding conductor must be identified. Covered or insulated equipment grounding conductors must have a continuous outer finish that is either green or green with one or more yellow stripes. As with the grounded conductor, if the conductor is larger than No. 6, some alternative marking methods are permitted. See Figures 33(A), (B) and (C) for examples of branch circuit layouts for these three conductors.

Figure 32 – Distribution panelboard

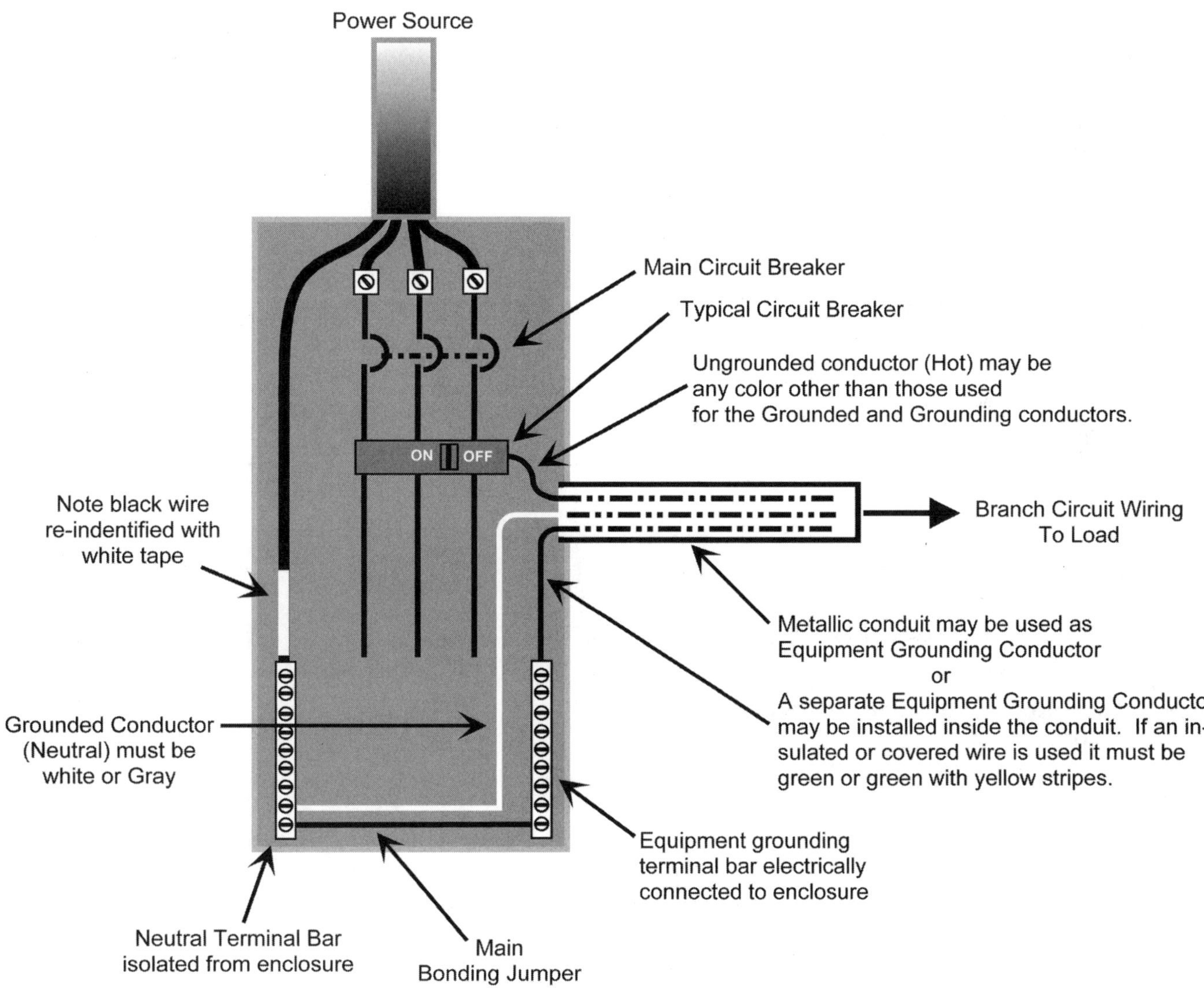

(a)(2) **Polarity of connections.** No grounded conductor may be attached to any terminal or lead so as to reverse designated polarity.

Polarity of Connections

When plugs, receptacles, and connectors are used in an electrical branch circuit, correct polarity between the ungrounded conductor, the grounded conductor, and the grounding conductor must be maintained.

Reversed polarity results when the identified circuit conductor (the grounded conductor) is incorrectly connected to the ungrounded terminal of a plug, receptacle, or other type of connector. A correctly wired circuit is shown in Figure 33(A). If the identified conductor is reversed with the ungrounded conductor, the current path downstream of the improper connection is reversed. Figure 33(B) illustrates reversed polarity. Under these conditions motor windings of tools and appliances and the screw shells of lampholders remain energized even after the equipment is turned off. Simultaneous contact with these parts and any ground will cause a shock, or electrocution, and the only place to turn off the power would be at the branch circuit disconnect.

Figure 33A - Correct wiring of duplex receptacles

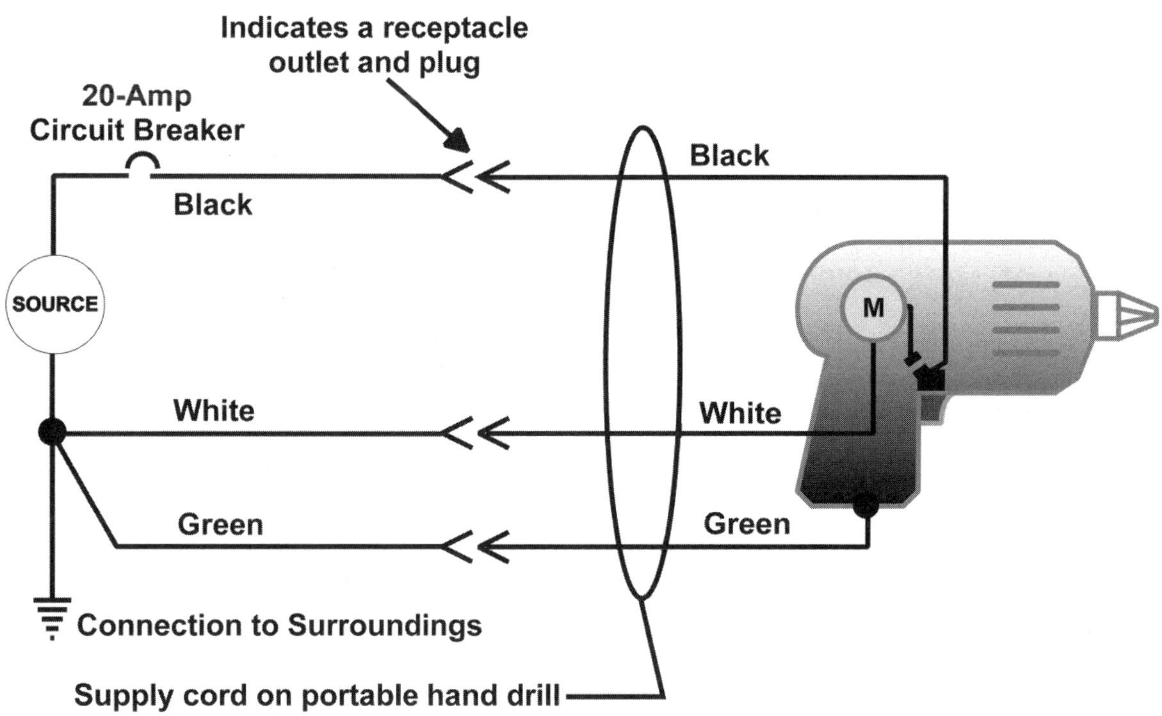

(a)(3) **Use of grounding terminals and devices.** A grounding terminal or grounding-type device on a receptacle, cord connector, or attachment plug may not be used for purposes other than grounding.

Use of Grounding Terminals and Devices

Although the use of grounding terminals had been in the previous edition of the OSHA standard it was not part of the requirements applicable to all installations regardless of when they were installed. OSHA stated in the rulemaking record that "for example, using a grounding terminal as the attachment point for a circuit conductor can energize the frame of equipment used by employees. If an employee was to touch such miswired equipment and a grounded surface at the same time, he or she would receive an

.electric shock and possibly die of electrocution." Since the use of grounding terminals for purposes other than grounding has never been permitted by the NEC© and such a condition would be an imminent danger electrocution hazard, OSHA has made this requirement mandatory regardless of when the installation was installed.

Figures 33(C) and (D) illustrate two other possible incorrect wiring conditions that are prohibited by paragraph (a)(3). The grounding terminal may be used only for attachment of the grounding conductor. The example in Figure 33(C) would not be hazardous with a single receptacle. However, under certain conditions, such as that receptacle placed in a circuit following a receptacle with reversed polarity, a hazard would exist. The example in Figure 33(D) is the most hazardous condition shown, with the ungrounded circuit conductor connected to the drill housing. Correct polarity is achieved when the grounded conductor is connected to the corresponding grounded terminal and the ungrounded conductor is connected to the corresponding ungrounded terminal. The reverse of the designated polarity is prohibited. Figure 34 illustrates a duplex receptacle correctly wired. Terminals are designated and identified to avoid confusion. An easy way to remember the correct polarity is "white to light" -- The white wire should be connected to the light or nickel colored terminal; "black to brass" -- the black or multicolored wire should be connected to the brass terminal; and "green to green," the green or bare equipment grounding wire should be connected to the green hexagonal head terminal screw.

(b) **Branch circuits.**

(b)(1) **Identification of multiwire branch circuits.** Where more than one nominal voltage system exists in a building containing multiwire branch circuits, each ungrounded conductor of a multiwire branch circuit, where accessible, shall be identified by phase and system. The means of identification shall be permanently posted at each branch-circuit panelboard

Figure 33B – Incorrect wiring of duplex receptacles. The illustration represents "reversed polarity," a situation where the ungrounded and grounded conductors are interchanged.

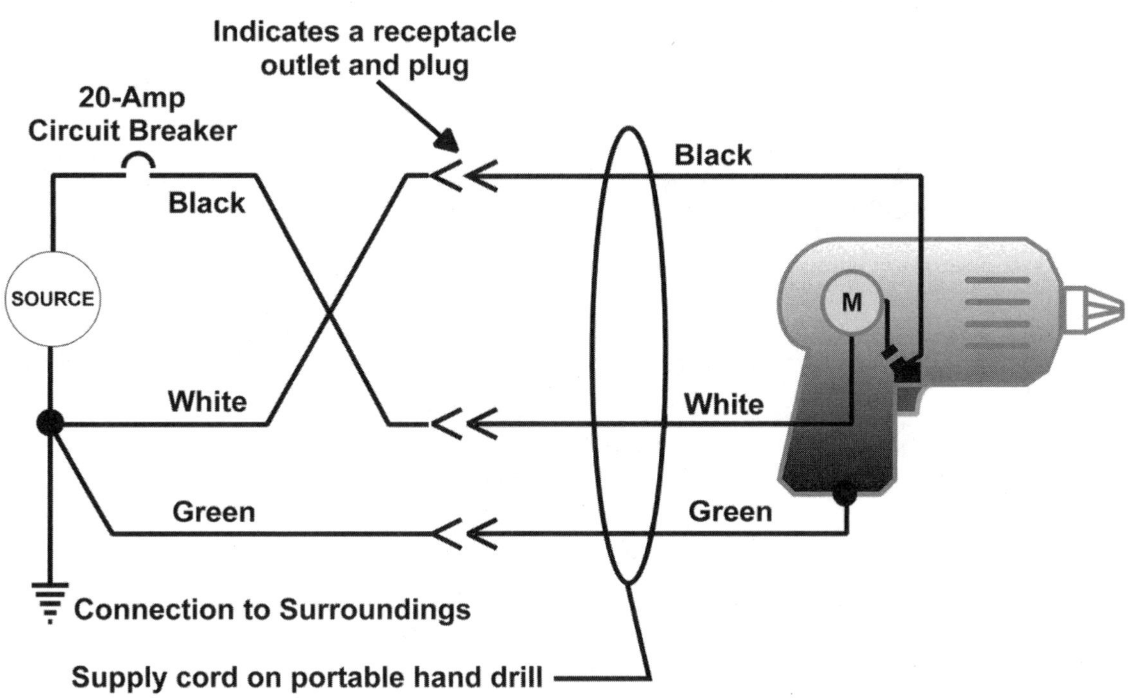

Figure 33C – Incorrect wiring of duplex receptacles. In the figure, the grounded and grounding conductors are interchanged, which allows current flow through the equipment grounding conductor.

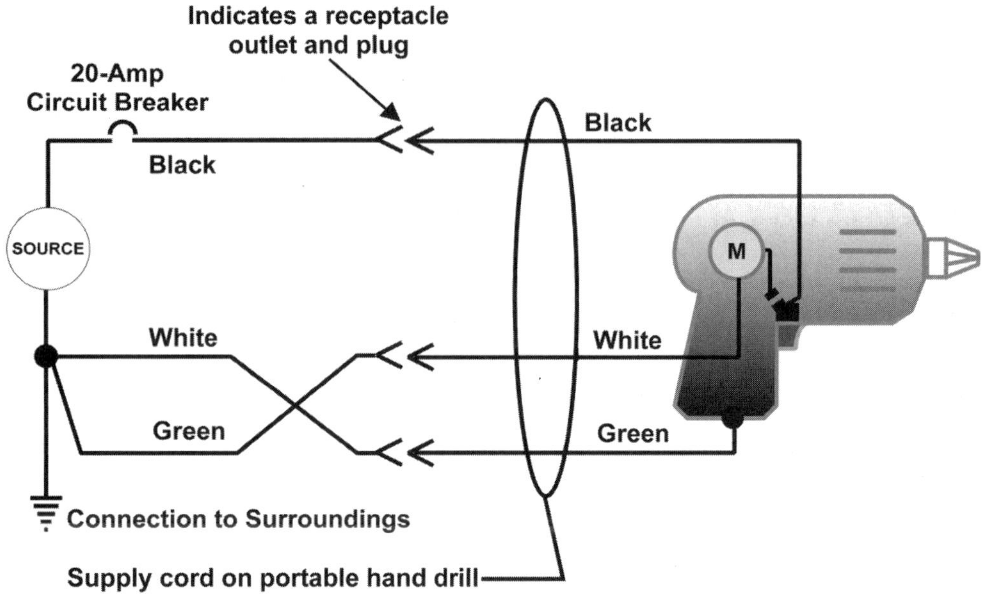

Figure 33D – Incorrect wiring of duplex receptacles: Note in the figure that the ungrounded and grounding conductors have been interchanged, forcing current to flow over the housing of the equipment, in this case a hand-held drill.

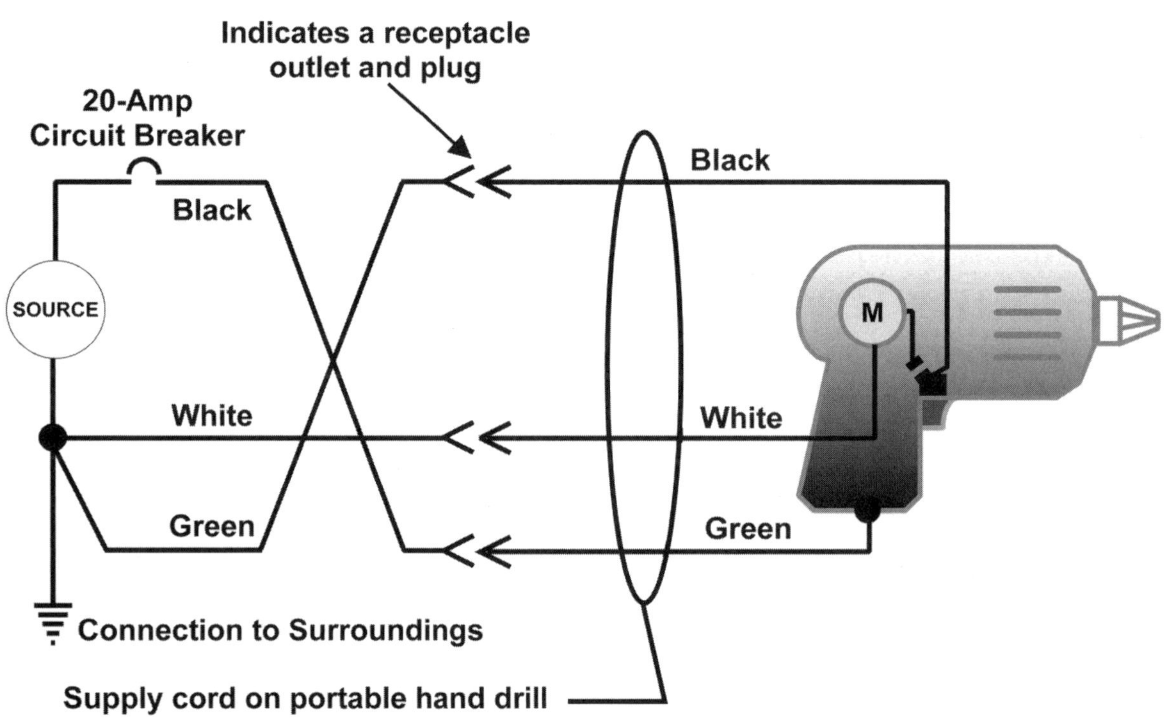

1910.304

Title: Wiring design and protection.
Subpart: S
SubPart Title: Electrical - Design Safety Standards for Electrical Systems

Figure 34 - Duplex receptacle correctly wired for designated terminals.

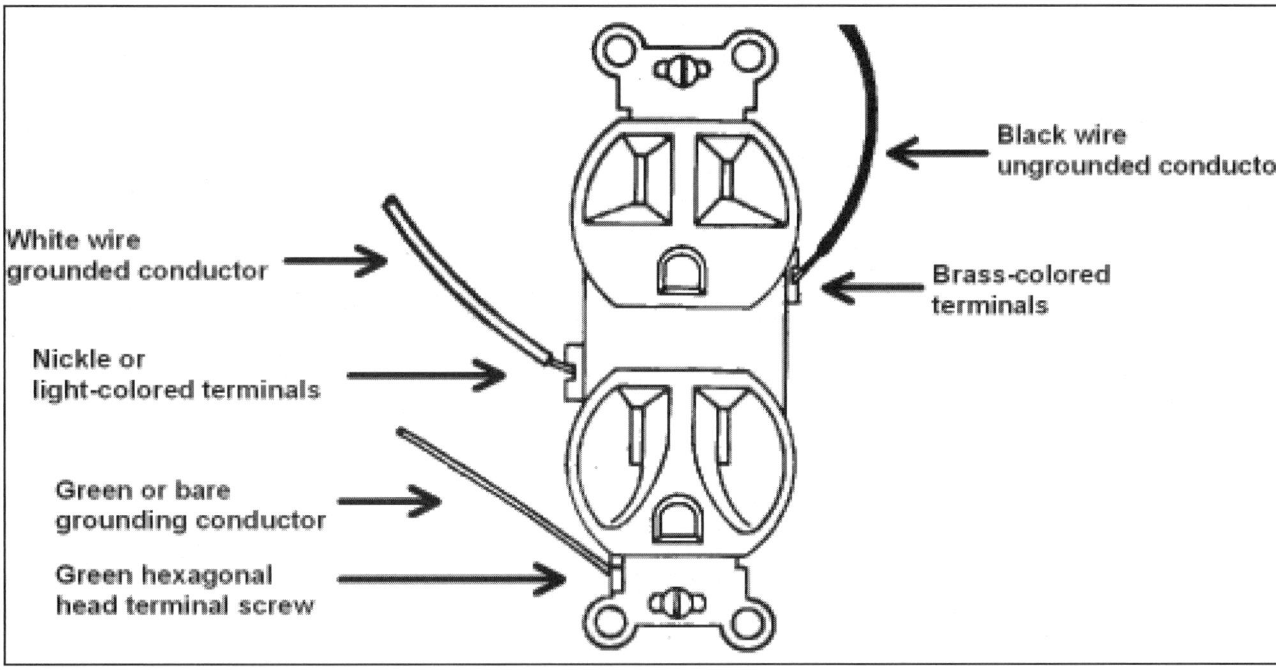

(b)(2)	**Receptacles and cord connectors.**
(b)(2)(i)	Receptacles installed on 15- and 20-ampere branch circuits shall be of the grounding type except as permitted for replacement receptacles in paragraph (b)(2)(iv) of this section. Grounding-type receptacles shall be installed only on circuits of the voltage class and current for which they are rated, except as provided in Table S–4 and Table S–5.
(b)(2)(ii)	Receptacles and cord connectors having grounding contacts shall have those contacts effectively grounded except for receptacles mounted on portable and vehicle-mounted generators in accordance with paragraph (g)(3) of this section and replacement receptacles installed in accordance with paragraph (b)(2)(iv) of this section.
(b)(2)(iii)	The grounding contacts of receptacles and cord connectors shall be grounded by connection to the equipment grounding conductor of the circuit supplying the receptacle or cord connector. The branch circuit wiring method shall include or provide an equipment grounding conductor to which the grounding contacts of the receptacle or cord connector shall be connected.

The importance of equipment grounding to prevent electrocution is further emphasized by the addition of 1910.304(b)(2), requiring that all branch circuit receptacles and cord connectors be connected.

(b)(2)(iv)	Replacement of receptacles shall comply with the following requirements:
(b)(2)(iv)(A)	Where a grounding means exists in the receptacle enclosure or a grounding conductor is installed, grounding-type receptacles shall be used and shall be connected to the grounding means or conductor;
(b)(2)(iv)(B)	Ground-fault circuit-interrupter protected receptacles shall be provided where replacements are made at receptacle outlets that are required to be so protected elsewhere in this subpart; and
(b)(2)(iv)(C)	Where a grounding means does not exist in the receptacle enclosure, the installation shall comply with one of the following provisions:
(b)(2)(iv)(C)(1)	A nongrounding-type receptacle may be replaced with another nongrounding-type receptacle; or

(b)(2)(iv)(C)(2) A nongrounding-type receptacle may be replaced with a ground-fault circuit-interrupter-type of receptacle that is marked "No Equipment Ground;" an equipment grounding conductor may not be connected from the ground-fault circuit-interrupter-type receptacle to any outlet supplied from the ground-fault circuit-interrupter receptacle; or

(b)(2)(iv)(C)(3) A nongrounding-type receptacle may be replaced with a grounding-type receptacle where supplied through a ground-fault circuit-interrupter; the replacement receptacle shall be marked "GFCI Protected" and "No Equipment Ground;" an equipment grounding conductor may not be connected to such grounding-type receptacles.

(b)(2)(v) Receptacles connected to circuits having different voltages, frequencies, or types of current (ac or dc) on the same premises shall be of such design that the attachment plugs used on these circuits are not interchangeable.

Although the equipment grounding requirements of Subpart S are found under 1910.302(b)(1), Extent of Application, as applying "to all electrical installations and utilization equipment, regardless of when they were designed or installed," OSHA did integrate allowances for the replacement of receptacle outlets, as found in the NEC®. Where equipment grounding does not exist, GFCIs may be used to alleviate the electrocution hazard, but must be appropriately labeled; receptacle outlets protected by GFCIs must be labeled "GFCI Protected", and if no equipment grounding conductor is in place, the outlets must also be labeled "No Equipment Ground". These labels are typically included with the GFCI receptacle or circuit breaker. Where replacements are made, all current requirements for GFCIs must be met.

(b)(3) **Ground-fault circuit interrupter protection for personnel.**

(b)(3)(i) All 125-volt, single-phase, 15- and 20-ampere receptacles installed in bathrooms or on rooftops shall have ground-fault circuit-interrupter protection for personnel.

(b)(3)(ii) The following requirements apply to temporary wiring installations that are used during construction-like activities, including certain maintenance, remodeling, or repair activities, involving buildings, structures or equipment.

(b)(3)(ii)(A) All 125-volt, single-phase, 15-, 20-, and 30-ampere receptacle outlets that are not part of the permanent wiring of the building or structure and that are in use by personnel shall have ground-fault circuit-interrupter protection for personnel.

Note 1 to paragraph (b)(3)(ii)(A) of this section: A cord connector on an extension cord set is considered to be a receptacle outlet if the cord set is used for temporary electric power.

Note 2 to paragraph (b)(3)(ii)(A) of this section: Cord sets and devices incorporating the required ground-fault circuit-interrupter that are connected to the receptacle closest to the source of power are acceptable forms of protection.

A new and expanded section on ground fault circuit interrupter protection for personnel appears in 1910.304 (b)(3). Although OSHA attempted to model the new standard after NFPA 70E, 2000 Edition (based on the NEC®), OSHA slightly modified the verbiage and used a new term "construction-like activities". As all the revisions are contained in 29CFR Part 1910, they apply only to general industry and do not apply to any construction activities, as covered by 29CFR Part 1926, most specifically Subpart K. The modification of the NFPA 70E verbiage combined with the new phrase "construction-like activities", which OSHA failed to define, has created much confusion regarding OSHA's intent. A clarification notice was published in the Federal Register of 10/29/2008, Number 73:64202-64205. OSHA has commented that it was meant to include activities that pose hazards similar to those associated with temporary receptacles on construction sites; in other words, OSHA felt that there is usually an elevated hazard due to more severe use or environmental conditions often encountered by employees using temporary wiring. Examples of such activities might include damage to a cord set from rough use, exposure to wet, damp, or conductive conditions, such as from working in an outdoor environment, and frequent reconfiguration and rearrangement of the electric equipment.

Some examples of "construction-like activity" were given in the preamble to the final rule, including clean up and disaster remediation. For example, if a storm blew over a tree on a factory's premises and temporary wiring was employed to power a chainsaw and other clean-up equipment, such remediation activity would be "construction-like." Other examples of construction-like activities might include employees engaged in a minor building repair using a temporary power outlet unit; manufacturing prefabricated housing, in which houses or portions of houses are assembled in a manufacturing plant; performing heat exchanger tubing water-blasting (hydro-cleaning) using a temporary electrical power outlet unit, where conditions similar to construction sites might exist; setting up and taking down of concert or carnival events. Such work is not construction work (as defined in 1910.12 and 1926.32(g)), but typically involves the use of temporary wiring and conductive work locations such as those found on a construction site. GFCI protection can be provided in numerous ways, but is often easily provided through the use of a portable device such as those pictured in Figures 35A and B.

Figure 35(A) – Example of portable GFCI device.

Figure 35(B) – HD-Pro 480V portable GFCI by TRC.

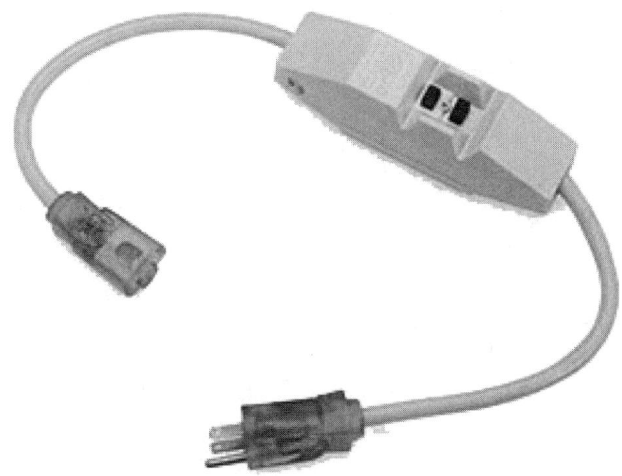

(b)(3)(ii)(B) Receptacles other than 125 volt, single-phase, 15-, 20-, and 30-ampere receptacles that are not part of the permanent wiring of the building or structure and that are in use by personnel shall have ground-fault circuit-interrupter protection for personnel.

(b)(3)(ii)(C) Where the ground-fault circuit-interrupter protection required by paragraph (b)(3)(ii)(B) of this section is not available for receptacles other than 125-volt, single-phase, 15-, 20-, and 30-ampere, the employer shall establish and implement an assured equipment grounding conductor program covering cord sets, receptacles that are not a part of the building or structure, and equipment connected by cord and plug that are available for use or used by employees on those receptacles. This program shall comply with the following requirements:

(b)(3)(ii)(C)(1) A written description of the program, including the specific procedures adopted by the employer, shall be available at the jobsite for inspection and copying by the Assistant Secretary of Labor and any affected employee;

(b)(3)(ii)(C)(2) The employer shall designate one or more competent persons to implement the program;

(b)(3)(ii)(C)(3) Each cord set, attachment cap, plug, and receptacle of cord sets, and any equipment connected by cord and plug, except cord sets and receptacles which are fixed and not exposed to damage, shall be visually inspected before each day's use for external defects, such as deformed or missing pins or insulation damage, and for indications of possible internal damage. Equipment found damaged or defective shall not be used until repaired;

(b)(3)(ii)(C)(4) The following tests shall be performed on all cord sets and receptacles which are not a part of the permanent wiring of the building or structure, and cord- and plug-connected equipment required to be grounded:

(b)(3)(ii)(C)(4)(i) All equipment grounding conductors shall be tested for continuity and shall be electrically continuous;

(b)(3)(ii)(C)(4)(ii) Each receptacle and attachment cap or plug shall be tested for correct attachment of the equipment grounding conductor. The equipment grounding conductor shall be connected to its proper terminal; and

(b)(3)(ii)(C)(4)(iii) All required tests shall be performed before first use; before equipment is returned to service following any repairs; before equipment is used after any incident which can be reasonably suspected to have caused damage (for example, when a cord set is run over); and at intervals not to exceed 3 months, except that cord sets and receptacles which are fixed and not exposed to damage shall be tested at intervals not exceeding 6 months;

(b)(3)(ii)(C)(5) The employer shall not make available or permit the use by employees of any equipment which has not met the requirements of paragraph (b)(3)(ii)(C) of this section; and

(b)(3)(ii)(C)(6) Tests performed as required in paragraph (b)(3)(ii)(C) of this section shall be recorded. This test record shall identify each receptacle, cord set, and cord- and plug-connected equipment that passed the test and shall indicate the last date it was tested or the interval for which it was tested. This record shall be kept by means of logs, color coding, or other effective means and shall be maintained until replaced by a more current record. The record shall be made available on the jobsite for inspection by the Assistant Secretary and any affected employee.

Where GFCI protection is not available for other than 125 volt, single-phase, 15-, 20-, and 30-ampere receptacles, an assured equipment grounding program is permitted to be used. A written program, testing and documentation are required to be maintained, and it is generally found that the cost and resources required outweight the convienence of using permanently installed or portable GFCIs, such as the 480V portable GFCI pictured in Figure 35B.

(b)(4) **Outlet devices.** Outlet devices shall have an ampere rating not less than the load to be served and shall comply with the following provisions:

(b)(4)(i) Where connected to a branch circuit having a rating in excess of 20 amperes, lampholders shall be of the heavy-duty type. A heavy-duty lampholder shall have a rating of not less than 660 watts if of the admedium type and not less than 750 watts if of any other type; and

1910.304

Title: Wiring design and protection.

Subpart: S

SubPart Title: Electrical - Design Safety Standards for Electrical Systems

The original intent of this requirement in the NEC® is to limit flourescent lighting branch circuits to 20 amperes, as most lampholders manufactured for use with flourescent fixtures are not rated as heavy-duty (660W or 750W, depending on the type).

(b)(4)(ii) Receptacle outlets shall comply with the following provisions:

(b)(4)(ii)(A) A single receptacle installed on an individual branch circuit shall have an ampere rating of not less than that of the branch circuit;

(b)(4)(ii)(B) Where connected to a branch circuit supplying two or more receptacles or outlets, a receptacle may not supply a total cord- and plug-connected load in excess of the maximum specified in Table S-4; and

(b)(4)(ii)(C) Where connected to a branch circuit supplying two or more receptacles or outlets, receptacle ratings shall conform to the values listed in Table S-5; or, where larger than 50 amperes, the receptacle rating may not be less than the branch-circuit rating. However, receptacles of cord- and plug-connected arc welders may have ampere ratings not less than the minimum branch-circuit conductor ampacity.

(b)(5) **Cord connections.** A receptacle outlet shall be installed wherever flexible cords with attachment plugs are used. Where flexible cords are permitted to be permanently connected, receptacles may be omitted.

Table S-4
Maximum Cord- and Plug-Connected Load to Receptacle

Circuit rating (amperes)	Receptacle rating (amperes)	Maximum load (amperes)
15 or 20	15	12
20	20	16
30	30	24

OSHA has always had the general outlet ampacity reference under 1910.304(b)(4) but has expanded the section to include Table S-4, specifying the maximum cord and plug connected load to receptacles and Table S-5, showing receptacle ratings for various circuits. These tables have been in NFPA 70E and the NEC® for some time.

Table S-5
Receptacle Ratings For Various Size Circuits

Circuit rating (amperes)	Receptacle rating (amperes)
15	Not over 15
20	15 or 20
30	30
40	40 or 50
50	50

NOTE: Only sections 1910.302-.308 were revised by OSHA in 2007. 1910.332 and 1910.333 already contained tables identified as Table S-4 and Table S-5. OSHA inadvertently numbered these new tables as S-4 and S-5. It is expected that a correction notice will renumber these tables at some point.

(c) **Outside conductors, 600 volts, nominal, or less.** The following requirements apply to branch circuit, feeder, and service conductors rated 600 volts, nominal, or less and run outdoors as open conductors.

Figure 36 – Open conductors installed on insulators on the surface of a building.

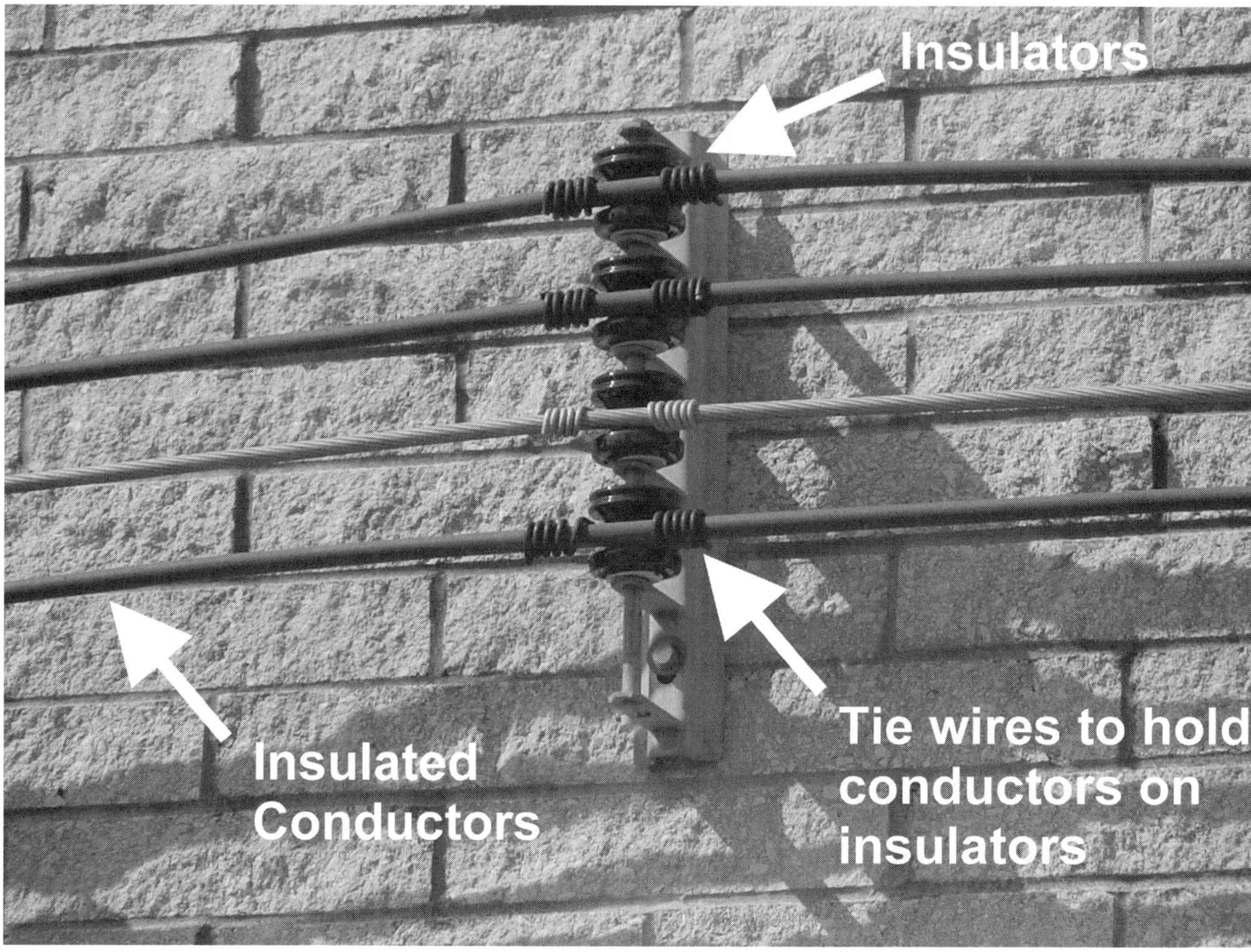

Outside Conductors.

Outside conductors run on buildings or structures must be installed in accordance with Article 225 of the National Electrical Code® which covers the requirements for Outside Branch Circuits and Feeders. Figure 36 illustrates open wiring on insulators, one of the wiring methods permitted by the National Electrical Code®. Open conductors are wires that are run as separate conductors in contrast to wires run through conduit, cables, or raceways. When run in this manner the installation must also meet the requirements of 1910.305(a)(4) and Article 398 of the National Electrical Code®, which covers additional requirements for Open Wiring on Insulators.

The following four paragraphs--(c)(1) conductors on poles, (c)(2) clearance from ground, (c)(3) clearance from building openings, and (c)(4) clearance over roofs--cover safety requirements regarding low voltage circuits (600 volts, nominal, or less) run outdoors as open conductors.

(c)(1)	**Conductors on poles.** Conductors on poles shall have a separation of not less than 305 mm (1.0 ft) where not placed on racks or brackets. Conductors supported on poles shall provide a horizontal climbing space not less than the following:
(c)(1)(i)	Power conductors below the communication conductors – 762 mm (30 in.);

(c)(1)(ii)	Power conductors alone or above communication conductors:
(c)(1)(ii)(A)	300 volts or less – 610 mm (24 in.),
(c)(1)(ii)(B)	Over 300 volts – 762 mm (30 in.);
(c)(1)(iii)	Communication conductors below power conductors – same as power conductors; and
(c)(1)(iv)	Communications conductors alone – no requirement.

Figure 37 – Minimum climbing spaces for power conductors located over communication conductors.

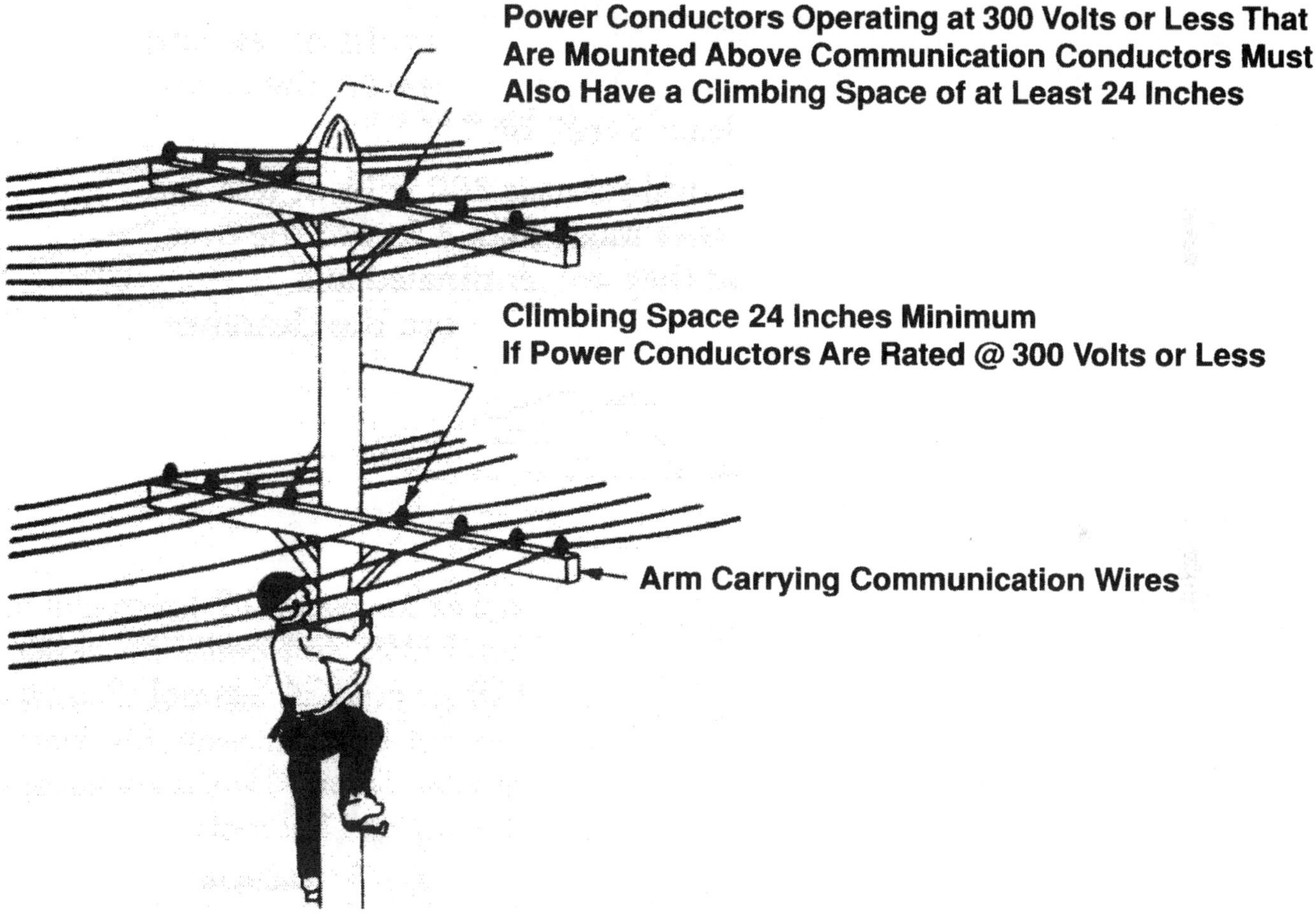

Conductors on Poles

Conductors on utility poles present a serious shock hazard to workers who are required to climb these poles. These conductors must be installed to allow a worker enough room to safely climb between them without making accidental contact with the conductor (see Figure 37). The minimum space that is allowed varies with the type conductor, the arrangement of conductors, and the voltage involved. Table 304-1 summarizes the minimum climbing spaces between conductors as described in the regulation.

(c)(2) **Clearance from ground.** Open conductors, open multiconductor cables and service-drop conductors of not over 600 volts, nominal, shall conform to the minimum clearances specified in Table S-6.

Table S-6
Clearances From Ground

Distance	Installations built before August 13, 2007		Installations built on or after August 13, 2007	
	Maximum voltage	Conditions	Voltage To ground	Conditions
3.05 m (10.0 ft)	< 600 V	Above finished grade or sidewalks, or from any platform or projection from which they might be reached. (If these areas are accessible to other than pedestrian traffic, then one of the other conditions applies).	< 150 V	Above finished grade or sidewalks, or from any platform or projection from which they might be reached. (If these areas are accessible to other than pedestrian traffic, then one of the other conditions applies).
3.66 m (12.0 ft)	< 600 V	Over areas, other than public streets, alleys, roads, and driveways, subject to vehicular traffic other than truck traffic.	< 300 V	Over residential property and driveways. Over commercial areas subject to pedestrian traffic or to vehicular traffic other than truck traffic. (This category includes conditions covered under the 3.05-m (10.0-ft) category where the voltage exceeds 150 V.)
4.57 m (15.0 ft)	< 600 V	Over areas, other than public streets, alleys, roads, and driveways, subject to truck traffic.	301 to 600 V	Over residential property and driveways. Over commercial areas subject to pedestrian traffic or to vehicular traffic other than truck traffic. (This category includes conditions covered under the 3.05-m (10.0-ft) category where the voltage exceeds 300 V.)
5.49 m (18.0 ft)	< 600 V	Over public streets, alleys, roads, and driveways.	< 600 V	Over public streets, alleys, roads, and driveways. Over commercial areas subject to truck traffic. Other land traversed by vehicles, including land used for cultivating or grazing and forests and orchards.

(c)(3) **Clearance from building openings.**

(c)(3)(i) Service conductors installed as open conductors or multiconductor cable without an overall outer jacket shall have a clearance of not less than 914 mm (3.0 ft) from windows that are designed to be opened, doors, porches, balconies, ladders, stairs, fire escapes, and similar locations. However, conductors that run above the top level of a window may be less than 914 mm (3.0 ft) from the window. Vertical clearance of final spans above, or within 914 mm (3.0 ft) measured horizontally of, platforms, projections, or surfaces from which they might be reached shall be maintained in accordance with paragraph (c)(2) of this section.

(c)(3)(ii) Overhead service conductors may not be installed beneath openings through which materials may be moved, such as openings in farm and commercial buildings, and may not be installed where they will obstruct entrance to these building openings.

(c)(4) **Above roofs.** Overhead spans of open conductors and open multiconductor cables shall have a vertical clearance of not less than 2.44 m (8.0 ft) above the roof surface. The vertical clearance above the roof level shall be maintained for a distance not less than 914 mm (3.0 ft) in all directions from the edge of the roof.

(c)(4)(i) The area above a roof surface subject to pedestrian or vehicular traffic shall have a vertical clearance from the roof surface in accordance with the clearance requirements of paragraph (c)(2) of this section.

(c)(4)(ii) A reduction in clearance to 914 mm (3.0 ft) is permitted where the voltage between conductors does not exceed 300 and the roof has a slope of 102 mm (4 in.) in 305 mm (12 in.) or greater.

(c)(4)(iii) A reduction in clearance above only the overhanging portion of the roof to not less than 457 mm (18 in.) is permitted where the voltage between conductors does not exceed 300 if:

(c)(4)(iii)(A) The conductors do not pass above the roof overhang for a distance of more than 1.83 m (6.0 ft), 1.22 m (4.0 ft) horizontally, and

(c)(4)(iii)(B) The conductors are terminated at a through-the-roof raceway or approved support.

(c)(4)(iv) The requirement for maintaining a vertical clearance of 914 mm (3.0 ft) from the edge of the roof does not apply to the final conductor span, where the conductors are attached to the side of a building.

Clearance Over Roofs.

Outside branch circuit, feeder, and service conductors which pass over roofs must be at least 8 feet above the highest outside point of the roof (See Figure 38).

Table 304-1 Minimum Climbing Spaces Between Low Voltage Conductors (600 volts or less)

Arrangement	Minimum Climbing Space Measured Horizontally	
	Between Power Conductors	**Between Communication Conductors**
Power conductors below communication conductors: 600 volts or less	30 inches	no requirement
Power conductors alone: 300 volts or less 301-600 volts	 24 inches 30 inches	 N/A N/A

Figure 38 – Required clearance between a roof and outside branch circuits, feeders, and service is 8 feet or more above a flat roof.

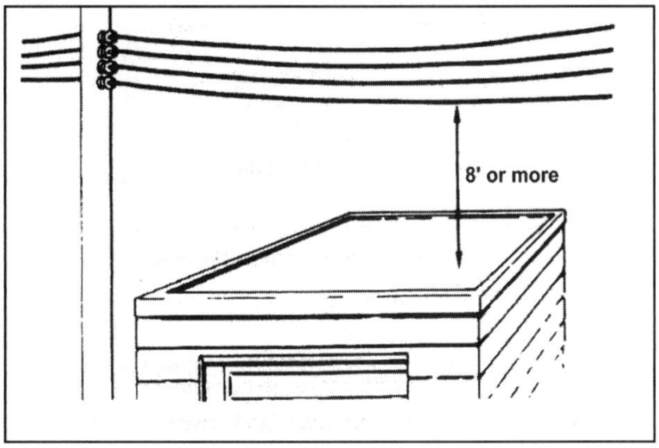

Figure 39 – Minimum distance of 3 feet from sloping roof to overhead wires.

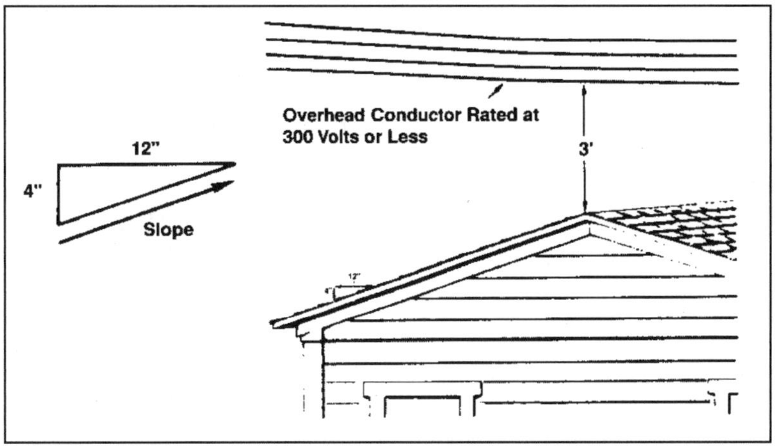

If the voltage between conductors is 300 volts or less and if the incline of the roof is at least 4 vertical inches for every 12 horizontal inches, these outside conductors can come no closer than 3 feet from the highest point of the roof (see Figure 39.) Employees are not likely to walk on roofs that have a greater pitch. Figures 40A and 40B show acceptable installation for sloping and flat roofs.

Figure 40A – Minimum clearance from roof to wires over slope roof overhang is shown.

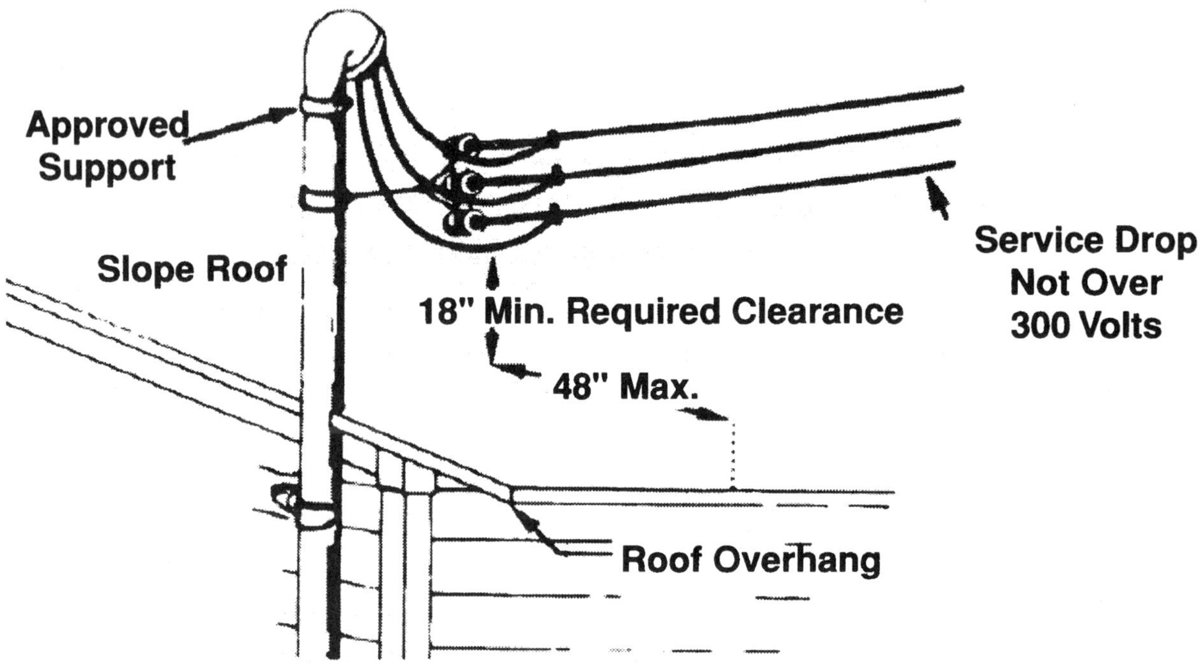

(d)	**Location of outdoor lamps.** Lamps for outdoor lighting shall be located below all live conductors, transformers, or other electric equipment, unless such equipment is controlled by a disconnecting means that can be locked in the open position or unless adequate clearances or other safeguards are provided for relamping operations.
(e)	**Services.**
(e)(1)	**Disconnecting means.**
(e)(1)(i)	Means shall be provided to disconnect all conductors in a building or other structure from the service-entrance conductors. The service disconnecting means shall plainly indicate whether it is in the open or closed position and shall be installed at a readily accessible location nearest the point of entrance of the service-entrance conductors.
(e)(1)(ii)	Each service disconnecting means shall simultaneously disconnect all ungrounded conductors.
(e)(1)(iii)	Each service disconnecting means shall be suitable for the prevailing conditions.

Service Disconnecting Means.

A readily accessible means of disconnecting conductors is required to be installed either outside a building or inside nearest the point of entrance of the service conductors, per the National Electrical Code®, Section 230.70(A)(1). The service point is the location where the utility distribution conductors connect to the premises wiring. The disconnecting means can be a switch or circuit breaker, and it must be capable of interrupting the current flow from the source of supply. This will disconnect the electrical equipment within the building from its source of supply in the event of an emergency or during normal servicing operations.

Three acceptable arrangements permitted by the National Electrical Code® for service-entrance disconnects are shown in Figure 41. The first is a single disconnect that opens all ungrounded conductors simultaneously. The second is a circuit breaker panel with up to six circuit breakers that will disconnect all ungrounded conductors to a building with no more than six movements of the hand. The third is a group of up to six individual switches located near where the service enters the building to disconnect all ungrounded conductors that supply a building. When more than one switch or circuit breaker is used, each must show whether it is in the on or off position and must simultaneously open all ungrounded conductors of the circuit it is intended to open.

Figure 40B – Minimum clearance from roof to wires over flat roof overhang.

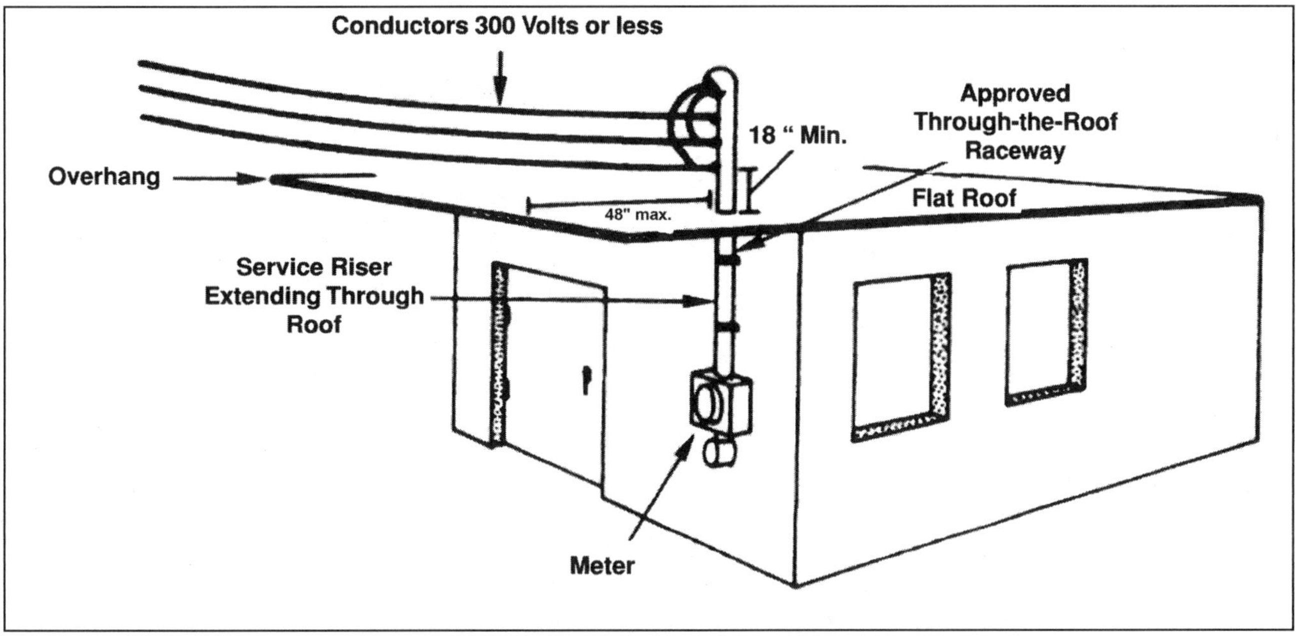

(e)(2)	**Services over 600 volts, nominal. The following additional requirements apply to services over 600 volts, nominal.**
(e)(2)(i)	Service-entrance conductors installed as open wires shall be guarded to make them accessible only to qualified persons.
(e)(2)(ii)	Signs warning of high voltage shall be posted where unqualified employees might come in contact with live parts.
(f)	**Overcurrent protection.**
(f)(1)	**600 volts, nominal, or less.** The following requirements apply to overcurrent protection of circuits rated 600 volts, nominal, or less.
(f)(1)(i)	Conductors and equipment shall be protected from overcurrent in accordance with their ability to safely conduct current.

1910.304(f)(1)(i) is a good example of how the "performance-oriented" standards of Subpart S are applied in the field. If OSHA Compliance Officers enforced this standard literally, ("Conductors and equipment shall be protected from overcurrent. . . ."), they could shut down the world! It is essential to understand that Subpart S states only the basic safety performance required. The National Electrical Code® describes recognized methods of providing such performance, including many justified exceptions.

Figure 41 – Three acceptable arrangements for service-entrance disconnecting means. (These arrangements are specified in the NEC®, with certain qualifying conditions.)

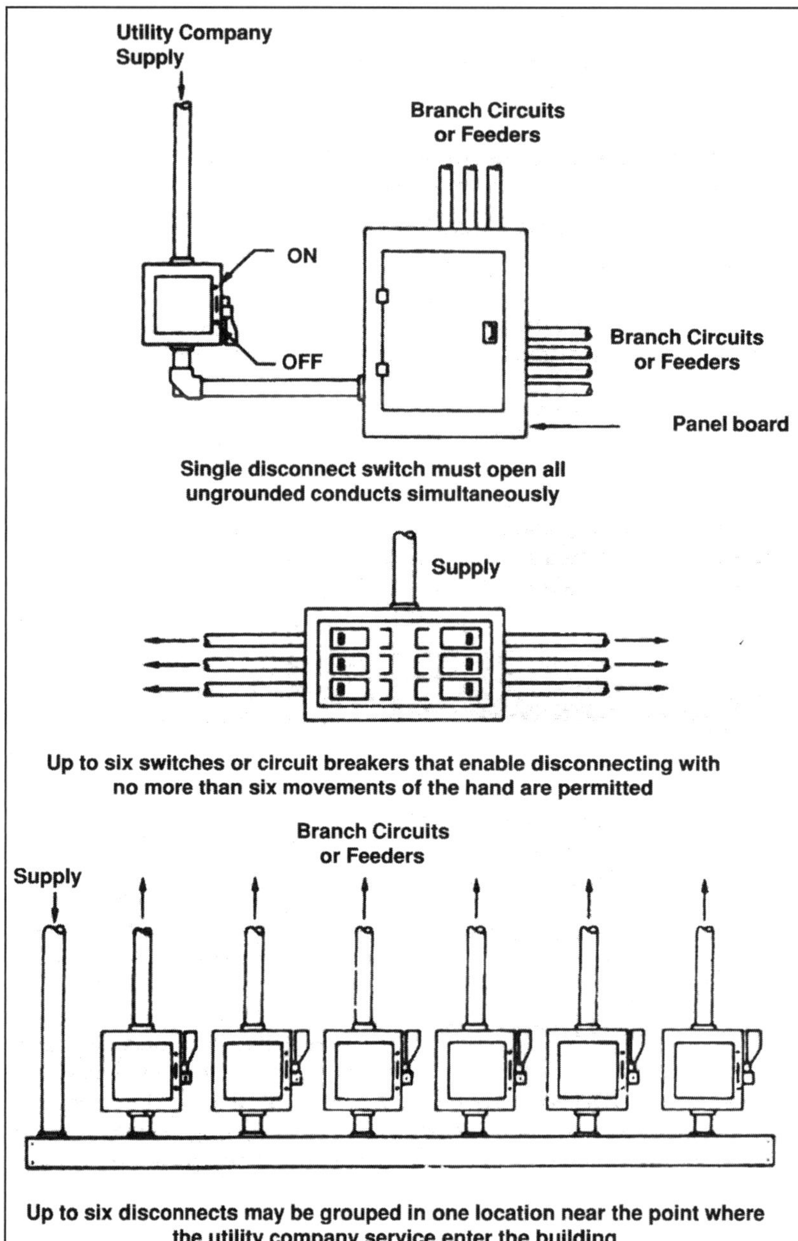

Protection of Conductors and Equipment

Electrical current is the flow of electrons through a conductor. The larger the wire, the more current can flow safely. If too much current flows through a conductor, excess heat is produced. If the circuit is not properly protected, the heat may continue to build and produce a temperature high enough to destroy the insulation and produce a fault and ultimately a fire.

Conductors and equipment are required to be protected from overcurrent conditions according to their ability to safely conduct electric current. Circuit breakers and fuses are protective devices designed to disconnect a circuit from its source of supply when a maximum allowable heat level is reached. A fuse functions to disconnect a circuit when an element inside the fuse body, a metal ribbon or link, melts from the heat produced when too much current passes through it. When this ribbon melts, the current flow through the circuit is stopped (See Figure 42).

A circuit breaker functions to disconnect a circuit when excessive flow causes a heat buildup in a bi-metal strip that acts as a temperature sensing device within the circuit breaker. When excessive current is sensed, the bi-metal strip moves and releases a spring-loaded switch that opens the circuit, thus stopping the current flow. (See Figure 43).

Either of these methods of circuit protection is acceptable. Circuit breakers can be reset when the handle is moved from the "trip" position to the "off" position and then moved to the "on" position to restore power to the circuit. By contrast, fuses must be replaced, thus requiring a supply of replacement fuses to be on hand. In cases of high voltage (over 600 volts), replacement of a fuse must be done only by qualified persons.

The National Electrical Code® specifies the allowable current flow permitted in certain-sized conductors. Ampacity is the term used to describe the current-carrying capacity of a conductor. The size of the circuit breaker or fuse required to provide protection is determined by the ampacity of the conductor in the circuit to be protected and the type of load that is in the circuit.

(f)(1)(ii) Except for motor running overload protection, overcurrent devices may not interrupt the continuity of the grounded conductor unless all conductors of the circuit are opened simultaneously.

(f)(1)(iii) A disconnecting means shall be provided on the supply side of all fuses in circuits over 150 volts to ground and cartridge fuses in circuits of any voltage where accessible to other than qualified persons so that each individual circuit containing fuses can be independently disconnected from the source of power. However, a current-limiting device without a disconnecting means is permitted on the supply side of the service disconnecting means. In addition, a single disconnecting means is permitted on the supply side of more than one set of fuses as permitted by the exception in § 1910.305(j)(4)(vi) for group operation of motors, and a single disconnecting means is permitted for fixed electric space-heating equipment.

Figure 42 – Common types of fuses: socket type (top) and cartridge.

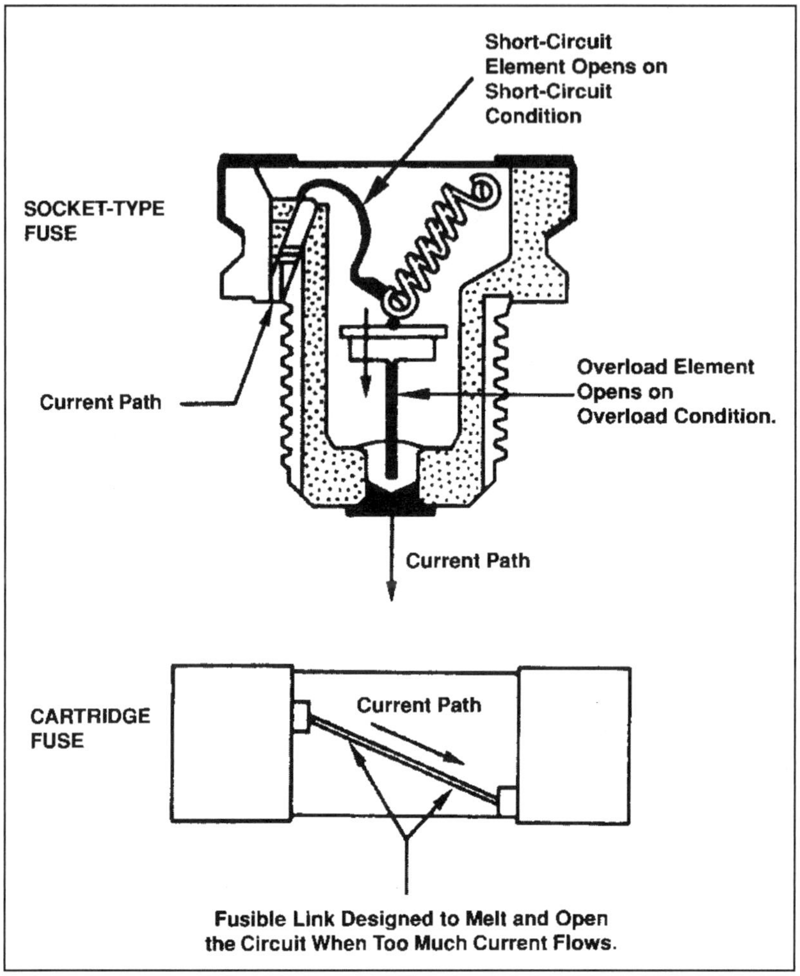

Disconnection of Fuses and Thermal Cutouts

All cartridge-type circuit fuses that are accessible to unqualified employees must be equipped with a disconnecting means to allow the fuse to be serviced safely. In addition, where cartridge-type fuses are located in systems that operate over 150 volts to ground, a disconnecting means is always required regardless of the qualifications of the persons having access to the fuses. The disconnect switches must be installed so that only the circuits that are protected by or otherwise related to the overcurrent device will be shut off. In this way power to other circuits and equipment in a system will not be unnecessarily disrupted when one overcurrent device in the system must be serviced or replaced. For example in Figure 44 if process motor No. 3 developed an overcurrent condition and blew a fuse, then disconnect No. 3 would be the only switch actuated to service the fuse. These provisions do not apply to service fuses.

Figure 43 – Cutaway view of a circuit breaker.

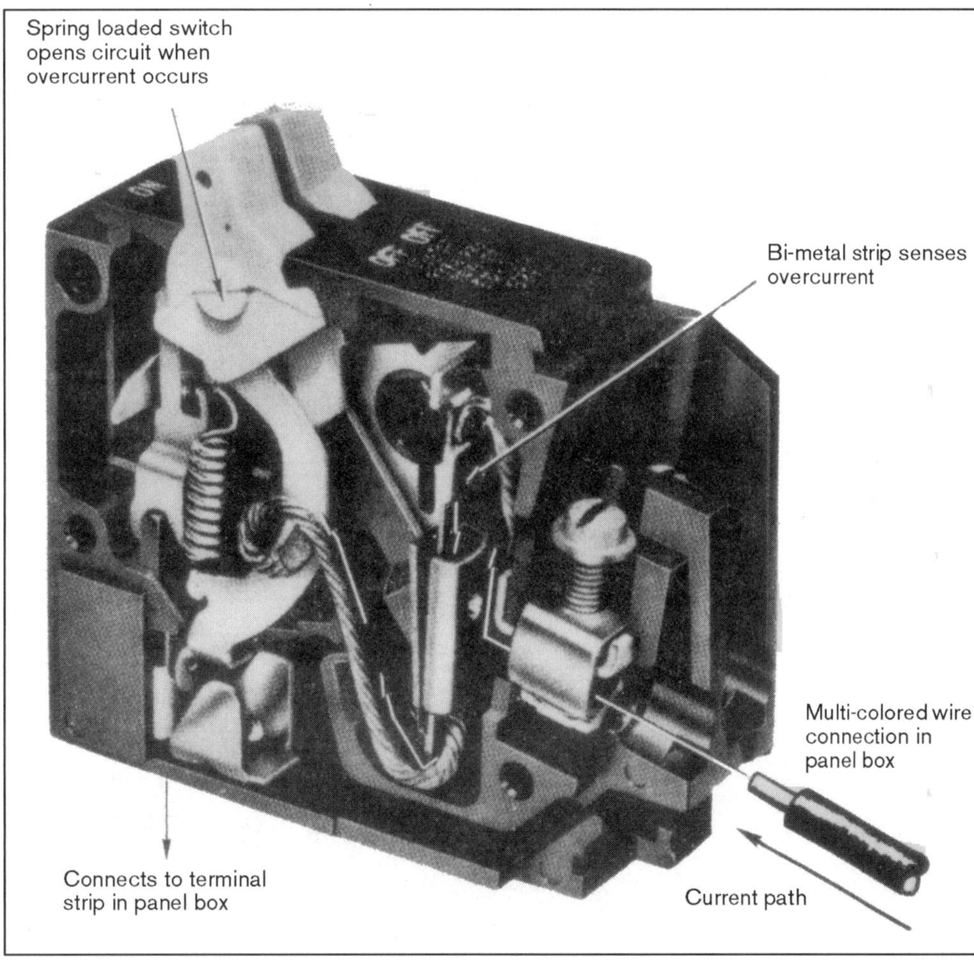

Figure 44 – Disconnects for overcurrent devices. This is a "one-line diagram," a form of shorthand that is helpful in visualizing electrical systems. Each single line (and each fuse) represents the total number of wires and fuses (1, 2 or 3) that circuit requires. In the diagram, each of the fuses shown might represent the three fuses of a three-phase, three-wire system

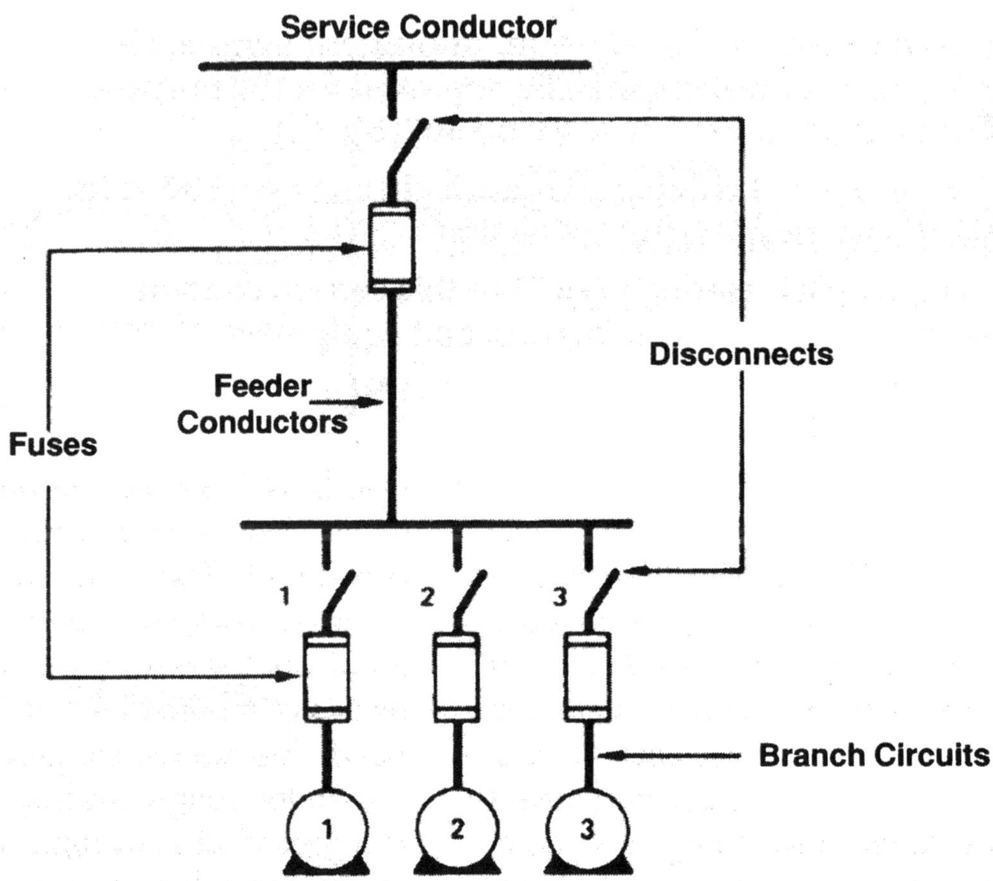

(f)(1)(iv) Overcurrent devices shall be readily accessible to each employee or authorized building management personnel. These overcurrent devices may not be located where they will be exposed to physical damage nor in the vicinity of easily ignitable material.

Overcurrent devices need only be readily accessible to "authorized and qualified" employees. In some cases it is necessary for these devices to be in locked rooms or enclosures to ensure that only authorized and qualified employees will operate them. As long as the authorized and qualified employee has access to the keys, the devices would be considered not only accessible but readily accessible. This is further emphasized by 110.26(G) of the NEC®, added for 2008, which states, "Electrical equipment rooms or enclosures housing electrical apparatus that are controlled by a lock(s) shall be considered **accessible** to qualified persons."

(f)(1)(v) Fuses and circuit breakers shall be so located or shielded that employees will not be burned or otherwise injured by their operation. Handles or levers of circuit breakers, and similar parts that may move suddenly in such a way that persons in the vicinity are likely to be injured by being struck by them, shall be guarded or isolated.

(f)(1)(vi) Circuit breakers shall clearly indicate whether they are in the open (off) or closed (on) position.

(f)(1)(vii) Where circuit breaker handles on switchboards are operated vertically rather than horizontally or rotationally, the up position of the handle shall be closed (on) position.

(f)(1)(viii) Circuit breakers used as switches in 120-volt and 277-volt, fluorescent lighting circuits shall be listed and marked "SWD."

Circuit breakers used as switches for fluorescent lighting, such as the one pictured in Figure 45, are required to be marked "SWD" as a minimum. If circuit breakers are used as switches for high-intensity discharge lighting, they must also be marked "HID". Circuit breakers marked as "HID" are also rated "SWD". The circuit breaker pictured here is marked with both.

(f)(1)(ix) A circuit breaker with a straight voltage rating, such as 240 V or 480 V, may only be installed in a circuit in which the nominal voltage between any two conductors does not exceed the circuit breaker's voltage rating. A two-pole circuit breaker may not be used for protecting a 3-phase, corner-grounded delta circuit unless the circuit breaker is marked 1Φ-3Φ to indicate such suitability. A circuit breaker with a slash rating, such as 120/240 V or 480Y/277 V, may only be installed in a circuit where the nominal voltage of any conductor to ground does not exceed the lower of the two values of the circuit breaker's voltage rating and the nominal voltage between any two conductors does not exceed the higher value of the circuit breaker's voltage rating.

Markings such as 480Y/277V and 120/240V indicate that the circuit breakers have been tested for use on circuits with higher phase-to-phase voltage and lower phase-to-ground voltage. A circuit breaker labeled 480Y/277V could not be used on a system with 480-volts to ground, but rather would require a circuit breaker marked either 480V or 600V.

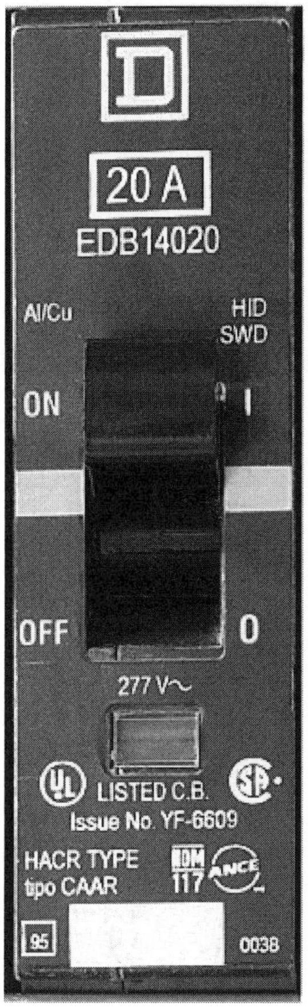

Figure 45 – SWD & HID markings on typical circuit breaker.

(f)(2)	**Feeders and branch circuits over 600 volts, nominal.** The following requirements apply to feeders and branch circuits energized at more than 600 volts, nominal:
(f)(2)(i)	Feeder and branch-circuit conductors shall have overcurrent protection in each ungrounded conductor located at the point where the conductor receives its supply or at a location in the circuit determined under engineering supervision;
(f)(2)(i)(A)	Circuit breakers used for overcurrent protection of three-phase circuits shall have a minimum of three overcurrent relays operated from three current transformers. On three-phase, three-wire circuits, an overcurrent relay in the residual circuit of the current transformers may replace one of the phase relays. An overcurrent relay, operated from a current transformer that links all phases of a three-phase, three-wire circuit, may replace the residual relay and one other phase-conductor current transformer. Where the neutral is not grounded on the load side of the circuit, the current transformer may link all three phase conductors and the grounded circuit conductor (neutral); and
(f)(2)(i)(B)	If fuses are used for overcurrent protection, a fuse shall be connected in series with each ungrounded conductor.
(f)(2)(ii)	Each protective device shall be capable of detecting and interrupting all values of current that can occur at its location in excess of its trip setting or melting point;
(f)(2)(iii)	The operating time of the protective device, the available short-circuit current, and the conductor used shall be coordinated to prevent damaging or dangerous temperatures in conductors or conductor insulation under short-circuit conditions; and
(f)(2)(iv)	The following additional requirements apply to feeders only:
(f)(2)(iv)(A)	The continuous ampere rating of a fuse may not exceed three times the ampacity of the conductors. The longtime trip element setting of a breaker or the minimum trip setting of an electronically actuated fuse may not exceed six times the ampacity of the conductor. For fire pumps, conductors may be protected for short circuit only; and
(f)(2)(iv)(B)	Conductors tapped to a feeder may be protected by the feeder overcurrent device where that overcurrent device also protects the tap conductor.
(g)	**Grounding**. Paragraphs (g)(1) through (g)(9) of this section contain grounding requirements for systems, circuits, and equipment.

Grounding

The subject of *grounding* has been, and continues to be, confusing. Neither the National Electrical Code® nor OSHA Subpart S was designed as an educational tool. The mystery and misunderstandings can be cleared up with just a little time invested in a review of a few fundamentals which govern how electricity does things—good or bad. These revealing fundamentals can be expressed in three simple and rather obvious statements:

1. Nothing happens unless current flows.
2. No current flows unless it finds a complete loop, from a source of voltage back to the same source.
3. Although current will flow through all possible loops, the only low impedance loop is one in which the outgoing and returning conductors are close together.

The classic analogy to illustrate electrical action has been that of water flowing through a pipe. Nothing happens in a water system either unless water flows through the pipe. But, the water analogy creates some false impressions, because the student visualizes a faucet turned on, and water flowing into the ground. Hence he or she may assume that electricity "seeks" the ground, and that equipment can be made electrically safe if we connect all of the enclosures to earth. This is not so, and this assumption has resulted in many casualties.

Figures 46-1A through 46-1D use a "marble analogy" that is a true and easily-visualized illustration of what happens. If the marbles obey the same rules that the electrons do in the circuit, they will be weightless, and no marble will move unless there is another coming along to take its place. The paddle wheel (generator) can drive marbles around the loop only as long as marbles are returning to the paddle wheel from the loop. Likewise, the voltage on the winding of a generator or transformer can only cause electrons (current) to flow if electrons are returning to the winding from the loop. Note that friction in the marble analogy acts like resistance in the electrical loop.

The third fundamental, one most frequently overlooked, is illustrated in Figures 46-2A through 46-2C, which shows how the magnetic reactance of an AC loop can be reduced by installing the outgoing and returning conductors close together. All electricians are aware that they must keep "L1" and "L2" close together (usually in the same conduit), but many of them do not realize that the ground-fault path must obey this same rule. The tendency to overlook this important effect in AC systems is surprising, because it represents the application to AC circuits of the basic electrical formula, Ohm's Law (see Table 304-2).

Figure 46-1A – No current flows unless there is a complete loop. The generator cannot force electrons to move through the tube unless they are returning to the generator.

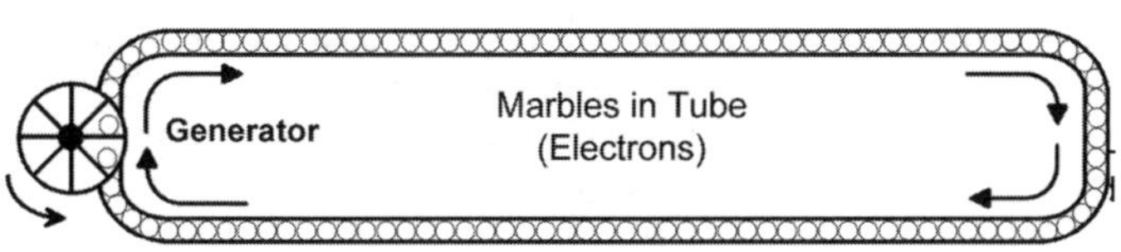

Figure 46-1B – If the loop (circuit) is not complete no electrons can return to the generator; therefore, electrons cease moving through the conductor.

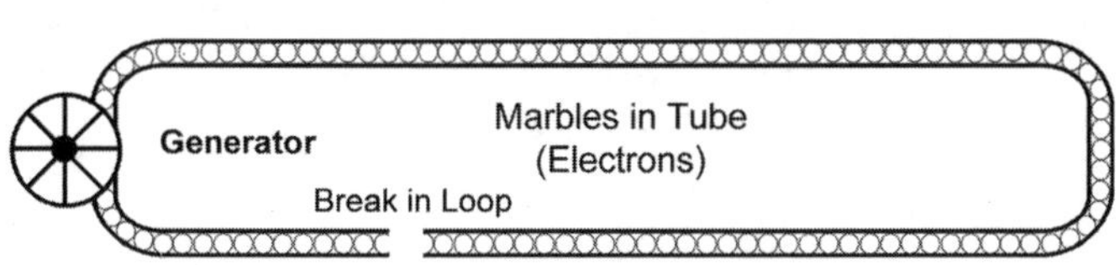

Figure 46-1C – Friction (resistance) inside of a conductor produces heat. The more electrons per second, the more heat will be produced.

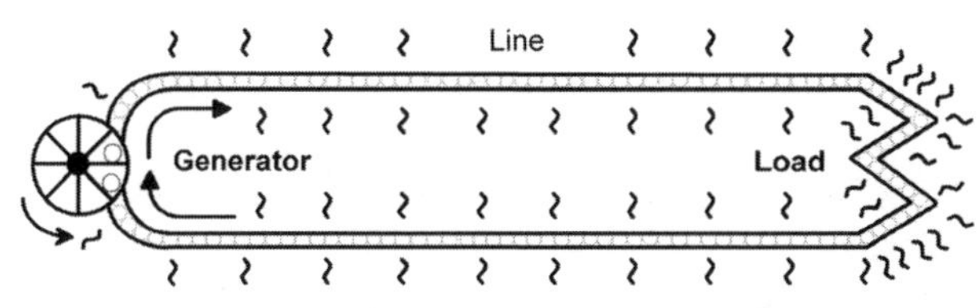

Figure 46-1D – A well-designed system has minimum resistance in lines, so that maximum effort is available to produce useful heat or work in the load

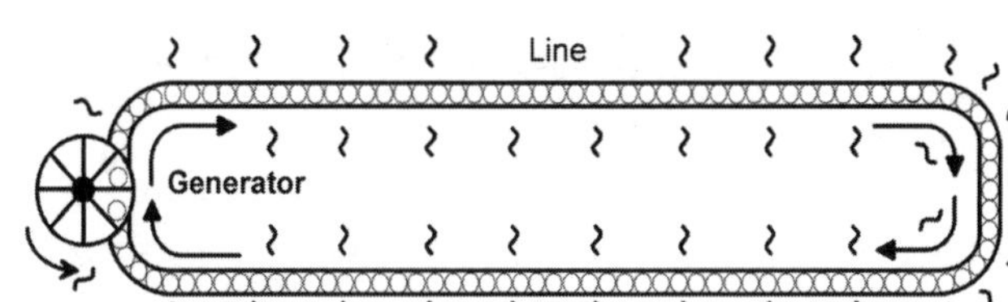

1910.304

Title: Wiring design and protection.

Subpart: S

SubPart Title: Electrical - Design Safety Standards for Electrical Systems

Table 304-2 Basic Electrical Fundamentals

FUNCTION	TERM	SYMBOL	UNIT OF MEASURE	ABBREVIATION
Force	Voltage	E	Volt	V
Result of Force	Current	I	Ampere	"A", "Amp"
Resists Current Flow Due to Physical Properties	Resistance	R	Ohm	Ω
Resists Current Flow Due to Magnetic Effect	Reactance	X_L	Ohm	Ω
Total Opposition to Current Flow in AC Systems	Impedance	Z	Ohm	Ω
Ohm's Law for ac Circuits, $I = E/Z$, $E = IZ$, $Z = E/I$				

The National Electrical Code® (henceforth referred to as the Code®) recognizes these fundamentals in Section 250.4(A)(5), which indicates that the effective ground fault path shall be permanent and continuous and shall have impedance sufficiently low to limit voltages between enclosures and to their surroundings, and to facilitate the operation of the circuit protective devices (fuses, etc.).

Figure 46-2A – When current flows through a conductor, a magnetic field is forced outside of the conductor. The current direction determines the direction of the magnetic field.

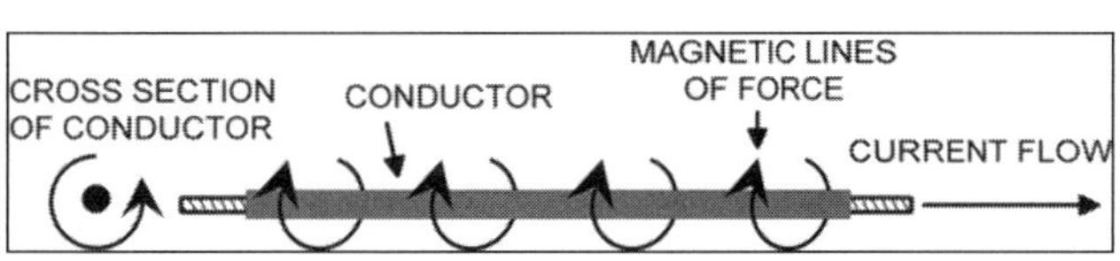

Figure 46-2B—In a conventional circuit, the current flowing in the returning line produces a field which equal and opposite to the field produced by the outing line.

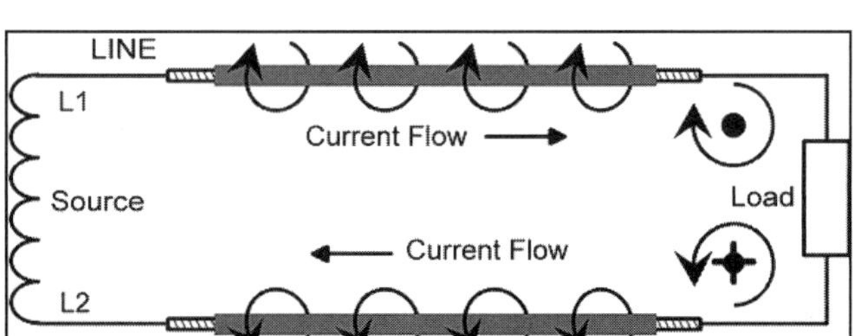

Figure 46-2C – If the outgoing and returning conductors of a circuit (loop) are close together, the two magnetic fields cancel each other. The magnetic "reactance" is then close to zero for that portion of the loop.

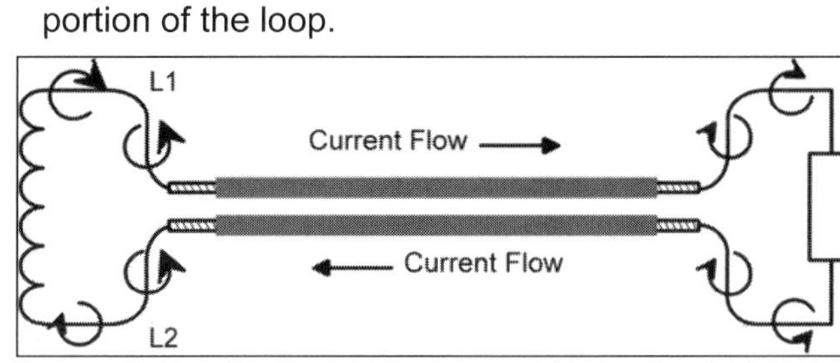

If we can predict we can control

Using these fundamentals, you can predict what will happen in a normal circuit, and also in a ground-fault circuit. However, it is first necessary to understand the language. Much of the confusion is created because the Code® uses the word "grounding" to mean three different things (see Table 304-3). The most important of the three grounding functions can be clarified if we think of it as "equipment bonding", which means connecting all electrical enclosures together.

When using the Code® or discussing code matters, we usually have to use Code® terminology, but in our own minds and in explaining the grounding functions to others, we can eliminate most of the confusion by using the less ambiguous terms suggested here. During our presentations of electrical safety seminars for inspectors and others, we have been told many times that this approach is of great value in design, installation, and inspection procedures. We are certain that it is helping to avoid ineffective grounding and arguments between inspectors and installers.

Why Grounding is Necessary

The normal functioning of an electrical circuit does not require a grounding or bonding conductor. The complete loop as shown in Figure 46-3 will carry current from the source, through the load and back to the source. Only when something goes wrong does the function of the grounding or bonding conductor enter the picture. The old approach of connecting the housing of the tool to a water pipe was only a partial solution for reasons which will be obvious as we proceed. The proper approach indicated by the diagram is to provide an assured low-impedance bonding path from the tool enclosure, bonding it to all other enclosures and eventually back to the main bonding jumper at the service entrance. This conductor is referred to in the Code® as the **equipment grounding conductor**. Its function is more clearly expressed if we refer to it as the equipment bonding conductor, inasmuch as connection to "ground" is not involved in the fault-clearing function of this conductor.

Figure 46-3 – Grounding (bonding) conductors must be carried with the supply conductors (NEC®250.134(B)).

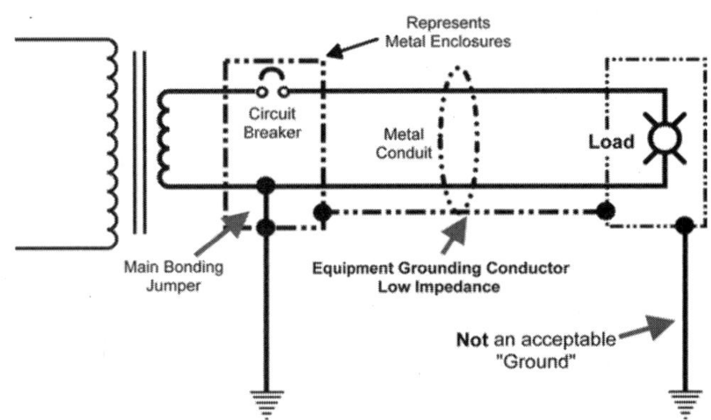

Figure 46-4 shows voltage and impedance values quite likely to be found in a real installation. Note that a good raceway or equipment bonding conductor is likely to have impedance (Z) of about 1/4 ohm, which will pass 240 amps of ground-fault current and instantly trip the circuit breaker. By contrast, a poor ground-fault path might have an impedance of, say, 5¾ ohms, which would not pass enough current to trip the 20 amp circuit breaker. Then the fault current would continue to flow indefinitely, causing a lethal voltage (E_G) of 115 volts to appear between the tool and ground. Because of the fact that accidents are still occurring due to improper grounding, OSHA and NEC® are increasingly requiring the use of Ground Fault Circuit Interrupters (GFCIs), which protect people even if the grounding fails to do so.

It is estimated that at least 95% of all hazards of this type would be eliminated if the equipment bonding conductor always provided a low-impedance ground-fault path. This assurance can be achieved only by proper installation, good maintenance, and routine testing of the ground loop impedance. There are instruments on the market for conveniently measuring this impedance.

Given the fact that the only low-impedance ground fault path is one which closely follows the supply conductors, it is not surprising that the Code® requires the equipment grounding conductor to be either a bonding conductor run along with the circuit conductors or a raceway containing the circuit conductors. The Code® expressly states that the structural metal frame of a building **shall not** be used as the required equipment-grounding conductor for AC equipment. There is a story in the trade about the homeowner who grounded his washing machine to a flower pot. But it's not funny when in reality some large installations depend on building steel as a ground-fault path—the flower pot would probably be about as effective. Again, the critical question is, "Have we provided a ground-fault path which will pass enough current to immediately operate the overcurrent protection device?" If not we are in violation of NEC® sections such as 250.4(A), 250.134, 250.136 and 250.138, and of common sense.

Table 304-3 – Grounding Terminology

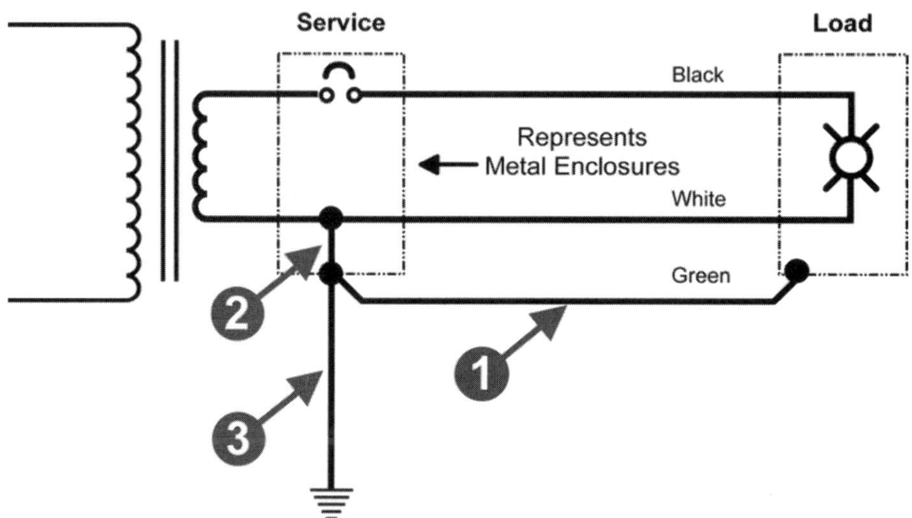

	CODE LANGUAGE	
	CONDUCTOR	FUNCTION
1	Equipment Grounding Conductor	Equipment Grounding
2	Main Bonding Jumper	System Grounding
3	Grounding Electrode Conductor	Grounding
	CLARIFICATION	
1	Equipment BONDING Conductor	Connects Enclosures TOGETHER (Not necessarily to ground)
2	System BONDING Jumper	Connects SYSTEM Conductor to Equipment Bonding Conductor
3	Earthing Conductor	Connects Equipment and the System to their SURROUNDINGS (Earth, Building Steel, Water Pipe).

Figure 46-4 – Ground impedance must be low.

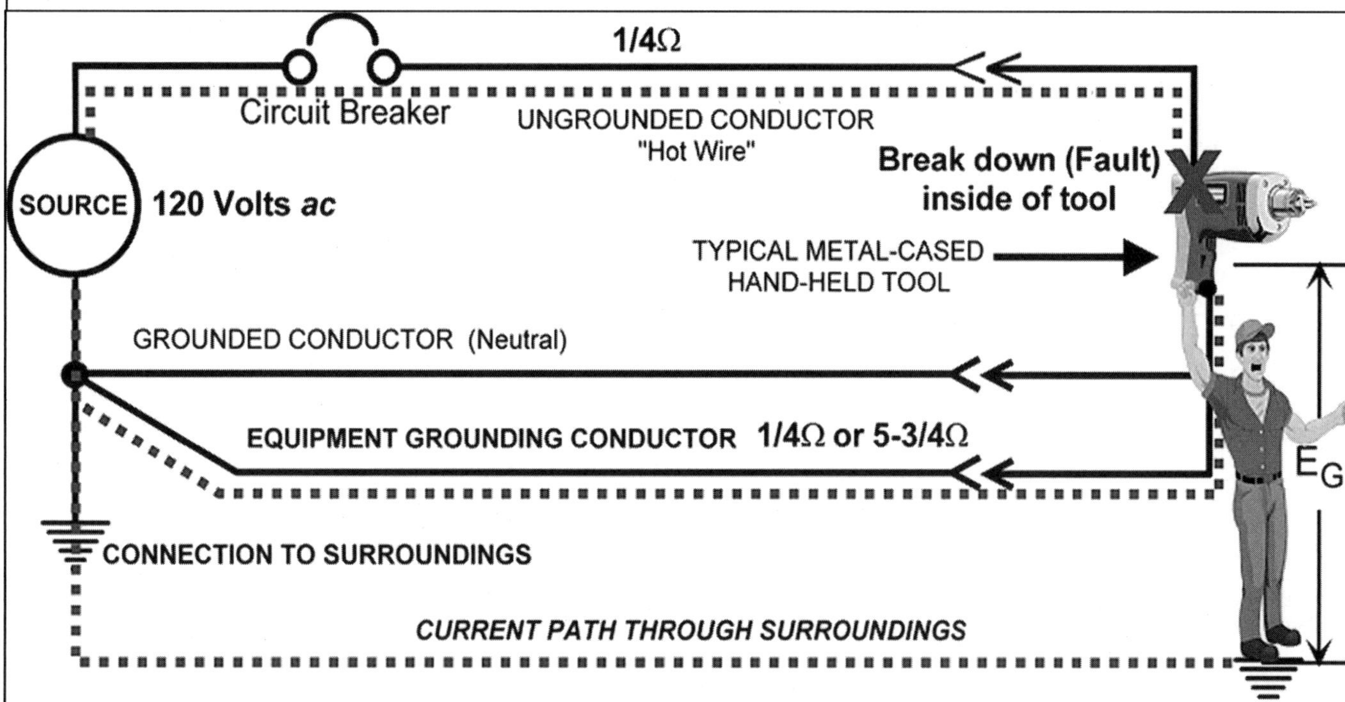

<u>If ground loop impedance is not low, the overcurrent protection device will not operate and dangerous voltages will exist.</u>

If the ground fault path is well maintained, then:

and

$I_{GF} = E/Z = 120v/.5\ \Omega = 240$ **amps**

$Z_{GF} = \frac{1}{4} + \frac{1}{4} = \frac{1}{2}\ \Omega$

Therefore circuit breaker <u>will trip instantaneously</u>.

But, if the ground fault path is not maintained, then:

$Z_{GF} = \frac{1}{4}\ \Omega + 5\frac{3}{4}\ \Omega = 6\ \Omega$

and

$I_{GF} = E/Z = 120v/6\ \Omega = 20$ **amps**

Therefore circuit breaker <u>will not trip</u>.

Also: There will be a voltage to ground (E_G) on the tool
$IZ = 20A \times 5\text{-}3/4\ \Omega = 115$ Volts

The grounding electrode conductor

If proper equipment bonding eliminates 95% of the hazards, what more can be done? We have established that any fault current from the transformer can only follow a loop which extends from one terminal to another terminal of the same transformer. Proper equipment grounding provides this loop and almost completely eliminates the dangers of ground-fault voltages or currents from this source. But in most installations, there is a remote possibility of voltages from some source other than the transformer secondary, in which case an additional precaution is necessary. Such "intruding" voltages will attempt to follow some unpredictable paths over which we have no control. Therefore, all we can do is to connect the entire electrical installation to its surroundings so that there can be no voltage between electrical enclosures and other grounded surfaces. This connection—to earth or some conductive body which serves in place of earth—is called the "grounding electrode conductor". (See Figures 46-5A and 46-5B.)

Don't confuse equipment grounding with system grounding

The electrical system consists of the supply source (usually a transformer) plus the actual conductors which carry current to and from all of the loads. In our discussion of grounding we have used systems with a "main bonding jumper" which connects one of the current-carrying conductors of the system to the electrical enclosures and to earth. These are defined as "grounded systems" and are the most common types of systems. They are specifically required for circuits supplying lighting and receptacles.

Certain systems are permitted to operate without a main bonding jumper, meaning that their electrical current-carrying conductors have no intentional connection to ground. Such systems are referred to as "ungrounded systems" and are frequently found in older occupancies. Note that all of the requirements for equipment grounding or bonding remain the same, whether or not the electric system is of the grounded type. The only difference is that the main bonding jumper is not used in an ungrounded system. The ungrounded system has advantages under certain circumstances, but it also has some disadvantages. One disadvantage is that the absence of the main bonding jumper means that the first ground fault will not clear, but will simply convert the system into a grounded system. Another disadvantage is that possible exposure to high voltages from some foreign source, such as transformer primary voltages, lightning, line surges, etc., may produce high voltages to ground on the ungrounded systems (see Figure 46-6A and 46-6B).

How do ungrounded systems clear faults?

Note that in Figure 46-7B, the first fault on an ungrounded system does not cause the operation of an overcurrent device, but if a second fault occurs on a different phase, fault current will flow between the two faults, thus causing operation of one or more fault-current protective devices to clear one of the faults. Also note that the ground-fault current which flows under these circumstances may have to follow a considerably longer path (in order to complete the loop) than would be the case if a main bonding jumper completed the loop (as in Figure 46-7A). For these reasons, it is even more important that the equipment grounding path be properly installed and maintained if the electrical system is to be operated ungrounded.

These explanations may require some study and thought by persons who have not "lived with" grounding problems for a number of years. But once the fundamentals are understood, much of the mystery about all aspects of grounding disappears. The effort is usually quite rewarding and will admit you to the somewhat exclusive company of those who really know what you have to do (and what you don't have to do) to provide successful protection from ground-fault hazards.

Figure 46-5A – The equipment grounding conductor protects against hazards of voltage between enclosures.

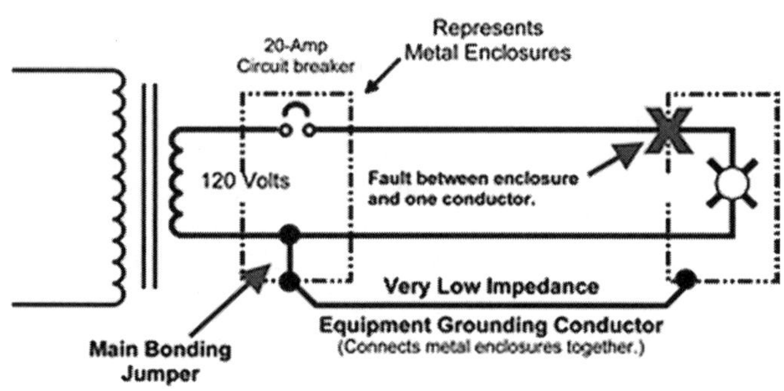

Figure 46-5B – The grounding-electrode conductor protects against hazards of voltage between enclosures and their surroundings.

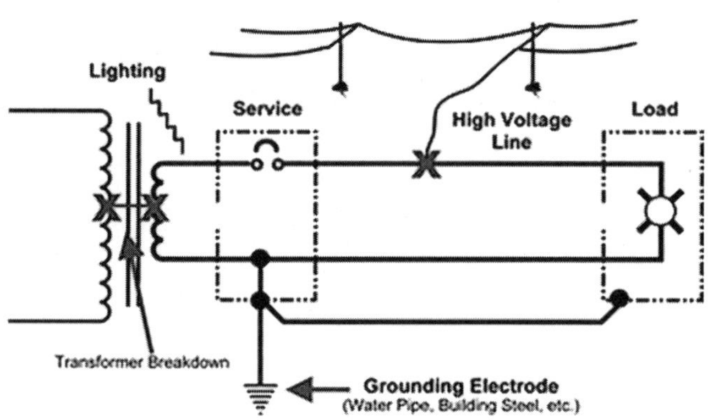

Figure 46-6A – A grounded system has an intentional connection from a current-carrying conductor to the equipment grounding conductor and to the grounding electrode.

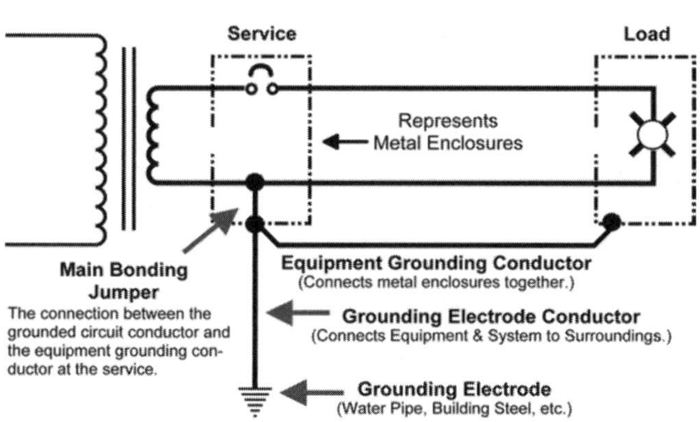

Figure 46-6B – An ungrounded *system has no main bonding jumper, but the equipment grounding conductors are required exactly as in the grounded system.*

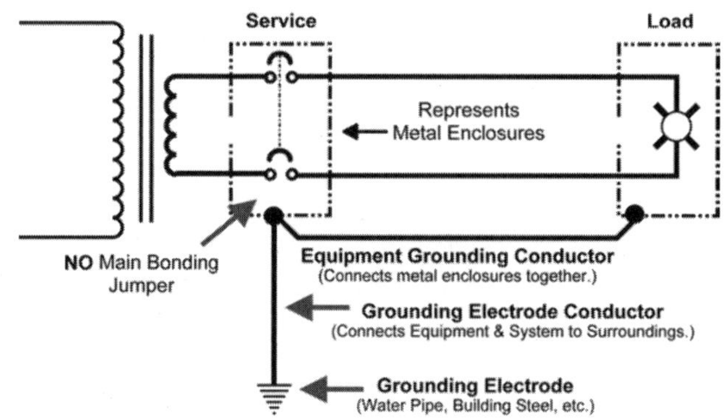

Figure 46-7A – In a ***grounded system*** each fault clears immediately, over a predictable path to the source.

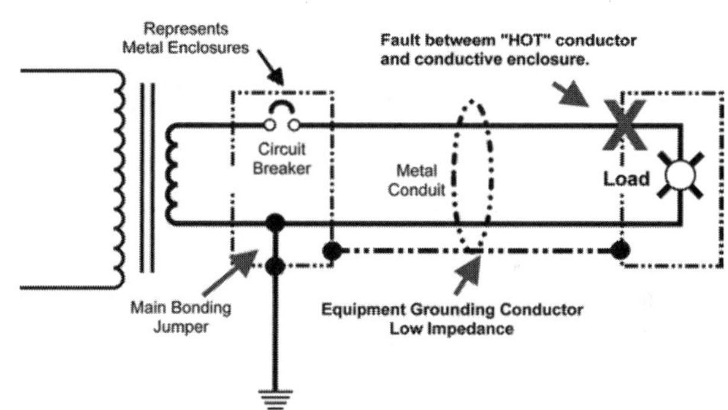

Figure 46-7B – In an ***ungrounded system*** fault current flows between faults over unpredictable paths.

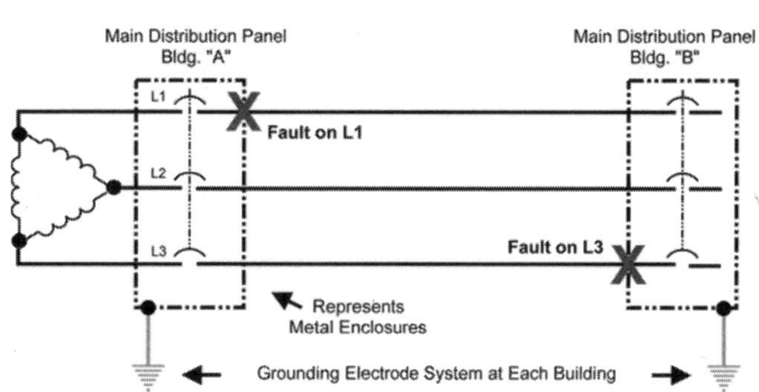

Figure 46-8 – System and equipment grounding

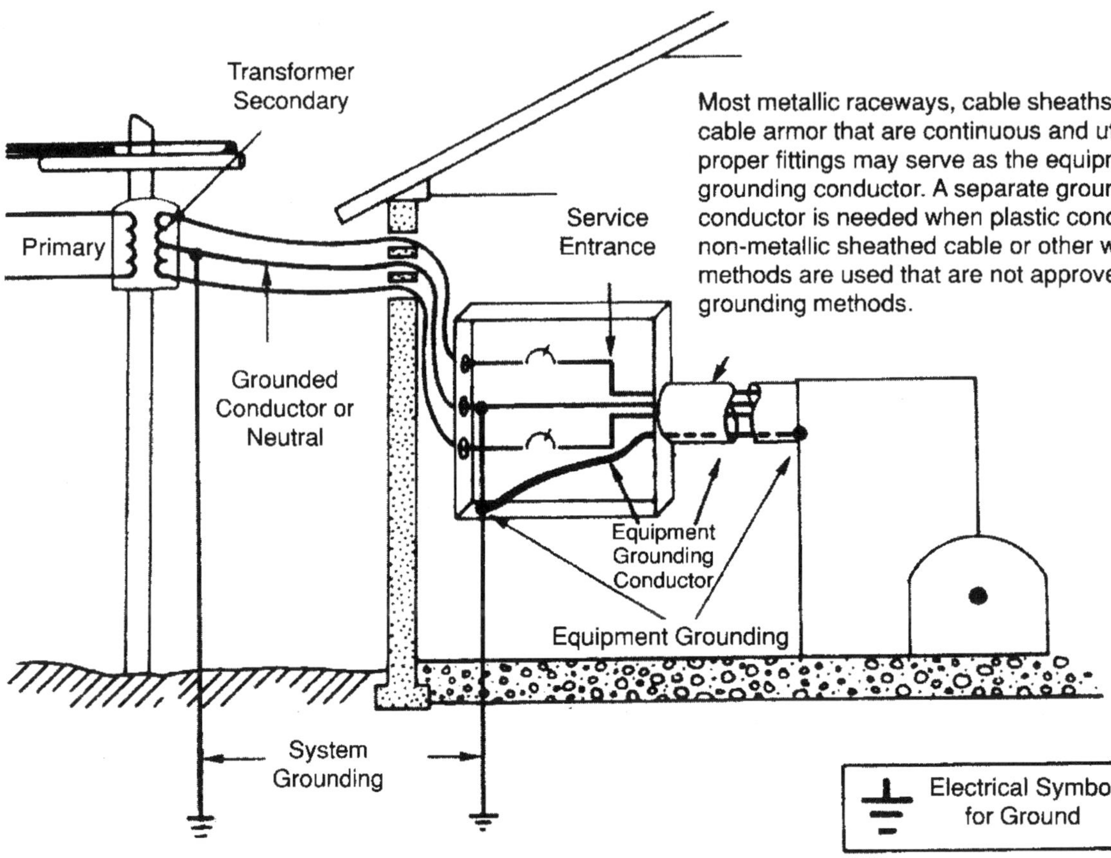

Discussion of Requirements

Grounding of electrical circuits and electrical equipment is required to protect employees against electrical shock, safeguard against fire, and protect against damage to electrical equipment.

There are two types of grounding: electrical circuit or system grounding, and electrical equipment grounding. Electrical system grounding is accomplished when one conductor of the circuit is intentionally connected to earth. This is done to protect the circuit should lightning strike or other high voltage contact occur. Grounding a system also balances the voltage in the system so expected voltage levels are not exceeded under normal conditions. Equipment grounding is accomplished when all metal frames of equipment and enclosures containing electrical equipment or conductors are grounded by means of a permanent and continuous connection or bond. The equipment grounding conductor provides a path for dangerous fault current to return to the system ground at the supply source of the circuit should an insulation failure take place. If installed properly, the equipment grounding conductor is the current path that enables protective devices, such as circuit breakers and fuses, to operate when a fault occurs. Figure 46-8 illustrates both types of grounding.

(g)(1)	**Systems to be grounded.** The following systems which supply premises wiring shall be grounded as **follows**:
(g)(1)(i)	All 3-wire dc systems shall have their neutral conductor grounded;

3-Wire DC System

A 3-wire DC system is generally used in heavy industrial applications to provide for smooth starting of motors and speed control. Three-wire DC motors are commonly used to operate large mills and extrusion machines. With a 3-wire circuit, they can run at half speed or at full speed, depending on the voltage selected. A 250 volt DC 3-wire system, for example, can supply both 125 volts DC and 250 volts DC to a motor. Figure 47 illustrates a schematic diagram of a 3-wire DC system that runs three motors, two single speed motors (one at 125 volts DC, the other at 250 volts DC) and a motor with two speed ranges (low speed at 125 volts DC and high speed at 250 volts DC). The neutral conductor is the conductor that requires a ground connection to earth. This connection must be made at a point near the DC source of supply.

Figure 47 – 250-volt, three-wire DC system. The neutral conductor is grounded at the source.

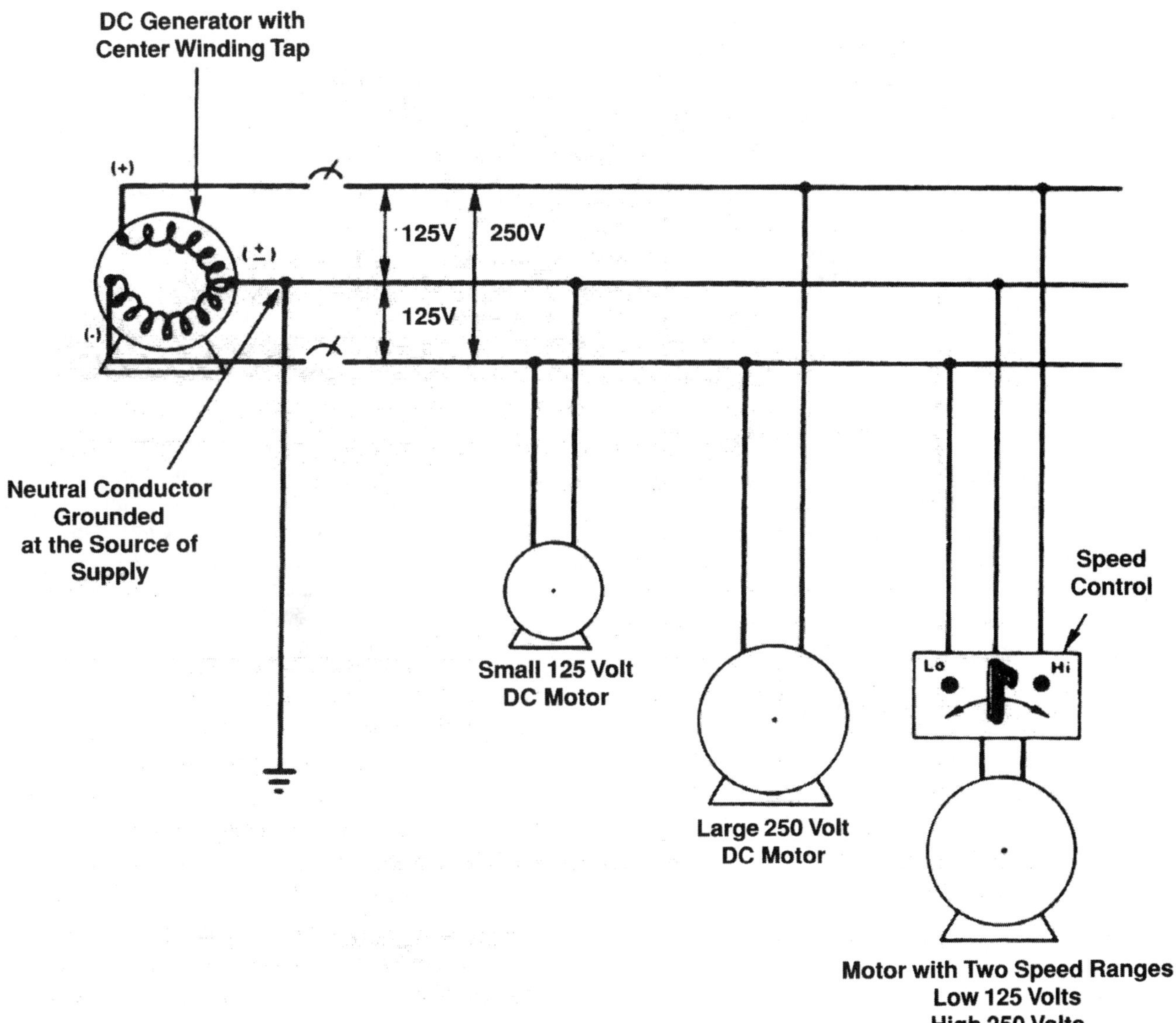

Figure 48 – Grounded 2-wire DC system supplying an electroplating operation.

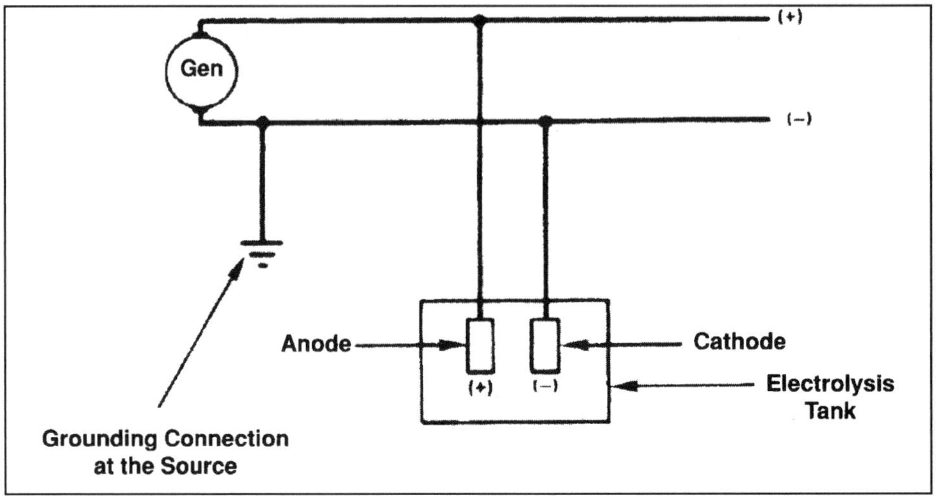

2-Wire DC Systems

(g)(1)(ii)	Two-wire dc systems operating at over 50 volts through 300 volts between conductors shall be grounded unless:
(g)(1)(ii)(A)	They supply only industrial equipment in limited areas and are equipped with a ground detector; or
(g)(1)(ii)(B)	They are rectifier-derived from an AC system complying with paragraphs (g)(1)(iii), (g)(1)(iv), and (g)(1)(v) of this section; or
(g)(1)(ii)(C)	They are fire-alarm signaling circuits having a maximum current of 0.030 amperes.

2-wire DC systems that operate between 50 volts and 300 volts and service premises must be grounded. Common 2-wire systems include battery charging, electroplating, and some crane operations. Grounding is accomplished by making an electrical connection to ground at the source as shown in Figure 48.

There are three exceptions to this requirement. First, if a DC system powers only industrial equipment in a limited area and is equipped with a sensor to detect any current leaks to ground, then no grounding is required. In such cases, a ground fault, resulting from an insulation failure, can be easily detected, located, and corrected before someone is injured. Second, if a DC system is derived from a rectifier that is an AC to DC converter, and the AC system complies with paragraphs *(g)(1)(iii), (g)(1)(iv), (g)(1)(v),* then no grounding is required in the DC system. Rectifiers also service limited areas; when the AC system which supplies the rectifier is properly grounded, a ground fault can be easily located and corrected. These systems must be properly maintained to assure that ground faults do not occur. Third, fire protective signaling circuits that operate under 30 milliamps do not require grounding because of the low current flow, which means that no serious shock hazard exists. Figure 48 illustrates a DC-powered electroplating operation with the grounding connection at the source.

Figure 49 – AC circuits under 50 volts requiring grounding.

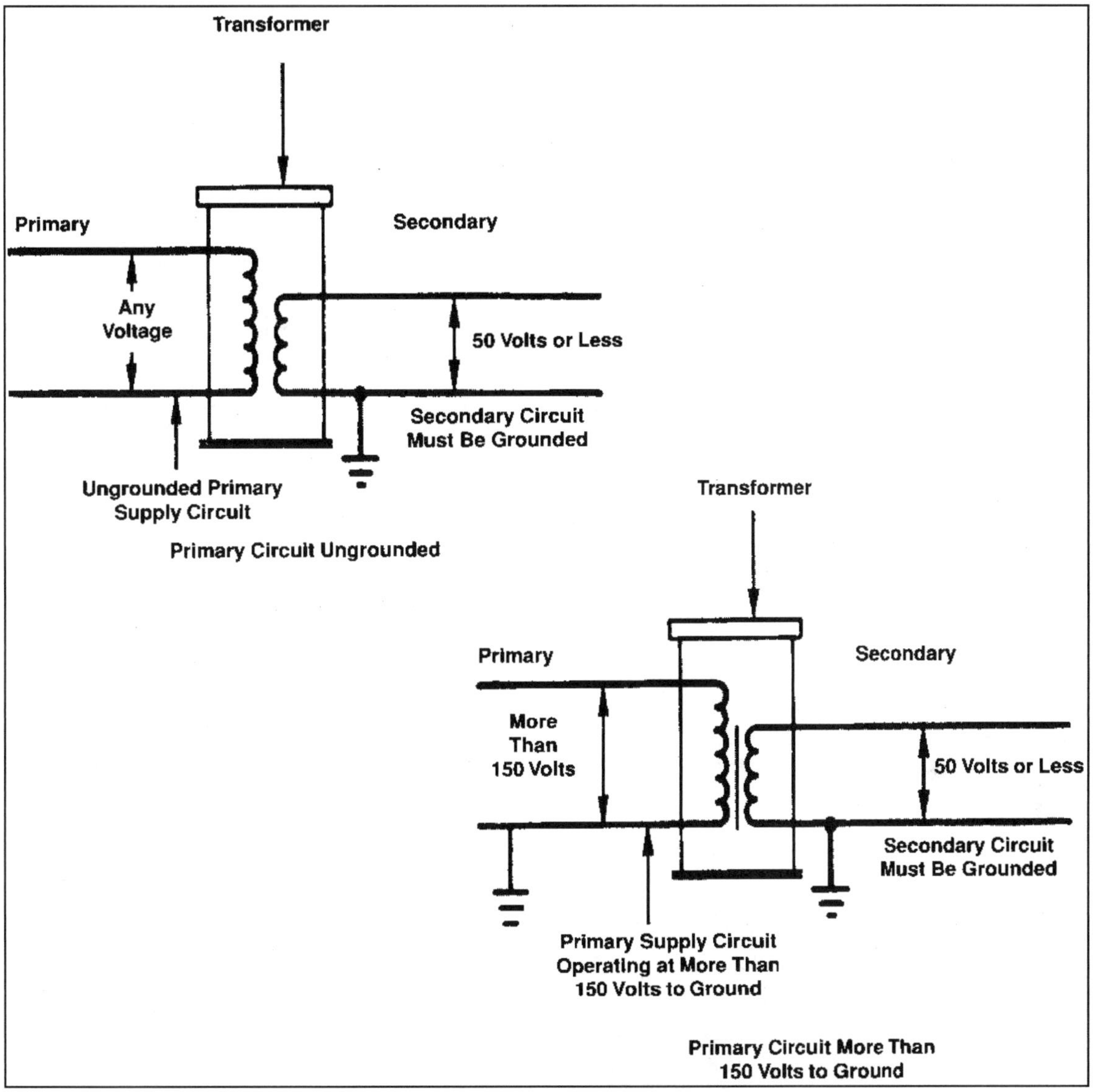

(g)(1)(iii) AC circuits of less than 50 volts shall be grounded if they are installed as overhead conductors outside of buildings or if they are supplied by transformers and the transformer primary supply system is ungrounded or exceeds 150 volts to ground.

Grounding of Systems and Circuits Less than 50V AC

Only under certain conditions are low-voltage AC systems required to be grounded. These are conditions where there is a likelihood of foreign higher voltage -- such as lightning or higher primary voltages -- being conducted across damaged transformer windings, creating a serious shock hazard to someone working with the low voltage system. In each case, grounding is required at the secondary side of the low voltage transformer. The first condition requires grounding when the conductors are run as overhead conductors outdoors. In this case, the low voltage system must be protected against high voltage surges resulting from lightning strikes and against contact with high voltage overhead lines. The second condition requires grounding when low voltage systems are supplied by transformers that are supplied by an ungrounded primary system of any voltage or when the primary voltage to ground exceeds 150 volts. Figure 49 illustrates examples of circuits operating at less than 50 volts that are required to be grounded by the second condition. Grounding of the low-voltage system on the secondary side of the transformer will prevent energizing of the secondary circuit with the higher primary voltages if the transformer windings are damaged.

(g)(1)(iv)	**AC systems of 50 volts to 1000 volts shall be grounded under any of the following conditions, unless exempted by paragraph (g)(1)(v) of this section:**
(g)(1)(iv)(A)	**If the system can be so grounded that the maximum voltage to ground on the ungrounded conductors does not exceed 150 volts;**
(g)(1)(iv)(B)	If the system is nominally rated three-phase, four-wire wye connected in which the neutral is used as a circuit conductor;
(g)(1)(iv)(C)	If the system is nominally rated three-phase, four-wire delta connection in which the midpoint of one phase is used as a circuit conductor; or
(g)(1)(iv)(D)	If a service conductor is uninsulated;

Grounding of Systems and Circuits 50-1000V AC

The most common AC systems in use in industry include several single- and three-phase systems. Grounding of these systems is required to provide for the prompt operation of over-current devices, to allow for the quick isolation of faults, to provide for positive lightning protection, and to afford protection for the equipment and for personnel. Table 304-4 summarizes the requirements for the following systems:

1. 240/120 volt single-phase 3-wire systems used most often for general light and power circuits;
2. 208Y/120 volt, 3-phase, 4-wire systems used to power commercial and small industrial equipment, lighting and receptacles;
3. 480Y/277 volt, 3-phase, 4-wire systems typically used in industrial applications;
4. 240/120 volt, 3-phase, 4-wire delta connected systems.

In the three-phase systems above (2, 3 and 4), note that the first voltage specified in each title is the 3-phase voltage that appears between the three ungrounded conductors.

Each of these systems must be grounded. The 240/120 volt system and 208Y/120 volt system are required to be grounded because they are covered by condition (iv)(a). Its voltage to ground is less than 150 volts. The 480Y/277 volt system is required to be grounded because it is specifically mentioned in condition (iv)(b). The 3-phase, 4-wire delta connected system is not as common as the other systems described, but grounding is required as mentioned in condition (iv)(c). In this system there is no neutral conductor, and the conductor that must be grounded is the circuit conductor connected at the mid-point of one of the phase windings in the transformer.

(Many older plants have three-phase, three-wire delta connected systems that have no neutral and that are not used for lighting. These are either 240- or 480-volt systems, and are not required to be grounded.)

If the service conductors that provide power to a building are located overhead and are uninsulated, the system must be grounded. In this case, grounding is required to protect from foreign higher voltage conditions resulting from lightning or accidental contact with higher voltage wires.

(g)(1)(v)	AC systems of 50 volts to 1000 volts are not required to be grounded under any of the following conditions:
(g)(1)(v)(A)	If the system is used exclusively to supply industrial electric furnaces for melting, refining, tempering, and the like;

1910.304

Title: Wiring design and protection.

Subpart: S

SubPart Title: Electrical - Design Safety Standards for Electrical Systems

(g)(1)(v)(B) If the system is separately derived and is used exclusively for rectifiers supplying only adjustable speed industrial drives;

Table 304-4 AC Systems 50 to 1000 Volts Requiring Grounding.

System Schematic	Description	Common Uses
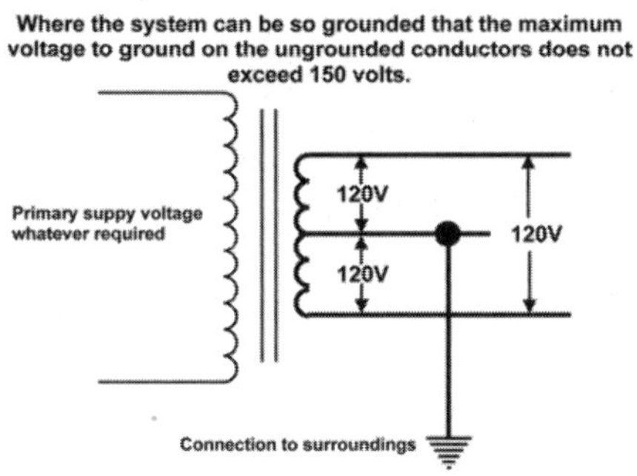 120/240 Volt, 1-Phase, 3-Wire System	**Voltage between ungrounded conductors is 240 volts. Voltage from an ungrounded conductor to the grounded conductor and/or ground (our surroundings) is 120 volts.**	**General lighting and receptacle loads. Primary system used in residential applications.** **Meets Condition 1910.304(g)(1)(iv)(A)**
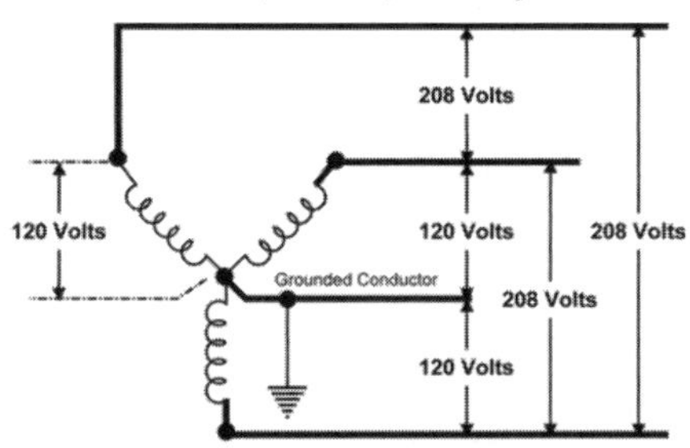 208Y/120 Volt, 3-Phase, 4-Wire System Wye Connected Transformer	**Voltage between ungrounded conductors is 208 volts. Voltage from an ungrounded conductor to ground (our surroundings) is 120 volts.**	**General lighting and receptacle loads, small machines, air-conditioning equipment.** **Meets Condition 1910.304(g)(iv)(A)**
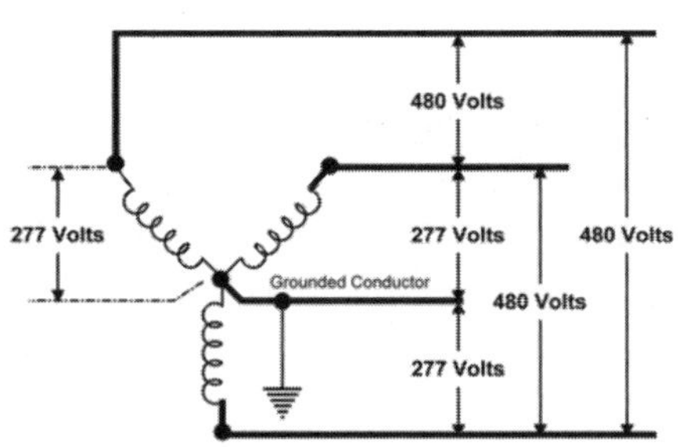 480Y/277 Volt, 3-Phase, 4-Wire System Wye Connected Transformer	**Voltage between ungrounded conductors is 480 volts. Voltage from an ungrounded conductor to ground (our surroundings) is 277 volts.**	**General lighting, special purpose welding receptacles, large air-conditioning equipment, and heavy industrial applications.** **Meets Condition 1910.304(g)(iv)(B)**
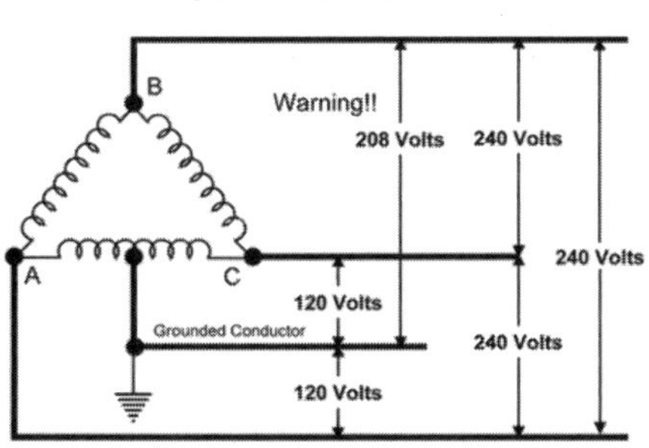 240Δ /120 Volt, 3-Phase, 4-Wire System Delta Connected Transformer	**Voltage between ungrounded conductors is 240 volts. Voltage from phase (A) or phase (C) to ground (our surroundings) is 120 volts. However the voltage from phase (B) to ground is 208 volts.**	**Older commercial and industrial facilities. Use varied greatly in different areas of the United States.** **Meets Condition 1910.304(g)(iv)(C)**

(g)(1)(v)(C)	If the system is separately derived and is supplied by a transformer that has a primary voltage rating less than 1000 volts, provided all of the following conditions are met:
(g)(1)(v)(C)(1)	The system is used exclusively for control circuits;
(g)(1)(v)(C)(2)	The conditions of maintenance and supervision assure that only qualified persons will service the installation;
(g)(1)(v)(C)(3)	Continuity of control power is required, and
(g)(1)(v)(C)(4)	Ground detectors are installed on the control system;
(g)(1)(v)(D)	If the system is an isolated power system that supplies circuits in health care facilities; or
(g)(1)(v)(E)	If the system is a high-impedance grounded neutral system in which a grounding impedance, usually a resistor, limits the ground-fault current to a low value for 3-phase ac systems of 480 volts to 1000 volts provided all of the following conditions are met:
(g)(1)(v)(E)(1)	The conditions of maintenance and supervision ensure that only qualified persons will service the installation;
(g)(1)(v)(E)(2)	Continuity of power is required;
(g)(1)(v)(E)(3)	Ground detectors are installed on the system; and
(g)(1)(v)(E)(4)	Line-to-neutral loads are not served.

No Grounding Required: Circuits 50-1000V AC

Under certain conditions AC systems that operate between 50 and 1,000 volts do not require grounding. Grounding is not required because they serve limited areas or are easily monitored for ground faults. They are closely supervised and must be properly maintained. If a ground fault occurs, it can be detected right away and corrected before a second ground fault occurs that may injure an employee.

(g)(2)	**Conductor to be grounded.** The conductor to be grounded for ac premises wiring systems required to be grounded by paragraph (g)(1) of this section shall be as follows:
(g)(2)(i)	One conductor of a single-phase, two-wire system shall be grounded;
(g)(2)(ii)	The neutral conductor of a single-phase, three-wire system shall be grounded;
(g)(2)(iii)	The common conductor of a multiphase system having one wire common to all phases shall be grounded;
(g)(2)(iv)	One phase conductor of a multiphase system where one phase is grounded shall be grounded; and
(g)(2)(v)	The neutral conductor of a multiphase system in which one phase is used as a neutral conductor shall be grounded.

Conductors to Be Grounded

In AC premises wiring systems (i.e., wiring used in servicing buildings, plants, and offices) required to be grounded, the identified conductor (the conductor with a white or gray outer cover) is the conductor required to be connected to the system ground.

(g)(3)	**Portable and vehicle-mounted generators.**
(g)(3)(i)	The frame of a portable generator need not be grounded and may serve as the grounding electrode for a system supplied by the generator under the following conditions:
(g)(3)(i)(A)	The generator supplies only equipment mounted on the generator or cord- and plug-connected equipment through receptacles mounted on the generator, or both; and
(g)(3)(i)(B)	The noncurrent-carrying metal parts of equipment and the equipment grounding conductor terminals of the receptacles are bonded to the generator frame.

The frame of a portable generator (one that is easily carried from one location to another) is not required to be connected to earth (or surroundings) if the generator supplies only outlets and/or equipment mounted to the generator, and the outlets of the generator have equipment grounding terminals bonded to the generator frame.

(g)(3)(ii) The frame of a vehicle need not be grounded and may serve as the grounding electrode for a system supplied by a generator located on the vehicle under the following conditions:

(g)(3)(ii)(A) The frame of the generator is bonded to the vehicle frame;

(g)(3)(ii)(B) The generator supplies only equipment located on the vehicle and cord- and plug-connected equipment through receptacles mounted on the vehicle;

(g)(3)(ii)(C) The noncurrent-carrying metal parts of equipment and the equipment grounding conductor terminals of the receptacles are bonded to the generator frame; and

(g)(3)(ii)(D) The system complies with all other provisions of paragraph (g) of this section.

If the vehicle-mounted generator has a neutral conductor, and is installed as a separately-derived system, the neutral conductor must be bonded to the generator and vehicle frame. The equipment grounding terminals of receptacle outlets and the noncurrent-carrying metal parts of equipment must be bonded to the generator frame.

(g)(3)(iii) A system conductor that is required to be grounded by the provisions of paragraph (g)(2) of this section shall be bonded to the generator frame where the generator is a component of a separately derived system.

(g)(4) **Grounding connections.**

(g)(4)(i) For a grounded system, a grounding electrode conductor shall be used to connect both the equipment grounding conductor and the grounded circuit conductor to the grounding electrode. Both the equipment grounding conductor and the grounding electrode conductor shall be connected to the grounded circuit conductor on the supply side of the service disconnecting means, or on the supply side of the system disconnecting means or overcurrent devices if the system is separately derived.

Grounding Electrode Conductor

When required, the electrical system's connection to ground must be made at the source of the electrical supply. (See paragraph (g)(3)(i)). For buildings, offices, plants and other premises, the source is at the service entrance, and the connection to ground must be made on the supply side of the service disconnect. The National Electrical Code® permits this connection to be made inside the service entrance equipment as shown in Figure 50. Typically, a main bonding jumper (A) is used to connect the grounded circuit conductor (B) (the white or gray wire, usually the neutral) to the equipment grounding conductor (C) (which is green or a bare wire or the metallic equipment housing and metal raceway system). Sometimes the bonding function of jumper wire (A) is accomplished by a screw furnished by the enclosure manufacturer that connects the grounded conductor's terminal to the equipment grounding conductor. It is important to note that this is the only connection between the grounded conductor and the grounding conductor that is allowed in the circuit.
The unspliced grounding electrode conductor (D) connects the building's electrical system to the grounding electrode system. The grounding electrode system (as defined by the NEC®) is composed of the following items (if present) bonded together: 1) metal underground water pipe, 2) metal frame of the building or structure, 3) concrete-encased electrode (also called an UFFer Ground), 4) ground ring and/or any of the approved electrodes described in Section 250.52(A)(5) thru (8) of the National Electrical Code®.

Figure 50 illustrates the typical grounding connections and components in a three-phase, 4-Wire utility supplied service. The grounded circuit conductor (neutral) (B) is connected to ground by means of a main bonding jumper (A) connected to the equipment grounding conductor terminal bar (C), which is bonded to the metal equipment enclosure. The grounding electrode conductor (D) is then connected to building steel and main water line which serves as the grounding electrode system (E).

Figure 50 – Typical connections for a GROUNDED 3-phase, 4-wire, service supplied system. These connections are made at the MAIN SERVICE DISCONNECT switch only. The grounding electrode system is discussed in the National Electrical Code®, Sections 250.50 and 250.52

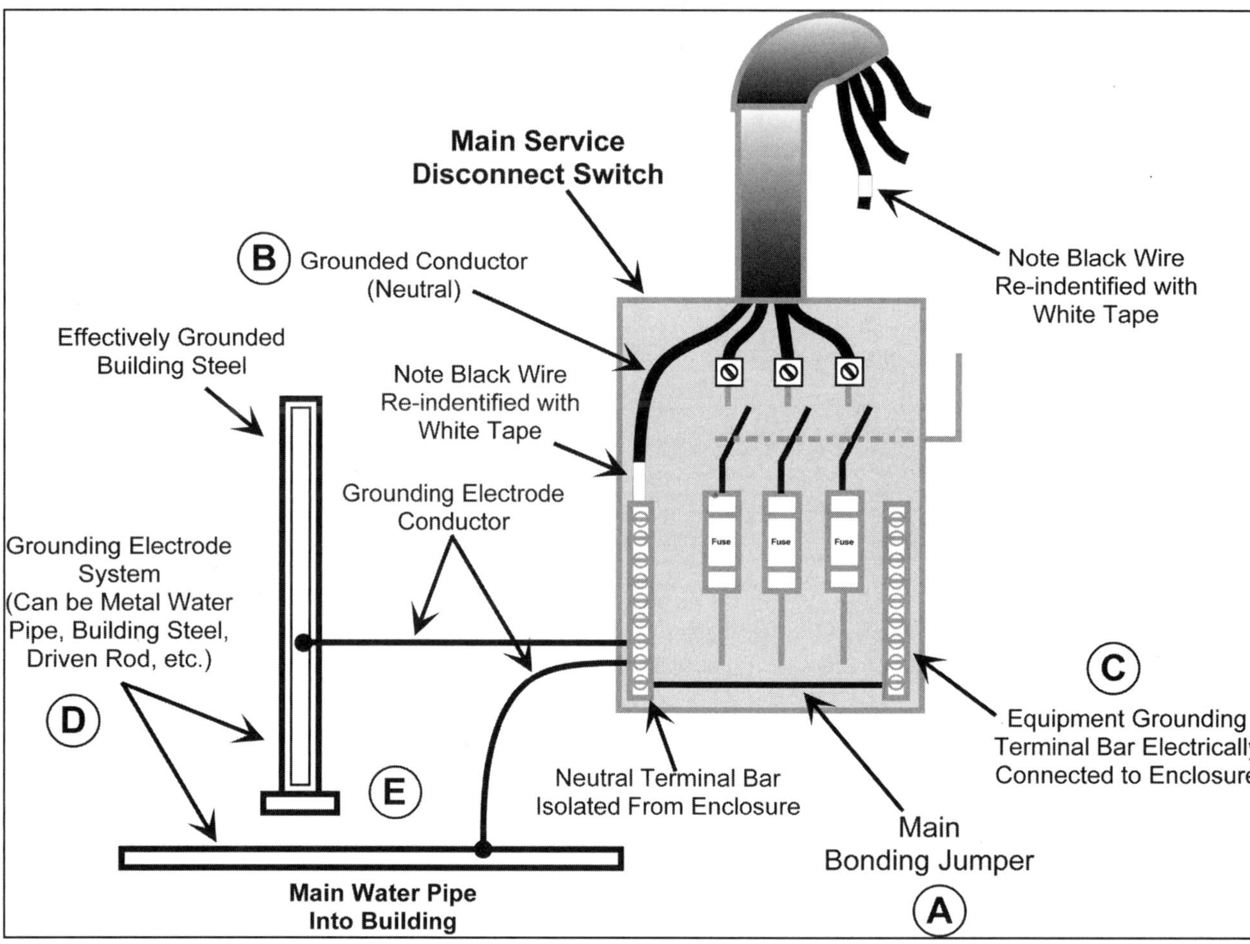

Grounding Connection for an Ungrounded Service-Supplied System

Where a facility is served by an ungrounded system, the connection for the equipment grounding conductor must be made at the source of electrical supply. Figure 51 illustrates where this connection is made. In this case, the main service disconnect switch at the service entrance for the premises.

Figure 51 - Typical connections for a UNGROUNDED 3-phase, 3-wire, service supplied system

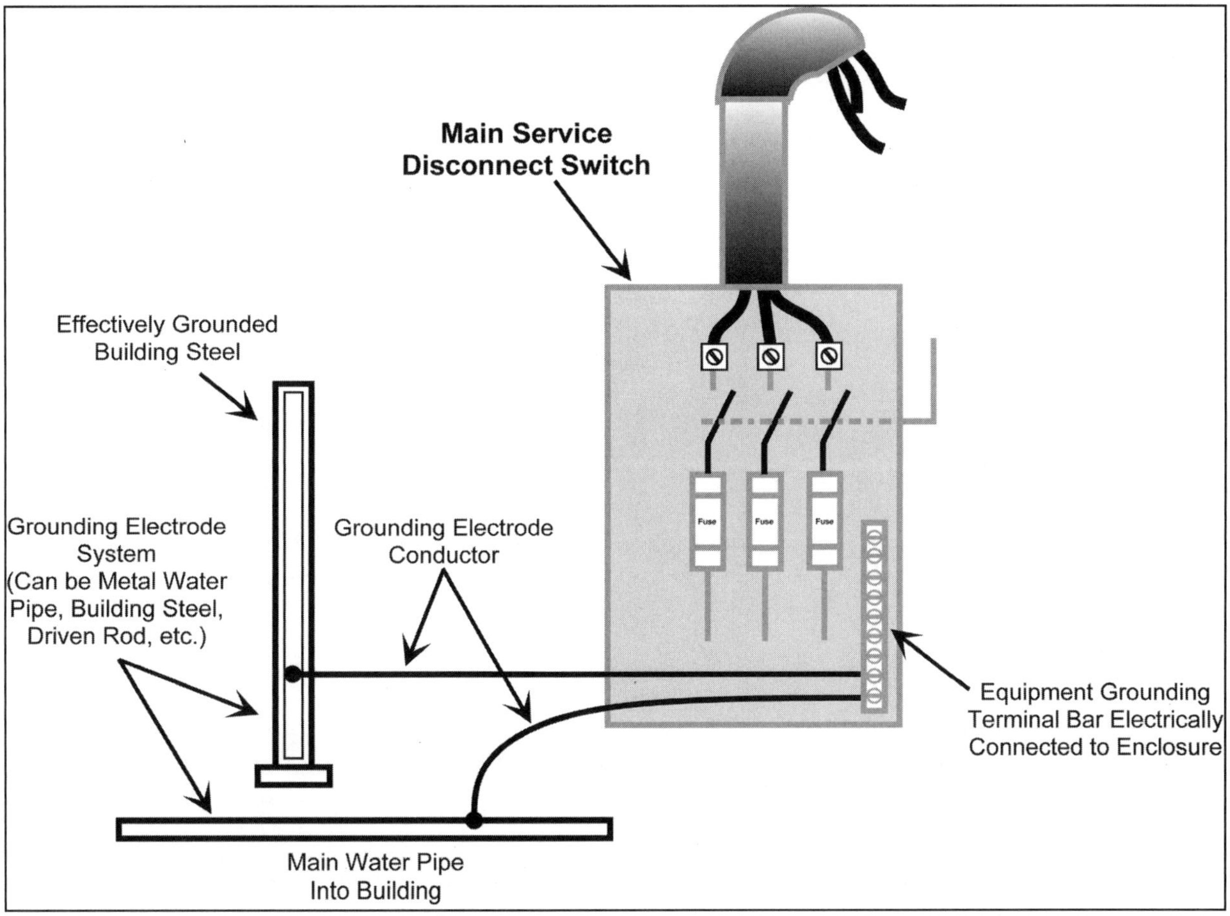

Major Difference Between "Grounded" and "Ungrounded" Supply Systems.

The major difference between the "grounded" and "ungrounded" supply system is the *main bonding jumper* (see Figures 52 and 53).. The main bonding jumper connects the equipment grounding conductors to the grounded conductor at the service or source of the separately-derived system.. As a result of this connection, the system is referred to as a "grounded" system. When the first fault occurs between an ungrounded (hot) conductor and a grounded metal enclosure the overcurrent device, protecting the circuit, will operate quickly.

In an ungrounded system, without a main bonding jumper, two faults must occur before an overcurrent device will operate. Although not specifically required by the OSHA regulations it is recommend that a "ground-fault indicator" device be installed on all ungrounded systems. Such a device will indicate when the first fault occurs allowing maintenance to correct the problem before the second fault results in a system outage.

Figure 52 – Simplified diagram of a GROUNDED system. Note the MAIN BONDING JUMPER.

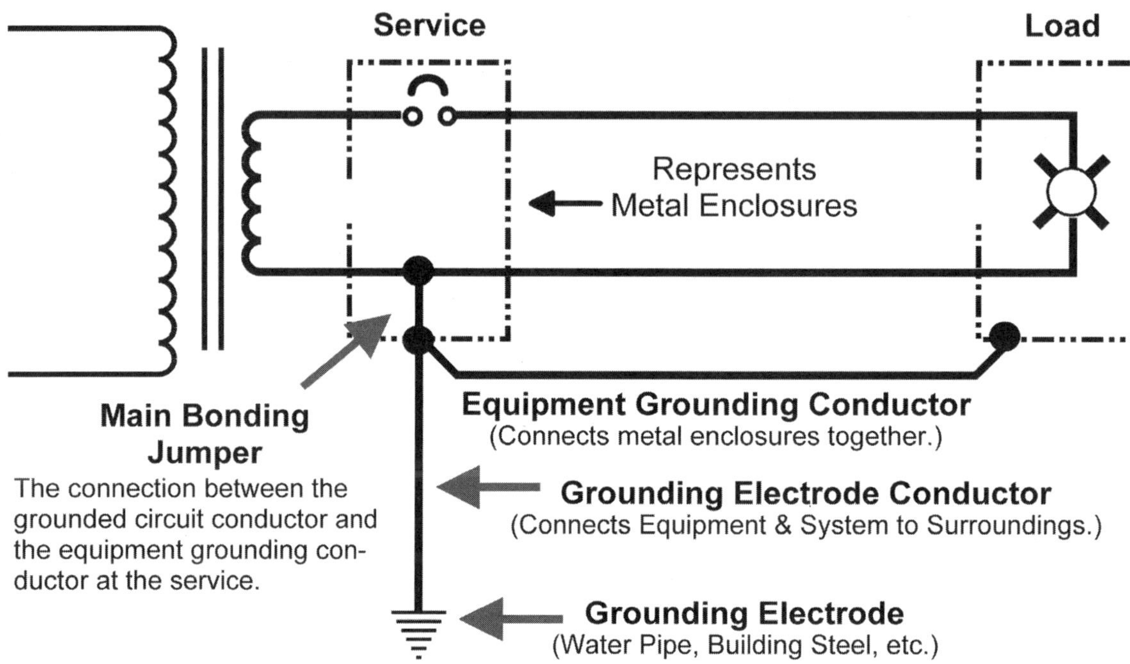

Figure 53 – Simplified diagram of an UNGROUNDED system. Note that there is NO MAIN BONDING JUMPER.

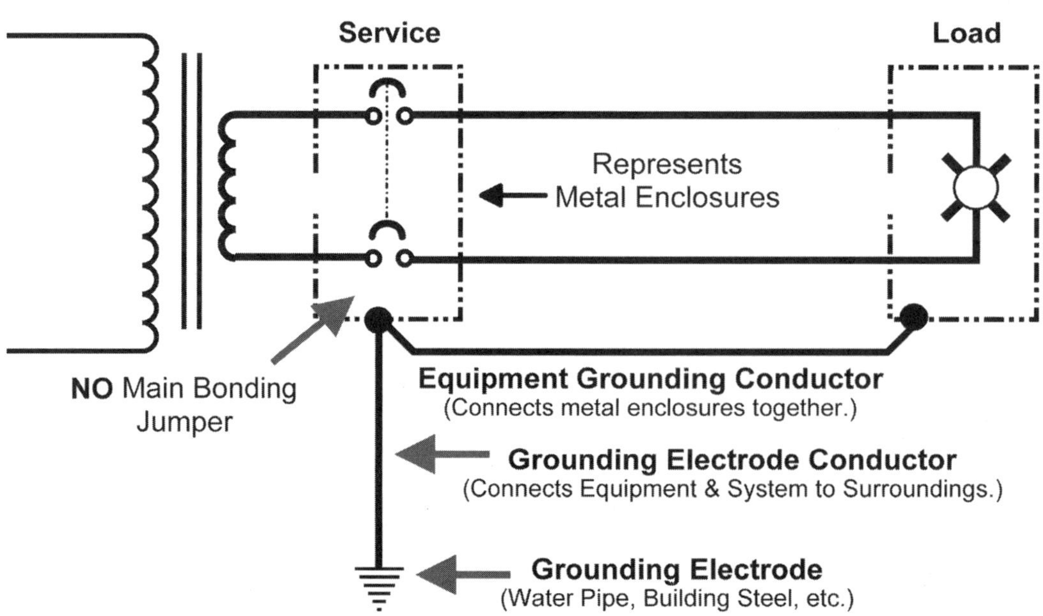

(g)(4)(iii) On extensions of existing branch circuits that do not have an equipment grounding conductor, grounding-type receptacles may be grounded to a grounded cold water pipe near the equipment if the extension was installed before August 13, 2007. When any element of this branch circuit is replaced, the entire branch circuit shall use an equipment grounding conductor that complies with all other provisions of paragraph (g) of this section.

In most instances, the requirements of the National Electrical Code® and NFPA 70E go hand-in-hand with the requirements of Subpart S. In this case however, OSHA chose, despite objections, to retain the allowance for the use of a cold water pipe for equipment grounding purposes in their revision to the regulations. The use of water piping for equipment grounding purposes was completely removed from the 1993 Edition of the NEC®, and such an installation would be currently be in violation of NEC® sections 250.52(A)(1), 250.134 and 300.3(B) as well as generally violating 1910.304(g)(5) and 1910.303(b)(5).

The 2002 Edition of the NEC® handbook, published by NFPA, explains in an editorial note under Article 250.130 that "because of the requirements of 250.52(A)(1) an interior metal water pipe more than 5 ft from the point of entrance of the water pipe into the building is no longer allowed to serve as a connection to the grounding electrode conductor." The 1990 NEC® Handbook published by NFPA provided additional clarification on the issue of permitting "the grounding conductor of a grounding type receptacle outlet to be grounded to a water pipe which is bonded in accordance with Section 250-80(a)." The editorial notes under 250-80(a), as written by NFPA engineers states, "The intent of the exception is to recognize the fact that increased use of nonmetallic water piping systems underground makes it virtually impossible to comply with the main rule of section 250-80 in the circumstances described."

It is recommended that the reader revisit the detailed explanation on grounding previously described in this book for the rationale of why water pipes provide inadequate equipment grounds. The detailed arguments for the proposed removal of this reference made to OSHA during the comment period of the standard rulemaking process, taken directly from the Subpart S preamble can be found in Appendix D.

(g)(5) **Grounding path.** The path to ground from circuits, equipment, and enclosures shall be permanent and continuous, and effective.

Article 250 in the National Electrical Code® has been extensively revised as compared to the 1978 edition that was referenced for the current version of Subpart S. The wording used by OSHA was originally found in Section 250-51. This section has been extensively reworded and relocated to Section 250.4(A) and (B) beginning with the 2002 Edition of the National Electrical Code®. However the intent of the requirement has remained the same.

Section 250-51 in the 1978 National Electrical Code® stated three important requirements for an effective ground fault current path: (1) be permanent and continuous; (2) have capacity to conduct safely any fault current likely to be imposed on it; and (3) have sufficiently low impedance to limit the voltage to ground and to facilitate the operation of the circuit protective devices in the circuit. As can be seen from the OSHA text in 1910.304(g)(5) OSHA omitted requirements (2) and (3). Contrary to some interpretations, this omission did not nullify these requirements. The omission was for simplification, as these requirements have been adequately covered by references such as 1910.303(b)(1), 1910.304(g)(8) and 1910.305(a)(1)(i). The addition of the word *"effective"* further underscores the import of a low impedance path to prevent electrocution by instantaneously tripping the overcurrent device in the event of a fault in the tool or equipment.

(g)(6) **Supports, enclosures, and equipment to be grounded.**

(g)(6)(i) *Supports and enclosures for conductors.* Metal cable trays, metal raceways, and metal enclosures for conductors shall be grounded, except that:

(g)(6)(i)(A) Metal enclosures such as sleeves that are used to protect cable assemblies from physical damage need not be grounded; and

(g)(6)(i)(B) Metal enclosures for conductors added to existing installations of open wire, knob-and-tube wiring, and nonmetallic-sheathed cable need not be grounded if all of the following conditions are met:

(g)(6)(i)(B)(1)	Runs are less than 7.62 meters (25.0 ft);
(g)(6)(i)(B)(2)	Enclosures are free from probable contact with ground, grounded metal, metal laths, or other conductive materials; and
(g)(6)(i)(B)(3)	Enclosures are guarded against employee contact.
(g)(6)(ii)	Metal enclosures for service equipment shall be grounded.
(g)(6)(iii)	Frames of electric ranges, wall-mounted ovens, counter-mounted cooking units, clothes dryers, and metal outlet or junction boxes that are part of the circuit for these appliances shall be grounded.
(g)(6)(iv)	Exposed non-current-carrying metal parts of fixed equipment that may become energized shall be grounded under any of the following conditions:
(g)(6)(iv)(A)	If within 2.44 m (8 ft) vertically or 1.52 m (5 ft) horizontally of ground or grounded metal objects and subject to employee contact.
(g)(6)(iv)(B)	If located in a wet or damp location and not isolated.
(g)(6)(iv)(C)	If in electrical contact with metal.
(g)(6)(iv)(D)	If in a hazardous (classified) location.
(g)(6)(iv)(E)	If supplied by a metal-clad, metal-sheathed, or grounded metal raceway wiring method.
(g)(6)(iv)(F)	If equipment operates with any terminal at over 150 volts to ground.
(g)(6)(v)	Notwithstanding the provisions of paragraph (g)(6)(iv) of this section, exposed noncurrent-carrying metal parts of the following types of fixed equipment need not be grounded:
(g)(6)(v)(A)	Enclosures for switches or circuit breakers used for other than service equipment and accessible to qualified persons only;
(g)(6)(v)(B)	Electrically heated appliances that are permanently and effectively insulated from ground;
(g)(6)(v)(C)	Distribution apparatus, such as transformer and capacitor cases, mounted on wooden poles, at a height exceeding 2.44 m (8.0 ft) above ground or grade level; and
(g)(6)(v)(D)	Listed equipment protected by a system of double insulation, or its equivalent, and distinctively marked as such.
(g)(6)(vi)	Exposed noncurrent-carrying metal parts of cord- and plug-connected equipment that may become energized shall be grounded under any of the following conditions:
(g)(6)(vi)(A)	If in hazardous (classified) locations (see § 1910.307);
(g)(6)(vi)(B)	If operated at over 150 volts to ground, except for guarded motors and metal frames of electrically heated appliances if the appliance frames are permanently and effectively insulated from ground;
(g)(6)(vi)(C)	If the equipment is of the following types:
(g)(6)(vi)(C)(1)	Refrigerators, freezers, and air conditioners;
(g)(6)(vi)(C)(2)	Clothes-washing, clothes-drying, and dishwashing machines, sump pumps, and electric aquarium equipment;
(g)(6)(vi)(C)(3)	Hand-held motor-operated tools, stationary and fixed motor-operated tools, and light industrial motor-operated tools;
(g)(6)(vi)(C)(4)	Motor-operated appliances of the following types: hedge clippers, lawn mowers, snow blowers, and wet scrubbers;
(g)(6)(vi)(C)(5)	Cord- and plug-connected appliances used in damp or wet locations, or by employees standing on the ground or on metal floors or working inside of metal tanks or boilers;
(g)(6)(vi)(C)(6)	Portable and mobile X-ray and associated equipment;
(g)(6)(vi)(C)(7)	Tools likely to be used in wet and conductive locations; and
(g)(6)(vi)(C)(8)	Portable hand lamps.

(g)(6)(vii) Notwithstanding the provisions of paragraph (g)(6)(vi) of this section, the following equipment need not be grounded:

(g)(6)(vii)(A) Tools likely to be used in wet and conductive locations if supplied through an isolating transformer with an ungrounded secondary of not over 50 volts; and

(g)(6)(vii)(B) Listed or labeled portable tools and appliances if protected by an approved system of double insulation, or its equivalent, and distinctively marked.

Figure 54 – Cord- and plug-connected equipment without a grounding conductor

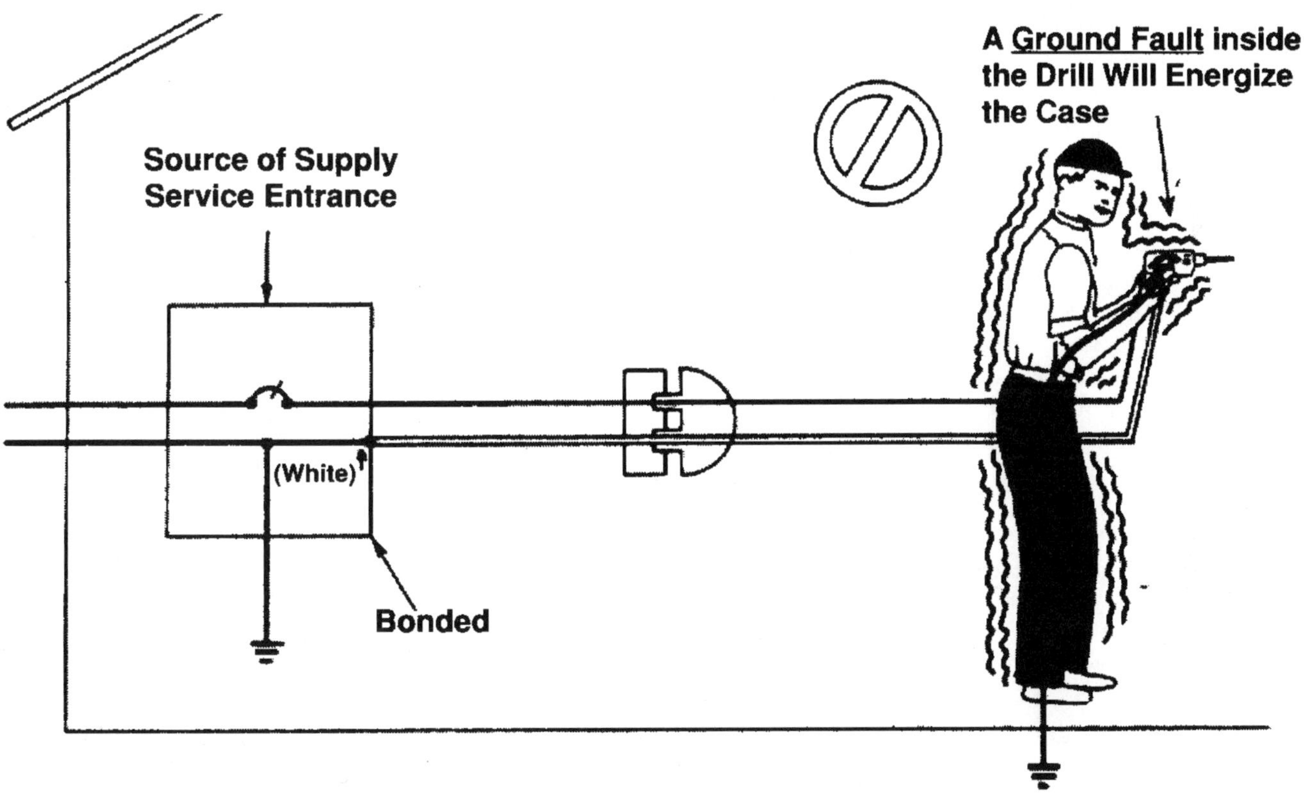

Equipment Connected by Cord and Plug

Under conditions described in paragraphs (g)(6)(vi)(A) through (C), exposed noncurrent-carrying metal parts of cord- and plug-connected equipment must be grounded. Grounding metal parts is not required if the equipment is supplied through an isolating transformer with an ungrounded secondary of not over 50 volts or if portable tools are protected by an approved system of double insulation. To ground cord- and plug-connected equipment, a third wire is commonly provided in the cord set and a third prong in the plug. The third wire serves as an equipment grounding conductor that is connected to the metal housing of a portable tool and a metal grounding bus inside the service entrance equipment. Usually the "third wire" (equipment grounding conductor) will be connected to other acceptable grounding conductors, such as metal conduit, which will eventually be connected to the grounding bus in the service entrance equipment. The service entrance equipment is located at the entrance point of the electric supply for a building or plant and contains (or serves other panelboards that contain) branch circuit protective devices such as fuses and circuit breakers. The third wire provides a path for fault current if an insulation failure should occur. In this manner dangerous fault current will be directed back to the source, the service entrance, and will enable circuit breakers or fuses to operate, thus opening the circuit and stopping the current flow. Figure 54 illustrates the potential shock hazard that exists when no third wire grounding conductor is used.

This accidental connection from the "hot" wire to the case of the drill provides a new path for current flow through the worker to some grounded surface, which probably continues the path back to the service entrance and the white wire. The normal current will continue to flow through the motor, but some additional current will follow the new path through the worker. It is appropriate to visualize the two symbols for ground as being connected together, as by earth, plumbing, etc.

Figure 55 illustrates the advantage of a properly connected equipment grounding conductor. It should be noted that properly bonded conduit and associated metal enclosures can serve as an equipment grounding conductor as described in paragraph (g)(8)(ii).

If a fault occurs, most of the current will follow the path of least resistance to ground. If the worker provides a path to ground as shown, some portion of the current will flow away from the grounded white conductor (neutral) and return to ground through the worker. The severity of the shock received will depend on the amount of current that flows through the worker.

Figure 55 – Cord- and plug-connected equipment with a grounding conductor. Dangerous fault current now is redirected along the equipment grounding conductor back to the source of electrical supply to operate overcurrent device.

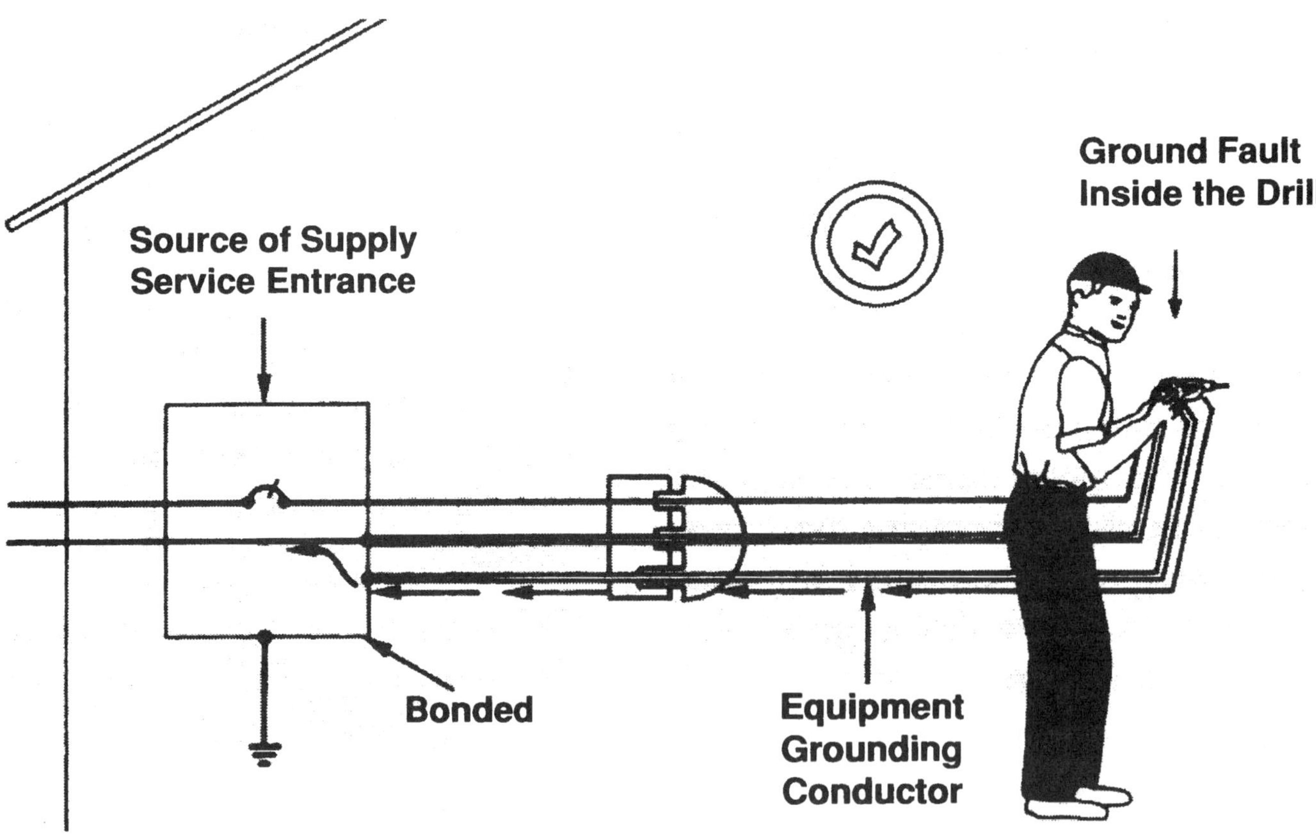

(g)(7) **Nonelectrical equipment.** The metal parts of the following nonelectrical equipment shall be grounded: frames and tracks of electrically operated cranes and hoists; frames of nonelectrically driven elevator cars to which electric conductors are attached; hand operated metal shifting ropes or cables of electric elevators; and metal partitions, grill work, and similar metal enclosures around equipment of over 750 volts between conductors.

(g)(8) **Methods of grounding fixed equipment.**

(g)(8)(i) Non-current-carrying metal parts of fixed equipment, if required to be grounded by this subpart, shall be grounded by an equipment grounding conductor which is contained within the same raceway, cable, or cord, or runs with or encloses the circuit conductors. For DC circuits only, the equipment grounding conductor may be run separately from the circuit conductors.

Grounding Non-Current-Carrying Metal Parts

The non-current-carrying metal parts of fixed electrical equipment that require grounding, i.e. metal enclosures, must be grounded by one of two methods that are acceptable for AC systems. Boxes, cabinets, and other metal enclosures can be grounded by an equipment-grounding conductor that is run with the circuit conductors serving the equipment or by using the metal cable armor, metal sheath, or the metal conduit that contains the power conductors. Proper fittings must be used between enclosures and conduits to assure that a good electrical connection is made so the grounding path is continuous.

Figure 56 illustrates examples of how equipment grounding conductors may be run. Figure 57 shows an acceptable arrangement for equipment grounding using equipment enclosures that are properly interconnected and fastened.

A separate equipment grounding conductor must be provided when plastic conduit is used. It must be contained in the same raceway or cable as the circuit of which it is part and properly connected to terminals and grounding muses inside panels and boxes.

DC systems, on the other hand, do not require the equipment grounding conductor to be in the same raceway as the circuit conductors. The NEC® describes raceway types that have proved effective for grounding. For example, flexible conduit and liquidtight flexible conduit are acceptable under certain limiting conditions.

Figure 57 shows a special case. Liquidtight flexible conduit and flexible metal conduit can be used in limited sizes and applications, as set forth in Section 250.118 of the NEC®.

(g)(8)(ii) Electric equipment is considered to be effectively grounded if it is secured to, and in electrical contact with, a metal rack or structure that is provided for its support and the metal rack or structure is grounded by the method specified for the noncurrent-carrying metal parts of fixed equipment in paragraph (g)(8)(i) of this section. Metal car frames supported by metal hoisting cables attached to or running over metal sheaves or drums of grounded elevator machines are also considered to be effectively grounded.

(g)(8)(iii) For installations made before April 16, 1981, electric equipment is also considered to be effectively grounded if it is secured to, and in metallic contact with, the grounded structural metal frame of a building. When any element of this branch circuit is replaced, the entire branch circuit shall use an equipment grounding conductor that complies with all other provisions of paragraph (g) of this section.

This "grandfather clause" permitting use of the grounded structural metal frame of a building as the ground-fault current path is an exception to the general rule that the grounding conductor shall closely follow the supply conductors. But it does require that the current path through the frame of the building will be sufficiently low impedance to prevent dangerous voltages and quickly operate the protective device. See 1910.304(g)(8) and the National Electrical Code® Section 250.4(A)(5) and 250.4(B)(4).

Figure 56 – Grounding methods for fixed equipment using a separate conductor or metal enclosures as the equipment grounding conductor.

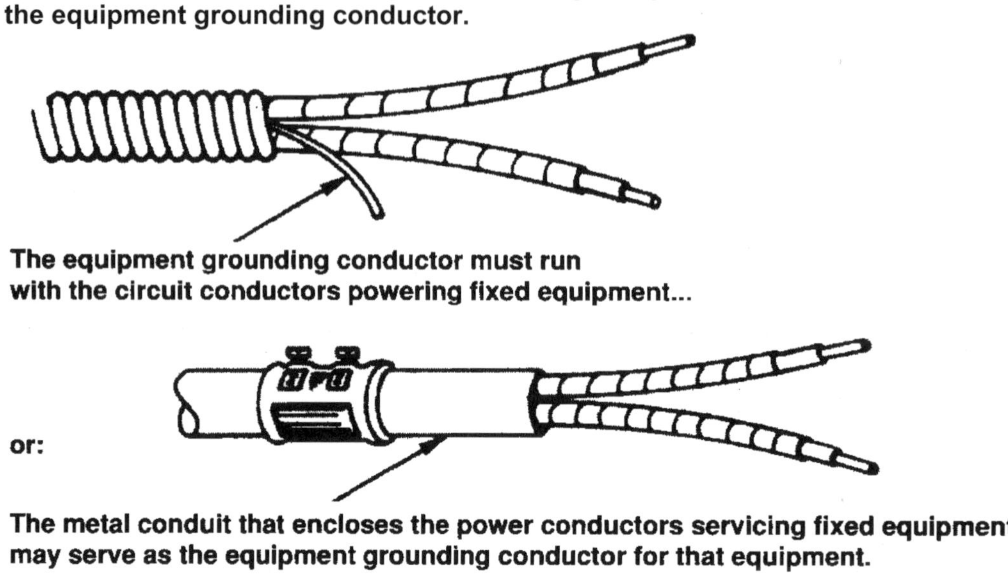

Figure 57 – Grounding method for fixed equipment using circuit conductor enclosures as equipment grounding conductor.

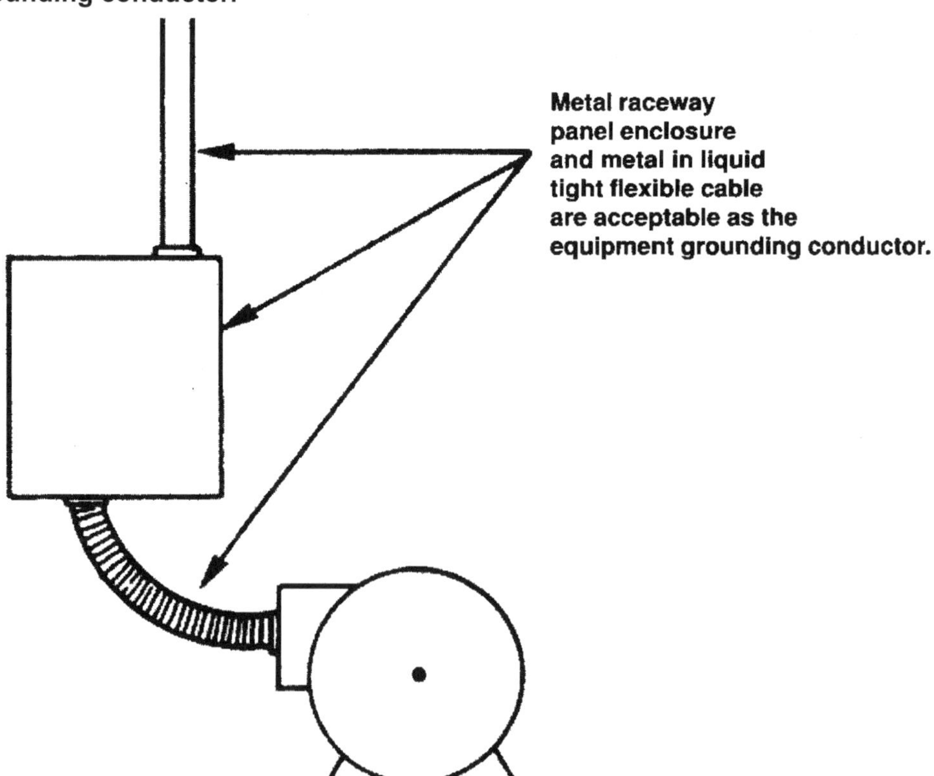

Figure 58 – Motor suitably grounded by its attachment to grounded structure.

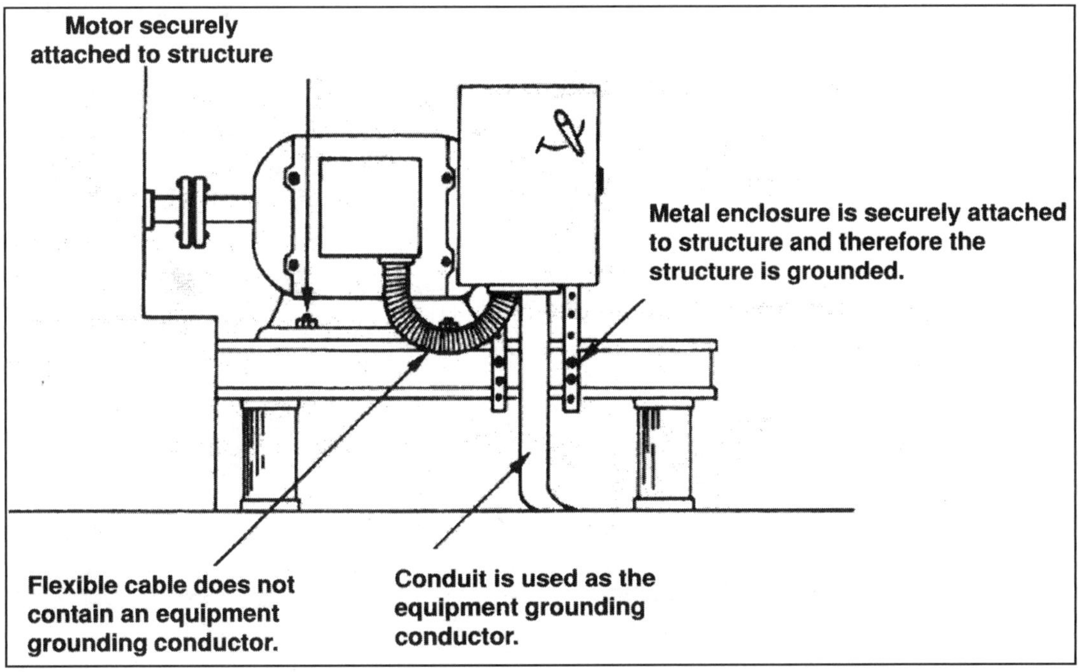

Structural Grounds

If electrical equipment that must be grounded is attached to a metal rack or structure that is grounded, and there is a good electrical connection between the equipment and the metal rack or structure, then the equipment is considered to be grounded. The structure must be effectively grounded by a grounded raceway or equipment grounding conductor run with the circuit conductors serving the equipment on the structure. Figure 58 shows a motor suitably grounded through its attachments to a grounded structure.

Metal elevator car frames are considered to be effectively grounded because there is continuous metal-to-metal contact between the car and the grounded elevator machine through the metal hoisting cables that are attached to both. Figure 59 shows metal hoisting cables for an elevator that are considered effectively grounded. Metal hoisting cables attached to or running over metal sheaves or drums of grounded elevator machines are considered grounded.

(g)(9) **Grounding of systems and circuits of 1000 volts and over (high voltage).** If high voltage systems are grounded, they shall comply with all applicable provisions of paragraphs (g)(1) through (g)(8) of this section as supplemented and modified by the following requirements:

It is important to notice that (g)(9) is a sudden change in subject from "equipment grounding" to "system and circuit grounding." Also, that (g)(9) covers high-voltage systems in general, whereas (g)(9)(i) covers only the very narrow subject of high-voltage systems supplying portable or mobile equipment.

(g)(9)(i) Systems supplying portable or mobile high voltage equipment, other than substations installed on a temporary basis, shall comply with the following:

(g)(9)(i)(A) The system shall have its neutral grounded through an impedance. If a delta-connected high voltage system is used to supply the equipment, a system neutral shall be derived.

Figure 59 – Equipment considered effectively grounded. The arrows in the figure indicate metal hoisting cables running over the metal sheave of a grounded elevator.

Grounding Portable High Voltage Systems

A delta-connected AC circuit, by its design, does not have a neutral conductor; therefore, when this type of high voltage system supplies power to portable equipment, a neutral is derived by means of a grounding transformer. This gives the system the advantages of a grounded system, including protection from accidental grounding. This is required because there is greater risk of accidental grounding on a portable system. (See Figure 60.)

Because of their current flows, high voltage systems that operate at 1000 volts or more cannot be grounded using low-impedance grounding conductors as is done on low-voltage systems. "Impedance" in an AC system is the total opposition to current flow. Portable high voltage equipment must be supplied with power from an electrical system that has its neutral conductor grounded through an impedance. An "impedance" is a device that, when it is placed in a circuit, will oppose current flow, thus aiding its control. The impedance controls the fault current so as not to damage the system. Figure 61 shows the location of an impedance in a grounding connection.

This explanation is specifically directed toward the requirements for high-voltage systems described in 1910.304(g)(9)(i), but the principles apply to grounding of other high-voltage systems as well. The intentional addition of impedance in a ground-fault path seems to be a contradiction of the requirement that effective grounding provide a low-impedance ground-fault path. The explanation is that the added impedance is high enough to control the fault current, but low enough to minimize the danger of voltages between accessible surfaces and to quickly operate the protective devices. In the high-voltage systems under discussion, the ground-fault path, even with the intentional impedance, will still be a fairly low-impedance path. And the protective devices used are more sensitive to ground fault current than are the fuses and circuit breakers used in low-voltage systems. (See 1910.304(g)(9)(i)(c).)

(g)(9)(i)(B) Exposed non-current-carrying metal parts of portable and mobile equipment shall be connected by an equipment grounding conductor to the point at which the system neutral impedance is grounded.

Figure 60 – Derived neutral for an ungrounded delta-connected system showing the neutral grounded through an impedance.

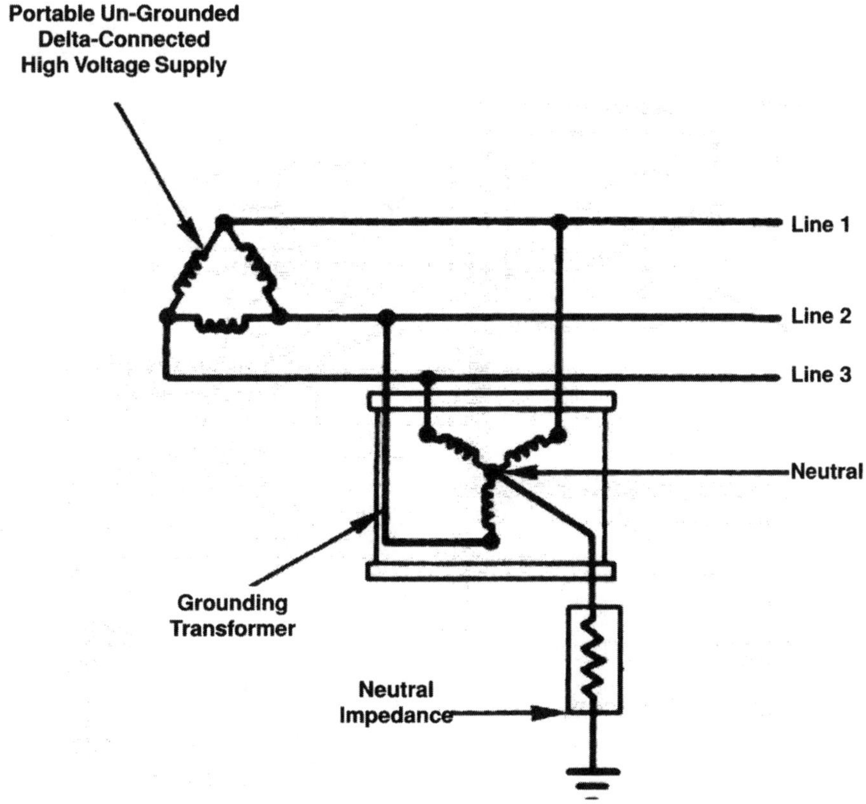

Figure 61 – Non-current-carrying metal parts connected.

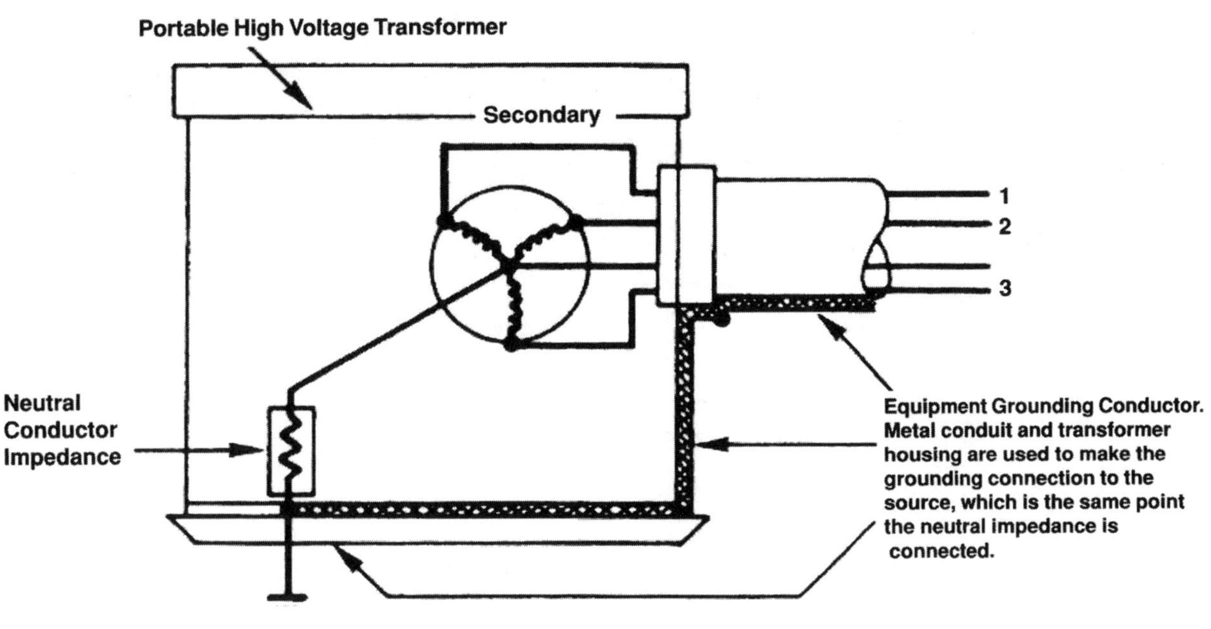

Title: Wiring design and protection.

Subpart: S

SubPart Title: Electrical - Design Safety Standards for Electrical Systems

Equipment Grounding Conductor

All exposed noncurrent-carrying metal parts and the metal enclosures, must be electrically interconnected with an equipment grounding conductor. Either the metal raceways, metal boxes, and other metal enclosures must be bonded together, or a separate conductor must be run with the circuit conductors and identified as the equipment grounding conductor. Figure 61 illustrates where the equipment grounding conductor is to be connected to ground. It should not be connected ahead of the neutral impedance. Making the grounding connection at the proper point facilitates the operation of ground-fault relays and protects the rest of the system from the damaging fault current.

(g)(9)(i)(C) Ground-fault detection and relaying shall be provided to automatically deenergize any high voltage system component that has developed a ground fault. The continuity of the equipment grounding conductor shall be continuously monitored so as to deenergize automatically the high voltage feeder to the portable equipment upon loss of continuity of the equipment grounding conductor.

Ground-Fault Detection and Relaying.

Portable high-voltage equipment must have a means of detecting faults and automatically disconnecting the power to the equipment if a ground fault develops. The detection system must continually monitor the system so that power will be disconnected as soon as a fault occurs.

In Figure 62, if a ground fault occurs at (A), the fault current will travel along the path of the equipment grounding conductor to the point where the neutral conductor impedance is grounded. It will then return to the center point of the transformer windings, to the neutral, passing through the ground-fault detector. Once the ground-fault detector senses fault current, a signal is transmitted to the ground fault relay, which will operate a disconnect to de-energize high voltage system components that have developed a fault. In addition, the continuity of the equipment grounding conductor must be continuously monitored. Should a break occur in the equipment grounding conductor, the circuit must automatically be de-energized.

Figure 63 is a photograph of a ground detector that monitors three legs of a 3-phase system.

Figure 62 – Ground fault detection and relaying

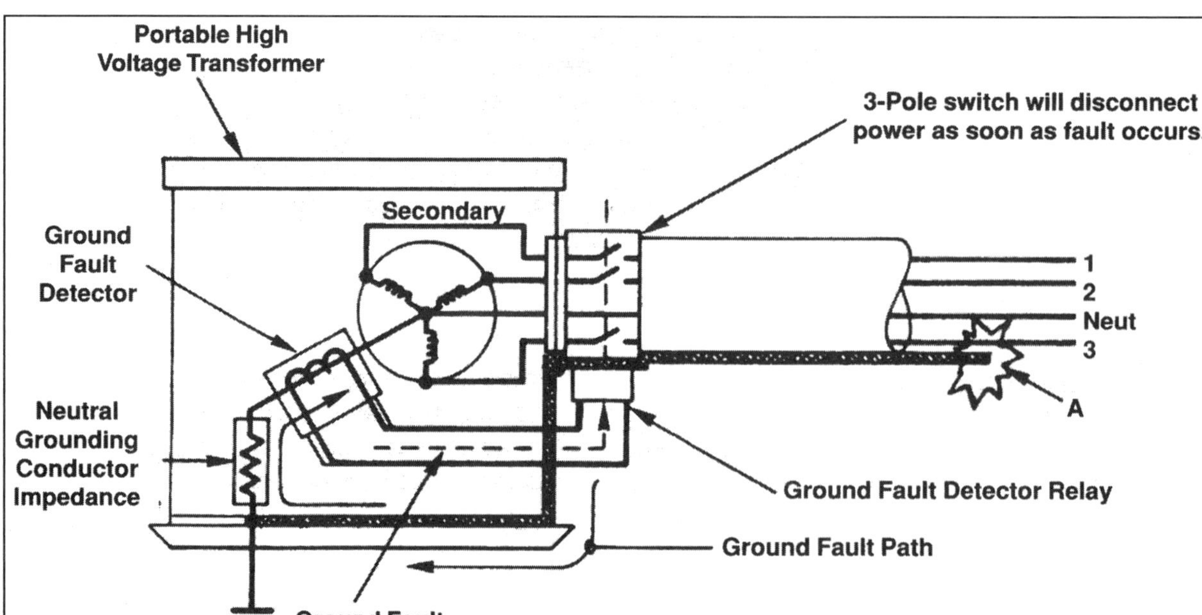

Figure 63 – Typical ground-fault indicator lights integrated into electrical equipment to indicate a ground-fault on a 3-phase delta system. Photo courtesy of Robert H Keis, retired chief electrical inspector. (www.bobkeis.com)

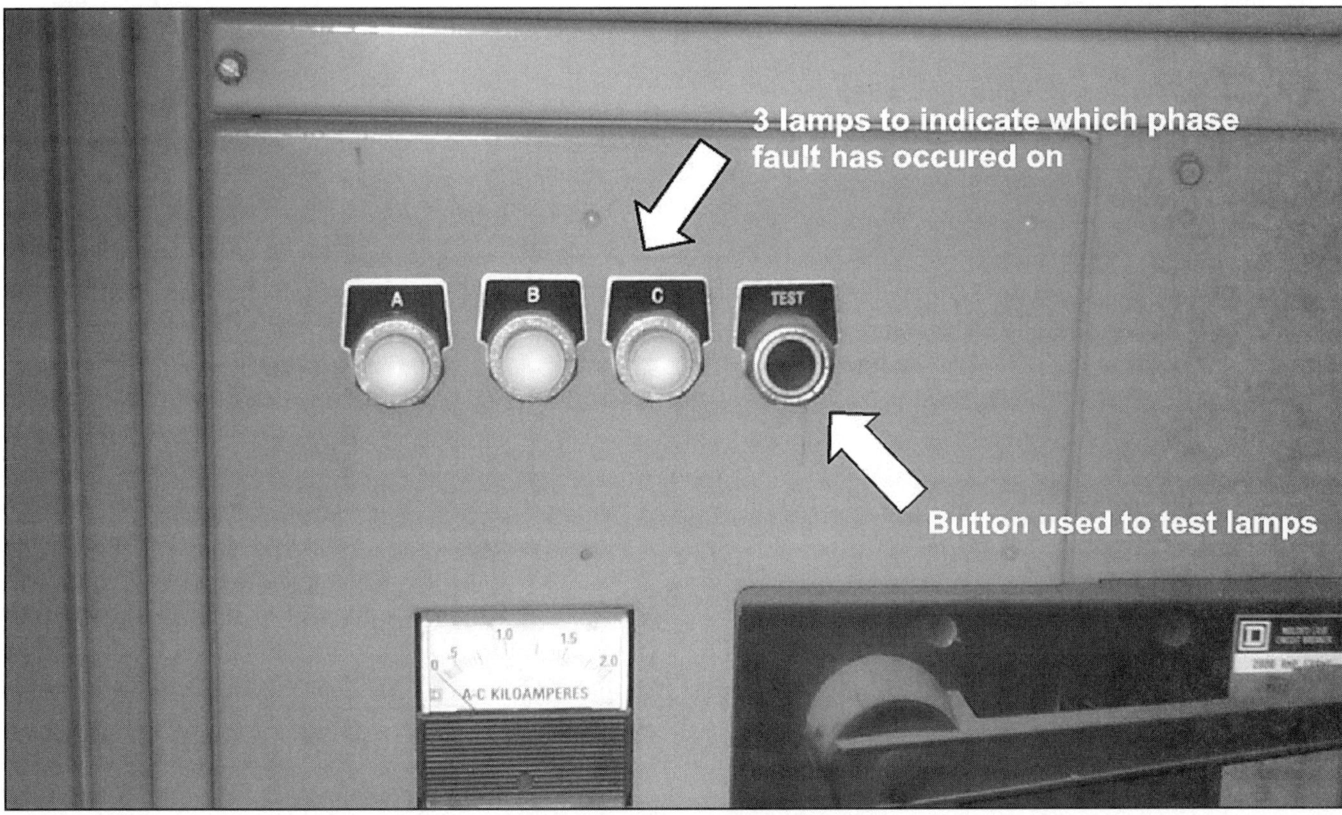

(g)(9)(i)(D) The grounding electrode to which the portable equipment system neutral impedance is connected shall be isolated from and separated in the ground by at least 6.1 m (20.0 ft) from any other system or equipment grounding electrode, and there shall be no direct connection between the grounding electrodes, such as buried pipe, fence, and so forth.

Grounding Electrodes

Grounding electrodes of systems supplying portable or mobile equipment shall be separated from any other electrical system or any other grounding electrodes. There must not be any direct electrical connection between these grounding electrodes and other objects such as buried pipes, fences, etc., to prevent interference between systems. See Figure 64.

(g)(9)(ii) All noncurrent-carrying metal parts of portable equipment and fixed equipment, including their associated fences, housings, enclosures, and supporting structures, shall be grounded. However, equipment that is guarded by location and isolated from ground need not be grounded. Additionally, pole-mounted distribution apparatus at a height exceeding 2.44 m (8.0 ft) above ground or grade level need not be grounded.

Figure 64 – Isolated grounding electrode.

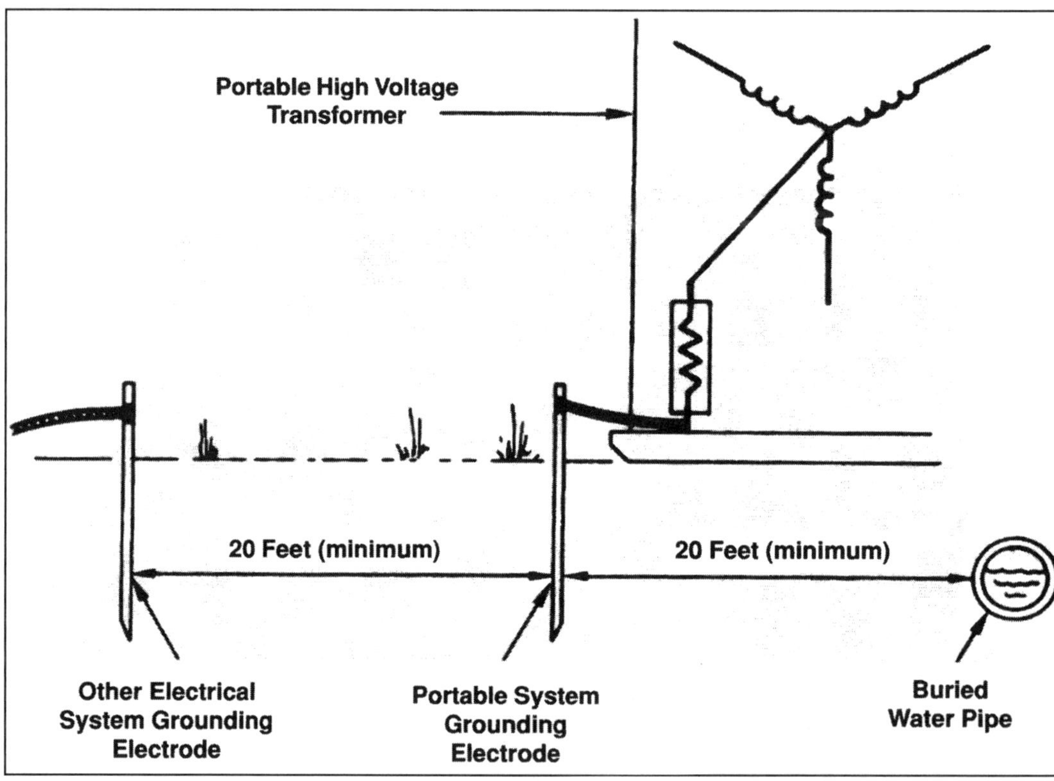

§ 1910.305 Wiring Methods, Components and Equipment for General Use

[46 FR 4056, Jan. 16, 1981; 46 FR 40185, Aug. 7, 1981; 72 FR 7201, Feb. 14, 2007]

(a) **Wiring methods.** The provisions of this section do not apply to the conductors that are an integral part of factory-assembled equipment.

(a)(1) **General requirements.**

(a)(1)(i) Metal raceways, cable trays, cable armor, cable sheath, enclosures, frames, fittings, and other metal noncurrent-carrying parts that are to serve as grounding conductors, with or without the use of supplementary equipment grounding conductors, shall be effectively bonded where necessary to ensure electrical continuity and the capacity to conduct safely any fault current likely to be imposed on them. Any nonconductive paint, enamel, or similar coating shall be removed at threads, contact points, and contact surfaces or be connected by means of fittings designed so as to make such removal unnecessary.

This is a basic requirement in both the OSHA and NEC® requirements. It emphasizes the importance of the low-impedance ground-fault path closely following the supply conductors. OSHA added additional wording consistent with the NFPA requirements to emphasize the importance of ensuring a effective connection between metal parts of electrical systems by removing paints, enamels or similar coatings on threads, contact points, and other contact surfaces.

(a)(1)(ii) Where necessary for the reduction of electrical noise (electromagnetic interference) of the grounding circuit, an equipment enclosure supplied by a branch circuit may be isolated from a raceway containing circuits supplying only that equipment by one or more listed nonmetallic raceway fittings located at the point of attachment of the raceway to the equipment enclosure. The metal raceway shall be supplemented by an internal insulated equipment grounding conductor installed to ground the equipment enclosure.

This section is consistent with NEC® 250.96(B), and allows sensitive electronic equipment to be isolated from the raceway to reduce electromagnetic interference, similar to how cord-and-plug-connected equipment would be. This option applies only to the metal equipment enclosure supplied by a branch circuit, and no other wiring, raceways, or equipment would be permitted beyond the insulated fitting. A fine print note (FPN) is included in the NEC and NFPA 70E which emphasizes that isolating the equipment grounding conductor does *not* relieve the requirement for grounding the raceway. The primary concern remains creating an effective ground fault path for the equipment.

(a)(1)(iii) No wiring systems of any type may be installed in ducts used to transport dust, loose stock, or flammable vapors. No wiring system of any type may be installed in any duct used for vapor removal or for ventilation of commercial-type cooking equipment, or in any shaft containing only such ducts.

(a)(2) **Temporary wiring.** Except as specifically modified in this paragraph, all other requirements of this subpart for permanent wiring shall also apply to temporary wiring installations.

It should be noted that the coverage of temporary wiring in this Part 1910 of the CFR applies to temporary wiring used for any purpose whatsoever. In the special case of temporary wiring used on construction sites, there are additional requirements in the Construction Standards of OSHA, CFR 1926, mostly in Subpart K.

Temporary Wiring

The requirements for temporary wiring for power and lighting purposes include provisions for wire connections, junction boxes, and overcurrent protection as well as the use of conductors. Temporary wiring does not have to meet all of the requirements that permanent wiring must meet. For example, an electrical meter on a construction site can be mounted on a pole or other temporary support, with open insulated conductors used to connect it. If the meteere part of a permanent installation, it would have to be mounted on a permanent structure, such as a building, and the conductors would normally be placed in conduit for permanent protection. Figure 65 shows a permissible temporary wiring arrangement.

1910.305

Title: Wiring methods, components, and equipment for general use.

Subpart: S

SubPart Title: Electrical - Design Safety Standards for Electrical Systems

One should visualize that an additional equipment grounding conductor is carried along with temporary feeders and branch circuits shown in Figure 65 and in Figures 66 and 67.

(a)(2)(i)	Temporary electrical power and lighting installations of 600 volts, nominal, or less may be used only as follows:
(a)(2)(i)(A)	During and for remodeling, maintenance, repair, or demolition of buildings, structures, or equipment, and similar activities;
(a)(2)(i)(B)	For a period not to exceed 90 days for Christmas decorative lighting, carnivals, and similar purposes; or
(a)(2)(i)(C)	For experimental or development work, and during emergencies.
(a)(2)(ii)	Temporary wiring shall be removed immediately upon completion of the project or purpose for which the wiring was installed.
(a)(2)(iii)	Temporary electrical installations of more than 600 volts may be used only during periods of tests, experiments, emergencies, or construction-like activities.

Figure 65 – Temporary wiring arrangement permissible for duration of construction, remodeling, maintenance, repair, or demolition of structures.

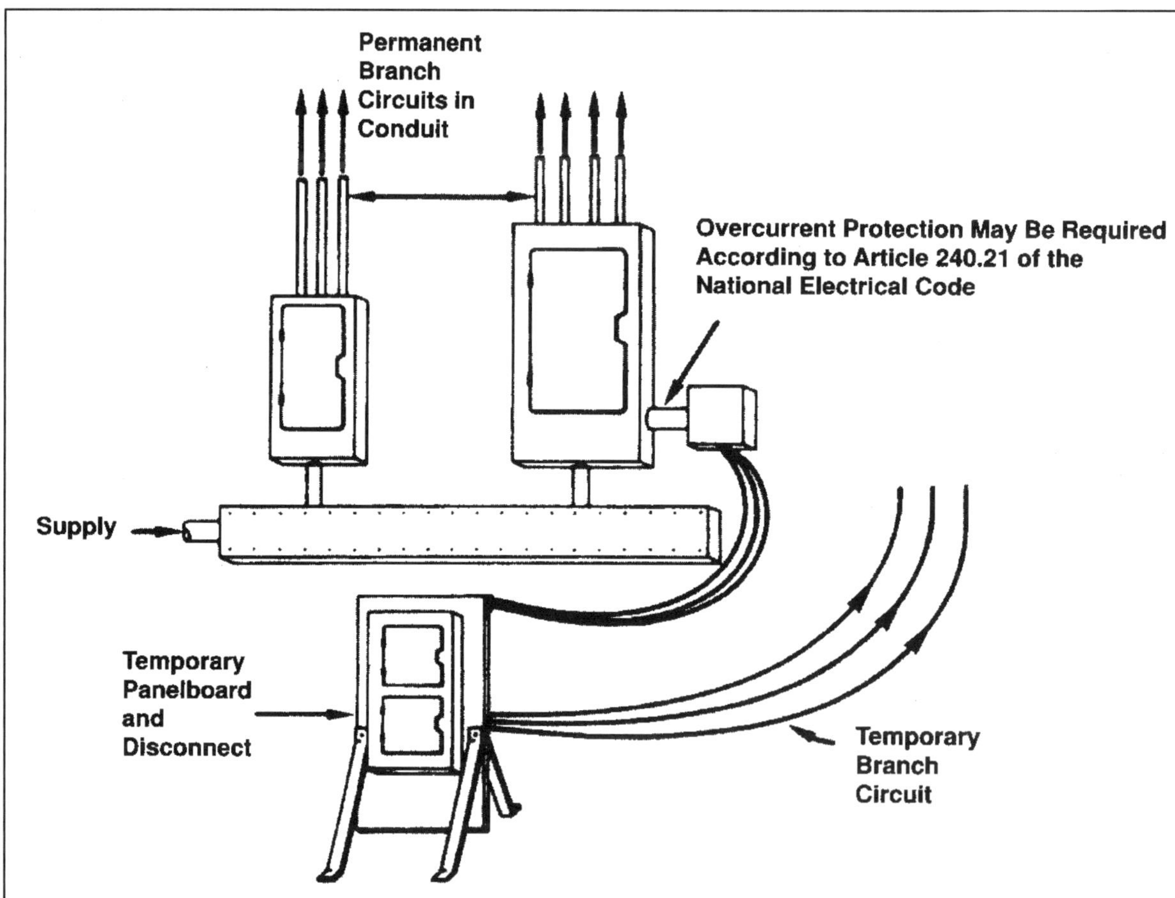

The wiring methods described in (a)(2) and in the National Electrical Code® Article 590 are used for temporary wiring from panelboards to fixed outlets. This is one of a very few places where flexible cord may be used as a substitute for the fixed wiring of a structure. When so used, the cord will generally be suspended from joists, etc., as it is not required to be flexed. When portable tools or appliances are plugged into these fixed outlets, flexible cord will most likely be used as extensions of the supply cord on the appliance. Extension cord sets used in this conventional, flexing manner are generally considered to be pieces of equipment, as described in the National Electrical Code® Article 400, not fixed wiring as described in the National Electrical Code® Article 590. (This is also true of the diagram in Figure 67.) There have been some misunderstandings regarding the plugging of one extension cord set into another extension cord set to extend the overall length. This is permitted by NFPA and not in violation of any standards.

OSHA also adopted an important requirement in 1910.305(a)(2)(ii). Temporary wiring must be disconnected and removed *immediately* upon completion of the project or purpose for which it was installed.

Feeder Circuits

Feeders (as defined in 1910.399) are conductors between the service equipment, or the generator switchboard of an isolated plant, and the final branch-circuit overcurrent device.

Feeders for temporary wiring must originate inside an approved distribution center, such as a panelboard, that is rated for the voltages and current the system is expected to carry. Some equipment is manufactured specifically for temporary use.

Feeders may be run as multi-conductor cords or cables, i.e., two or more conductors, each with their own insulation, run together in the same cord. When feeder conductors will not be subjected to physical damage, they can be installed as open conductors. However, they must be installed on insulating supports that are no more than 10 feet apart (see Figure 66).

Beginning with the 1996 edition of the National Electrical Code®, open conductors on insulators are permitted as feeders for emergencies & tests only. Although Federal OSHA is not enforcing this requirement, some state OSHA offices do, along with many local code enforcement agencies.

(a)(2)(v) The following requirements apply to branch circuits:

(a)(2)(v)(A) Branch circuits shall originate in an approved power outlet or panelboard.

(a)(2)(v)(B) Conductors shall be multiconductor cord or cable assemblies or open conductors. If run as open conductors, they shall be fastened at ceiling height every 3.05 m (10.0 ft).

(a)(2)(v)(C) No branch-circuit conductor may be laid on the floor.

(a)(2)(v)(D) Each branch circuit that supplies receptacles or fixed equipment shall contain a separate equipment grounding conductor if run as open conductors.

(a)(2)(iv) The following requirements apply to feeders:

(a)(2)(iv)(A) Feeders shall originate in an approved distribution center.

(a)(2)(iv)(B) Conductors shall be run as multiconductor cord or cable assemblies. However, if installed as permitted in paragraph (a)(2)(i)(C) of this section, and if accessible only to qualified persons, feeders may be run as single insulated conductors.

Figure 66 – Temporary feeders run as open conductors.

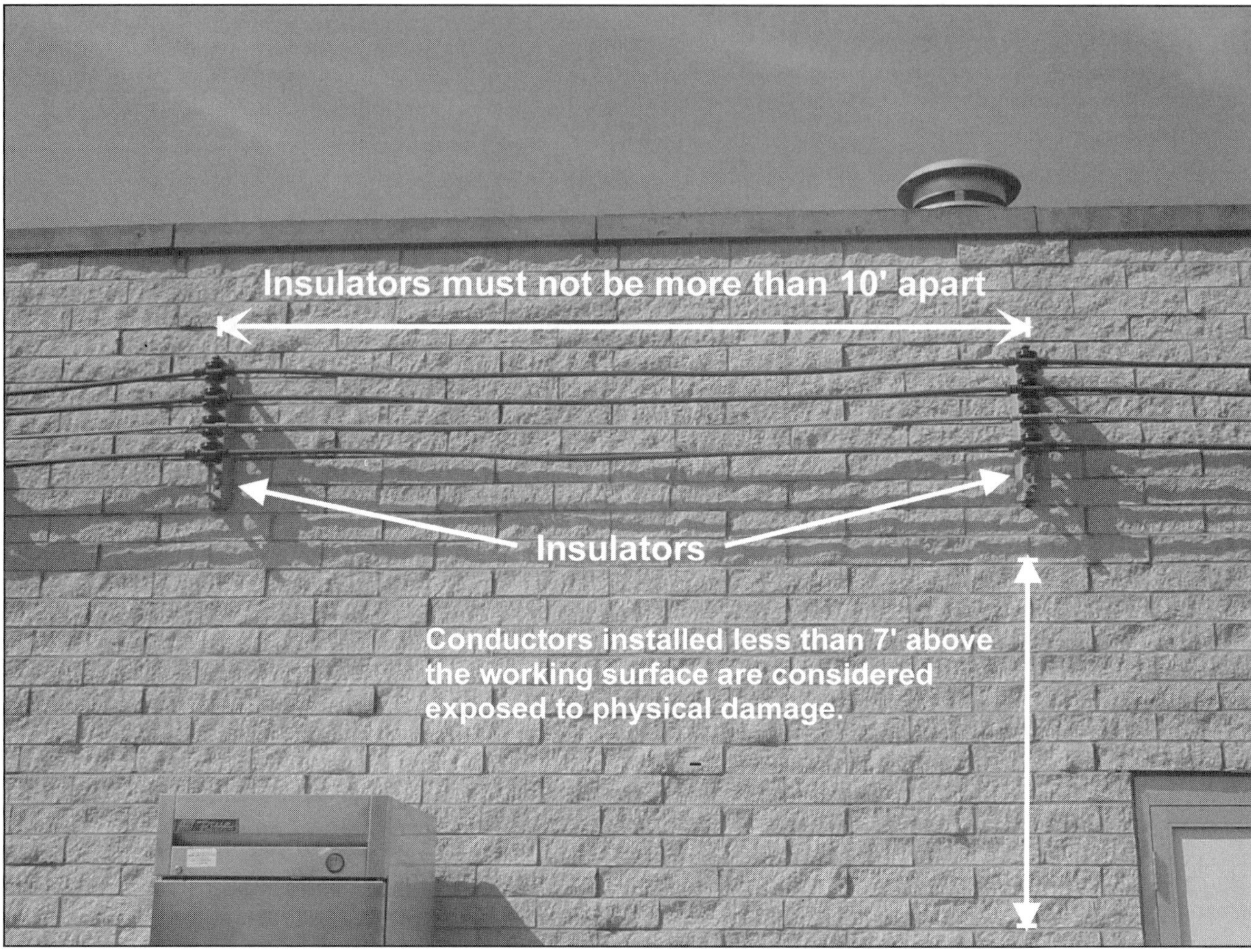

Branch Circuits

Branch circuits are the conductors between the last overcurrent device in an electrical system and the outlets, such as receptacles, lighting fixtures, and electrical equipment wired directly into a circuit. Branch circuits for temporary wiring must originate inside an approved panelboard or power outlet that is rated for the voltages and currents the system is expected to carry. As with feeders, branch circuit conductors can be contained within multi-conductor cord or cable assemblies, or they can be run as open conductors. If run as open conductors, they must be securely attached every 10 feet at ceiling height. The OSHA regulations state that the conductors must be mounted at "ceiling height" whereas the National Electrical Code® states that if the conductors are mounted less than 7 feet above the floor the conductors are considered exposed to physical damage and must be protected. To avoid physical damage, branch circuit conductors must never be placed on the floor (see Figure 67).

If a branch circuit supplies power to a receptacle or to fixed equipment through a system of open conductors, an additional conductor must be run with the circuit conductors to provide equipment grounding.

Figure 67 – Temporary branch circuit run as open wiring on insulators

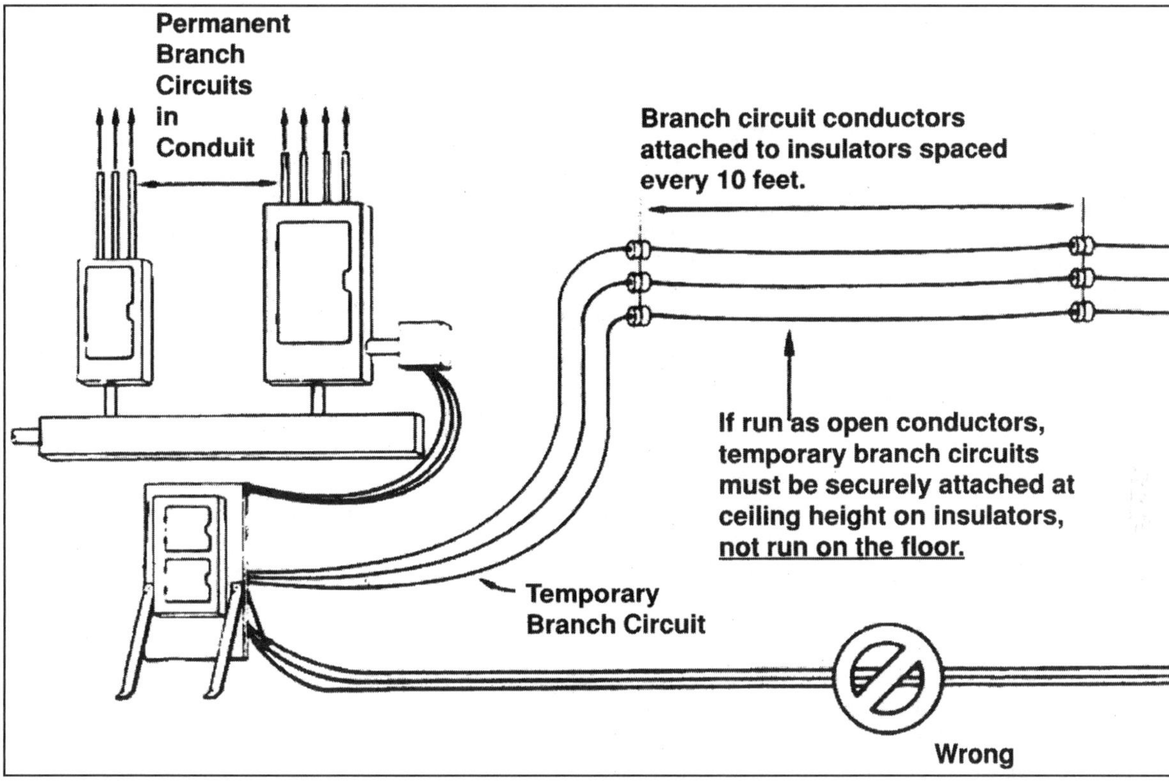

As stated above for feeders, beginning with the 1996 edition of the National Electrical Code®, open conductors on insulators are permitted as branch circuits for emergencies & tests only. Although Federal OSHA is not enforcing this requirement, some state OSHA offices do, along with many local code enforcement agencies.

Receptacles

Receptacles used in temporary wiring circuits must provide a connection for an equipment grounding conductor. Unless the receptacle is supplied by a metallic raceway that provides a continuous grounding path back to the source, a separate equipment grounding conductor must be placed in the branch circuit. There must be an effective electrical connection between the receptacle-grounding terminal and the equipment grounding conductor (see Figure 68).

(a)(2)(vi) Receptacles shall be of the grounding type. Unless installed in a continuous grounded metallic raceway or metallic covered cable, each branch circuit shall contain a separate equipment grounding conductor and all receptacles shall be electrically connected to the grounding conductor.

(a)(2)(vii) No bare conductors nor earth returns may be used for the wiring of any temporary circuit.

Figure 68 – Receptacle used for temporary wiring protected by a GFCI circuit breaker

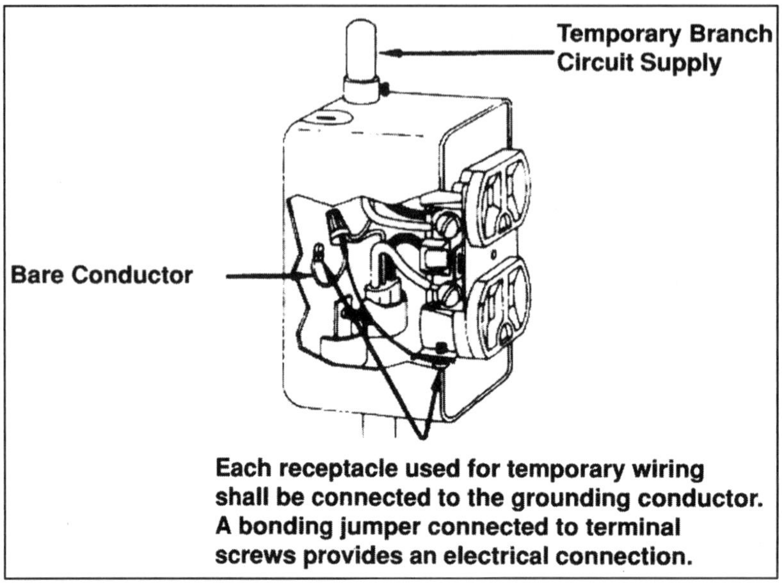

Bare Conductors

Bare conductors are conductors that do not have any coverings whatsoever. Bare conductors mustnot be used for wiring temporary circuits.

Earth Returns

Earth returns use the earth itself to provide a current path back to the supply source. This is done by implanting a grounding electrode at the equipment being served and connecting the equipment to the ungrounded conductor and to the grounding electrode. Since one side of the supply source is also connected to ground through a grounding electrode, a return path exists; however, its effectiveness is dependent on varying soil conditions. Earth returns must not be used for wiring temporary circuits because they are not effective and may present a serious hazard on temporary work sites. Figure 69 shows an earth return, which is not allowed, in contrast to a proper return system.

Disconnecting Means

(a)(2)(viii) Suitable disconnecting switches or plug connectors shall be installed to permit the disconnection of all ungrounded conductors of each temporary circuit. Multiwire branch circuits shall be provided with a means to disconnect simultaneously all ungrounded conductors at the power outlet or panelboard where the branch circuit originated.

Note to paragraph (a)(2)(viii) of this section. Circuit breakers with their handles connected by approved handle ties are considered a single disconnecting means for the purpose of this requirement.

Figure 69 – Earth returns.

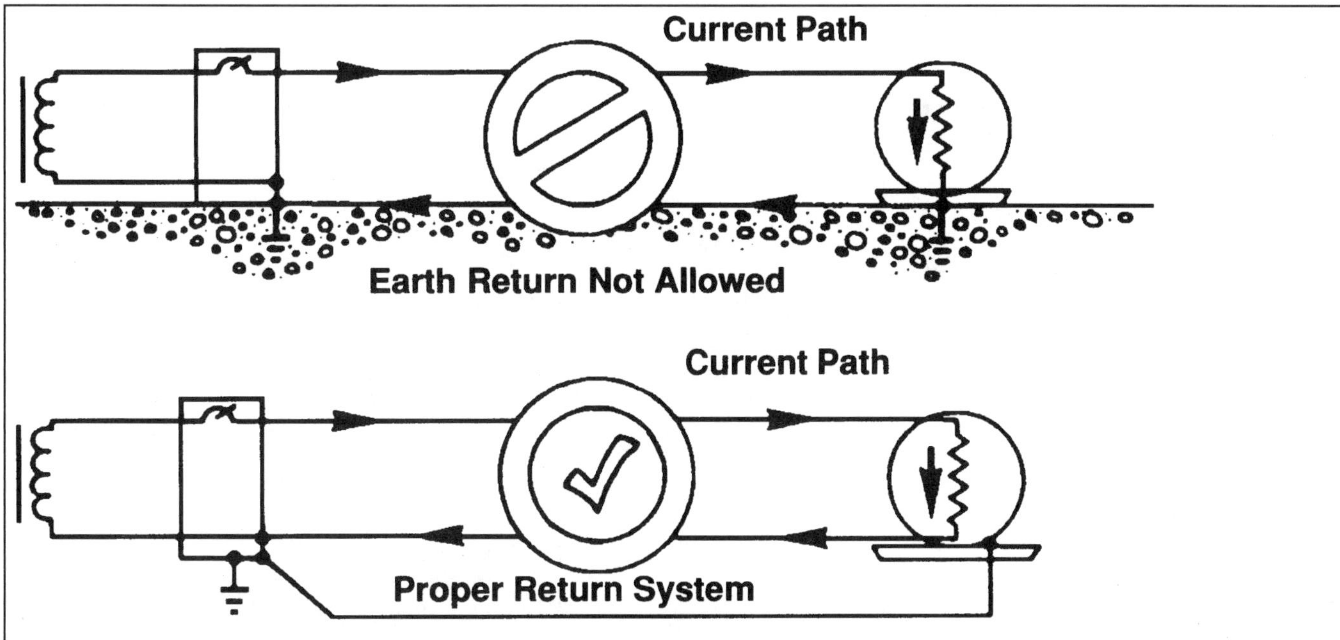

In addition, a separate equipment-grounding conductor must be used to provide a low-impedance path to the source. This path will allow sufficient current to flow to operate the circuit breaker when a fault occurs.

Suitable switches or plug connectors, such as those approved for use in damp locations, must be installed in all ungrounded, or hot, conductors used in temporary circuits. These switches allow power to the conductors to be quickly and safely disconnected in emergencies or for maintenance purposes. Figure 70 shows a plug connector that is acceptable for use with temporary circuits.

The "plug connector" shown is used to disconnect a portable tool from an extension cord, but the extension cord remains energized. The standard requires that a means be provided to also disconnect the extension cord. Generally the extension cord would be plugged into a fixed receptacle, which provides the disconnecting function. If a length of flexible cord is used for fixed temporary wiring, as when solidly wired from a panelboard to fixed receptacles, that fixed length of cord would be disconnected by opening a switch or circuit breaker in the panelboard.

Injuries and fatalities continue to occur in the industry on multi-wire branch circuits. OSHA took the opportunity to incorporate the NFPA requirement that multi-wire branch circuits have a means to disconnect all ungrounded conductors simultaneously.

Figure 70 – Plug connector for use with temporary circuits

Plug connectors like the one shown are permissible for use as disconnecting means for temporary circuits. Connectors like this provide an additional degree of protection in wet and damp locations.

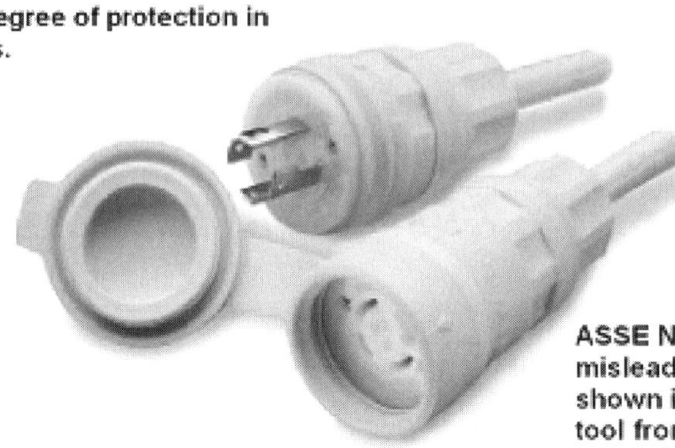

ASSE NOTE: The illustration may be misleading. The "plug connector" as shown is used to disconnect a portable tool from an extension cord, but the extension cord remains energized. The standard requires that a means be provided to also disconnect the extension cord. Generally, the extension cord would be plugged into a fixed receptacle, which provides the disconnecting function.
If a length of flexible cord is used for fixed temporary wiring, as when solidly wired from a panelboard to fixed receptacles, that fixed length of cord would be disconnected by opening a switch or circuit breaker in the panelboard.

(a)(2)(ix) All lamps for general illumination shall be protected from accidental contact or breakage by a suitable fixture or lampholder with a guard. Brass shell, paper-lined sockets, or other metal-cased sockets may not be used unless the shell is grounded.

(a)(2)(x) Flexible cords and cables shall be protected from accidental damage, as might be caused, for example, by sharp corners, projections, and doorways or other pinch points.

(a)(2)(xi) Cable assemblies and flexible cords and cables shall be supported in place at intervals that ensure that they will be protected from physical damage. Support shall be in the form of staples, cables ties, straps, or similar type fittings installed so as not to cause damage.

Unlike NEC® section 400.8(4), flexible cords are permitted to be attached to building **surfaces** where used for temporary wiring.

(a)(3) **Cable trays.**

(a)(3)(i) Only the following wiring methods may be installed in cable tray systems: armored cable; electrical metallic tubing; electrical nonmetallic tubing; fire alarm cables; flexible metal conduit; flexible metallic tubing; instrumentation tray cable; intermediate metal conduit; liquidtight flexible metal conduit; liquidtight flexible nonmetallic conduit; metal-clad cable; mineral-insulated, metal-sheathed cable; multiconductor service-entrance cable; multiconductor underground feeder and branch-circuit cable; multipurpose and communications cables; nonmetallic-sheathed cable; power and control tray cable; power-limited tray cable; optical fiber cables; and other factory-

	assembled, multiconductor control, signal, or power cables that are specifically approved for installation in cable trays, rigid metal conduit, and rigid nonmetallic conduit.
(a)(3)(ii)	In industrial establishments where conditions of maintenance and supervision assure that only qualified persons will service the installed cable tray system, the following cables may also be installed in ladder, ventilated-trough, or ventilated-channel cable trays:
(a)(3)(ii)(A)	Single conductor cable; the cable shall be No. 1/0 or larger and shall be of a type listed and marked on the surface for use in cable trays; where Nos. 1/0 through 4/0 single conductor cables are installed in ladder cable tray, the maximum allowable rung spacing for the ladder cable tray shall be 229 mm (9 in.); where exposed to direct rays of the sun, cables shall be identified as being sunlight resistant;
(a)(3)(ii)(B)	Welding cables installed in dedicated cable trays;
(a)(3)(ii)(C)	Single conductors used as equipment grounding conductors; these conductors, which may be insulated, covered, or bare, shall be No. 4 or larger; and
(a)(3)(ii)(D)	Multiconductor cable, Type MV; where exposed to direct rays of the sun, the cable shall be identified as being sunlight-resistant.
(a)(3)(iii)	Metallic cable trays may be used as equipment grounding conductors only where continuous maintenance and supervision ensure that qualified persons will service the installed cable tray system.
(a)(3)(iv)	Cable trays in hazardous (classified) locations may contain only the cable types permitted in such locations. (See § 1910.307.)
(a)(3)(v)	Cable tray systems may not be used in hoistways or where subjected to severe physical damage.

Cable tray requirements are more extensively covered by NEC® Article 392. Both the requirements stated here and those of the National Electrical Code® are commonly found violated in field installations. In 2005, as a response to a high number of violations and incidents recorded, OSHA posted a Fact Sheet on their web site specifically addressing the safety hazards of overloading cable trays, as shown in Figure 71. The fact sheet can be found at http://www.osha.gov/Publications/cable_trays_fs.html, or by searching for "Electrical Safety Hazards of Overloading Cable Trays Fact Sheet." The bulk of this section is dedicated to the types of wiring methods permitted in cable trays; it should be noted that flexible cords (multi-conductor power cables), which are listed for use in cable tray are permitted. As of this printing, no flexible cords currently marketed are listed for use in a cable tray.

(a)(4)	**Open wiring on insulators.**
(a)(4)(i)	Open wiring on insulators is only permitted on systems of 600 volts, nominal, or less for industrial or agricultural establishments, indoors or outdoors, in wet or dry locations, where subject to corrosive vapors, and for services.
(a)(4)(ii)	Conductors smaller than No. 8 shall be rigidly supported on noncombustible, nonabsorbent insulating materials and may not contact any other objects. Supports shall be installed as follows:
(a)(4)(ii)(A)	Within 152 mm (6 in.) from a tap or splice;
(a)(4)(ii)(B)	Within 305 mm (12 in.) of a dead-end connection to a lampholder or receptacle; and
(a)(4)(ii)(C)	At intervals not exceeding 1.37 m (4.5 ft), and at closer intervals sufficient to provide adequate support where likely to be disturbed.
(a)(4)(iii)	In dry locations, where not exposed to severe physical damage, conductors may be separately enclosed in flexible nonmetallic tubing. The tubing shall be in continuous lengths not exceeding 4.57 m (15.0 ft) and secured to the surface by straps at intervals not exceeding 1.37 m (4.5 ft).

Figure 71 – Overloaded multiple, ladder-type, cable trays.

(a)(4)(iv)	Open conductors shall be separated from contact with walls, floors, wood cross members, or partitions through which they pass by tubes or bushings of noncombustible, nonabsorbent insulating material. If the bushing is shorter than the hole, a waterproof sleeve of nonconductive material shall be inserted in the hole and an insulating bushing slipped into the sleeve at each end in such a manner as to keep the conductors absolutely out of contact with the sleeve. Each conductor shall be carried through a separate tube or sleeve.
(a)(4)(v)	Where open conductors cross ceiling joints and wall studs and are exposed to physical damage (for example, located within 2.13 m (7.0 ft) of the floor), they shall be protected.

Open wiring on insulators is an exposed wiring method, which is not permitted to be concealed. The National Electrical Code® covers Open Wiring on Insulators in detail in Article 398.

(b)	**Cabinets, boxes, and fittings.**
(b)(1)	**Conductors entering boxes, cabinets or fittings.**
(b)(1)(i)	Conductors entering boxes, cabinets, or fittings shall also be protected from abrasion, and openings through which conductors enter shall be effectively closed. Unused openings in cabinets, boxes, and fittings shall be effectively closed.
(b)(1)(ii)	Unused openings in cabinets, boxes, and fittings shall be effectively closed.
(b)(1)(iii)	Where cable is used, each cable shall be secured to the cabinet, cutout box, or meter socket enclosure. However, where cable with an entirely nonmetallic sheath enters the top of a surface-mounted enclosure through one or more nonflexible raceways not less than 457 mm (18 in.) or more than 3.05 m (10.0 ft) in length, the cable need not be secured to the cabinet, box, or enclosure provided all of the following conditions are met:
(b)(1)(iii)(A)	Each cable is fastened within 305 mm (12 in.) of the outer end of the raceway, measured along the sheath;
(b)(1)(iii)(B)	The raceway extends directly above the enclosure and does not penetrate a structural ceiling;
(b)(1)(iii)(C)	A fitting is provided on each end of the raceway to protect the cable from abrasion, and the fittings remain accessible after installation;

(b)(1)(iii)(D)	The raceway is sealed or plugged at the outer end using approved means so as to prevent access to the enclosure through the raceway;
(b)(1)(iii)(E)	The cable sheath is continuous through the raceway and extends into the enclosure not less than 6.35 mm (0.25 in.) beyond the fitting;
(b)(1)(iii)(F)	The raceway is fastened at its outer end and at other points as necessary; and
(b)(1)(iii)(G)	Where installed as conduit or tubing, the allowable cable fill does not exceed that permitted for complete conduit or tubing systems.

Conductors Entering Boxes, Cabinets, or Fittings

Since conductors can be damaged if they rub against the sharp edges of cabinets, boxes, or fittings, they must be protected from damage where they enter. To protect the conductors, some type of clamp or rubber grommet must be used. The device used must close the hole through which the conductor passes as well as provide protection from abrasion. If the conductor is in a conduit and the conduit fits tightly in the opening, additional sealing is not required.

The knockouts in cabinets, boxes, and fittings should be removed only if conductors are to be run through them. However, if a knockout is missing or if there is another hole in the box, the hole or opening must be closed (see Figures 72A and B). (See also 1910.303(b)(7))

(b)(2)	**Covers and canopies**.
(b)(2)(i)	All pull boxes, junction boxes, and fittings shall be provided with covers identified for the purpose. If metal covers are used, they shall be grounded. In completed installations, each outlet box shall have a cover, faceplate, or fixture canopy. Covers of outlet boxes having holes through which flexible cord pendants pass shall be provided with bushings designed for the purpose or shall have smooth, well-rounded surfaces on which the cords may bear.
(b)(2)(ii)	Where a fixture canopy or pan is used, any combustible wall or ceiling finish exposed between the edge of the canopy or pan and the outlet box shall be covered with noncombustible material.
(b)(3)	**Pull and junction boxes for systems over 600 volts, nominal.** In addition to other requirements in this section, the following requirements apply to pull and junction boxes for systems over 600 volts, nominal:
(b)(3)(i)	Boxes shall provide a complete enclosure for the contained conductors or cables.
(b)(3)(ii)	Boxes shall be closed by suitable covers securely fastened in place.
	Note to paragraph (b)(3)(ii) of this section: Underground box covers that weigh over 45.4 kg (100 lbs) meet this requirement.
(b)(3)(iii)	Covers for boxes shall be permanently marked "HIGH VOLTAGE." The marking shall be on the outside of the box cover and shall be readily visible and legible.
(c)	**Switches.**
(c)(1)	**Single-throw knife switches.** Single-throw knife switches shall be so placed that gravity will not tend to close them. Single-throw knife switches approved for use in the inverted position shall be provided with a locking device that will ensure that the blades remain in the open position when so set.
(c)(2)	**Double-throw knife switches.** Double-throw knife switches may be mounted so that the throw will be either vertical or horizontal. However, if the throw is vertical, a locking device shall be provided to ensure that the blades remain in the open position when so set.

Figure 72A – Unused openings in boxes, raceways, auxiliary gutters, cabinets, equipment cases, or housings shall be effectively closed to afford protection substantially equivalent to the wall of the equipment. Note the unused opening in the 4-Square box above.

Figure 72B – Various standard trade sizes of knockout covers in both plastic and metal are available to seal unused openings, such as that shown in Figure 72A.

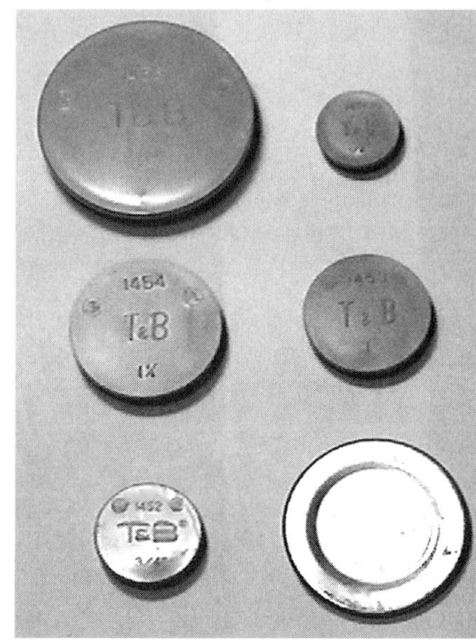

Knife Switches

Single-throw knife switches have one energized (closed or "on") position and one open (dead or "off") position. The switch must be designed so that when it is in the open position, the blades are not energized (i.e., the blades must be connected to the load side, not the supply side of the circuit). The switch must also be installed so that if it falls downward, it will not fall into its energized position. However, some single-throw knife switches are designed to be installed so that they open upward.

To be approved for this type of installation, they must have a latch or other locking device (such as a spring-loaded device) to secure the switch in the open position. The illustration in Figure 73A shows a single-throw knife switch connected so that the blades are dead when the switch is open. Additionally, Figure 73B shows a latch arrangement that holds the blade in the open position and will prevent gravity from pulling the switch closed.

Figure 73 – Knife switches (A and B)

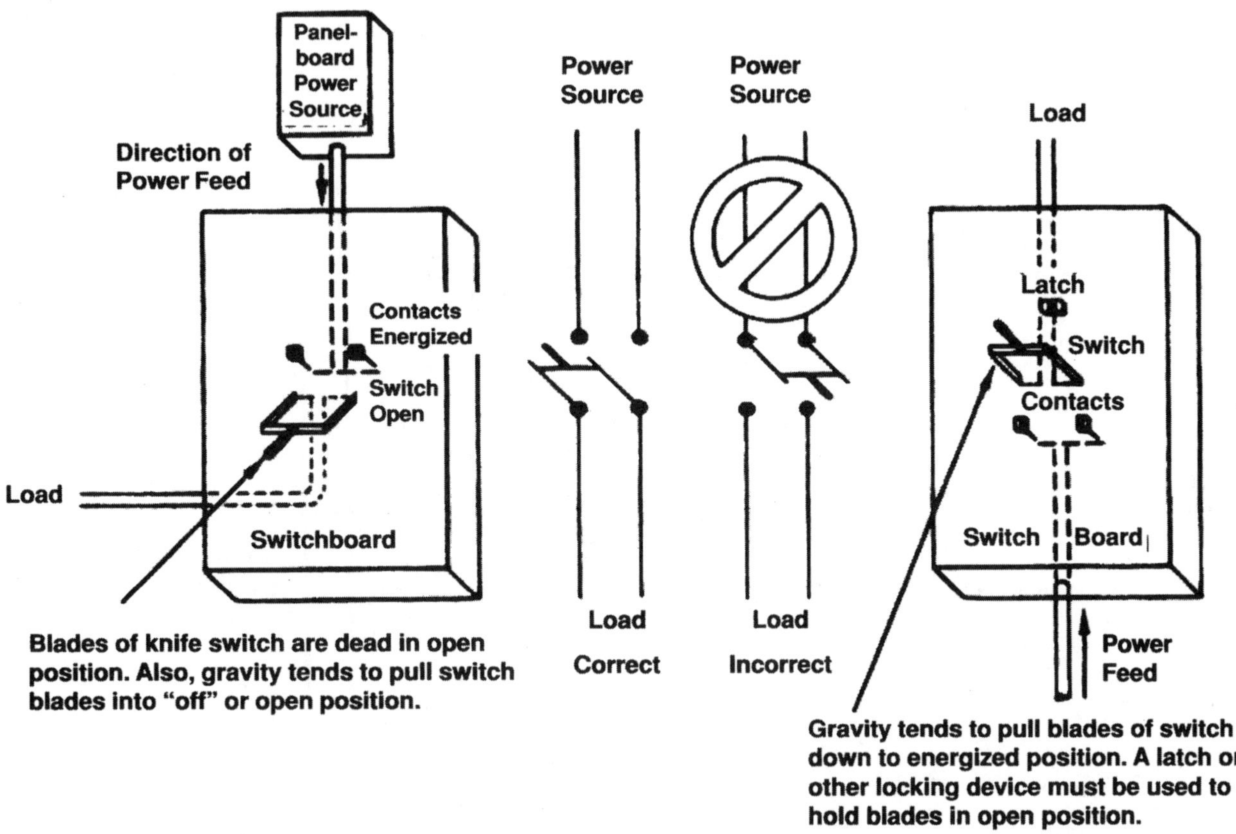

Double-Throw Knife Switches

Double-throw knife switches are knife switches that have two energized (closed or "on") positions and one open (dead or "off") position. These switches can be mounted vertically so that they are moved up and down, or horizontally so that they are moved back and forth. If switches are mounted vertically, they must have a locking device (such as a spring-loaded device) that will hold the switch blades in the open position (see Figure 74).

Figure 74 – Double-throw knife switches with locking device

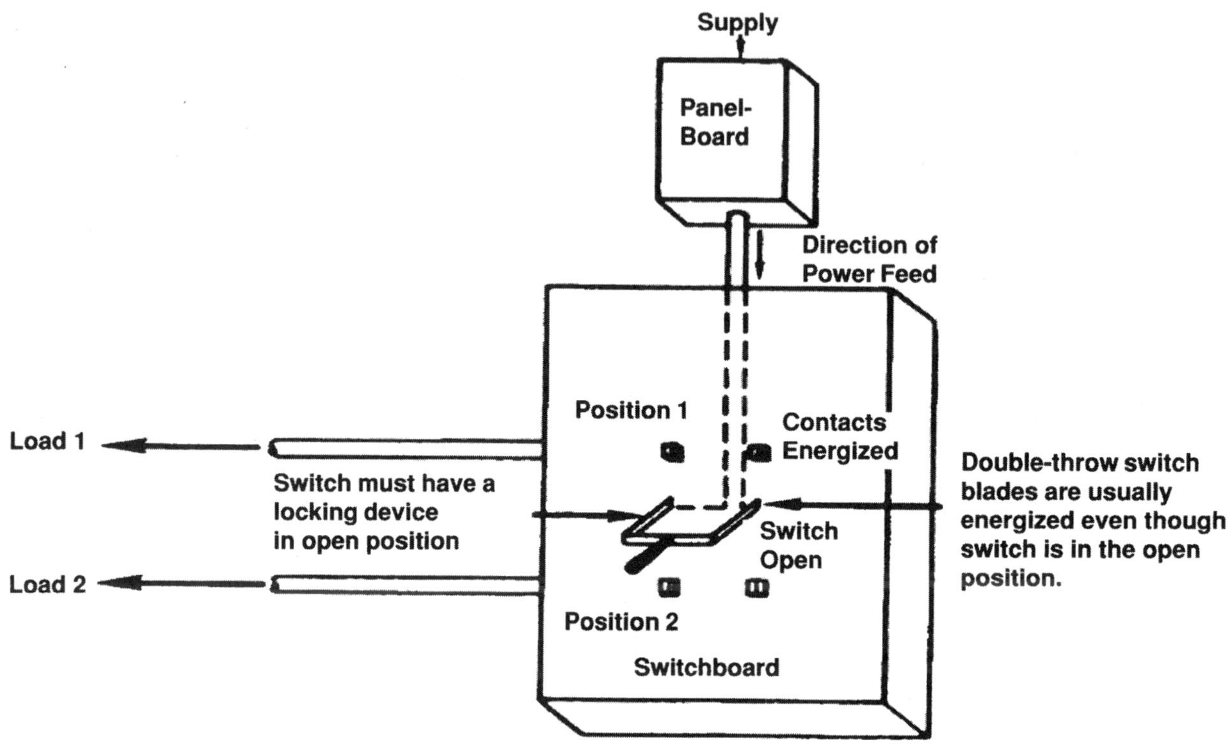

(c)(3) **Connection of switches.**

(c)(3)(i) Single-throw knife switches and switches with butt contacts shall be connected so that the blades are deenergized when the switch is in the open position.

(c)(3)(ii) Single-throw knife switches, molded-case switches, switches with butt contacts, and circuit breakers used as switches shall be connected so that the terminals supplying the load are deenergized when the switch is in the open position. However, blades and terminals supplying the load of a switch may be energized when the switch is in the open position where the switch is connected to circuits or equipment inherently capable of providing a backfeed source of power. For such installations, a permanent sign shall be installed on the switch enclosure or immediately adjacent to open switches that read, "WARNING - LOAD SIDE TERMINALS MAY BE ENERGIZED BY BACKFEED."

Switches that have energized blades when open must have barriers or other means to guard against accidental contact in order to protect against accidental contact when qualified persons are working. Typical sources of backfeed include, but are not limited to batteries, capacitors, generator systems, and double-ended switchgear. These installations could provide power to the load side of switches and circuit breakers, even when in the open position.

(c)(4) **Faceplates for flush-mounted snap switches.** Snap switches mounted in boxes shall have faceplates installed so as to completely cover the opening and seat against the finished surface.

(c)(5) **Grounding**. Snap switches, including dimmer switches, shall be effectively grounded and shall provide a means to ground metal faceplates, whether or not a metal faceplate is installed. However, if no grounding means exists within the snap-switch enclosure, or where the wiring method does not include or provide an equipment ground, a snap switch without a grounding connection is permitted for replacement purposes only. Such snap switches shall be provided with a faceplate of nonconducting, noncombustible material if they are located within reach of conducting floors or other conducting surfaces.

OSHA puts further emphasis on grounding by requiring that switches be effectively grounded. While the non-current-carrying metal parts of the switch itself is not typically exposed for contact by personnel, the main concern is the use of metallic faceplates, which could pose a shock or electrocution hazard if energized.

(d) **Switchboards and panelboards.**

(d)(1) **Switchboards with exposed live parts.** Switchboards that have any exposed live parts shall be located in permanently dry locations and shall be accessible only to qualified persons.

(d)(2) **Panel board enclosures.** Panelboards shall be mounted in cabinets, cutout boxes, or enclosures designed for the purpose and shall be dead front. However, panelboards other than the dead front externally-operable type are permitted where accessible only to qualified persons.

(d)(3) **Knife switches mounted in switchboards or panelboards.** Exposed blades of knife switches mounted in switchboards or panelboards shall be dead when open.

Switchboards and Panelboards

A switchboard that has exposed live parts must be located in an area that is not subject to wetness or dampness. One purpose of this regulation is to lessen the chance of severe shock if a worker accidentally came into contact with the live parts. Additionally, only qualified persons may have access to switchboards with exposed live parts. To limit access, the switchboard should be located in a locked room or within a locked cage or fenced area. Keys to the locks should be controlled to ensure that only properly trained personnel are allowed to enter the area.

Panelboards must be mounted in enclosures and must be rated for the voltages and currents involved. Panelboards, like switchboards, may have exposed live parts if only properly trained personnel will have access to them. If the panelboard has an exposed knife switch, the blades must be de-energized when the switch is open. (See also discussion of knife switches, previous pages.)

(e) **Enclosures for damp or wet locations.**

(e)(1) **Cabinets, cutout boxes, fittings, boxes, and panelboard enclosures.** Cabinets, cutout boxes, fittings, boxes, and panelboard enclosures in damp or wet locations shall be installed so as to prevent moisture or water from entering and accumulating within the enclosures and shall be mounted so there is at least 6.35-mm (0.25-in.) airspace between the enclosure and the wall or other supporting surface. However, nonmetallic enclosures may be installed without the airspace on a concrete, masonry, tile, or similar surface. The enclosures shall be weatherproof in wet locations.

(e)(2) **Switches, circuit breakers, and switchboards.** Switches, circuit breakers, and switchboards installed in wet locations shall be enclosed in weatherproof enclosures.

(f) **Conductors for general wiring.**

(f)(1) **Insulation.** All conductors used for general wiring shall be insulated unless otherwise permitted in this subpart.

(f)(2) **Type.** The conductor insulation shall be of a type that is approved for the voltage, operating temperature, and location of use.

(f)(3) **Distinguishable.** Insulated conductors shall be distinguishable by appropriate color or other suitable means as being grounded conductors, ungrounded conductors, or equipment grounding conductors.

1910.305

Title: Wiring methods, components, and equipment for general use.

Subpart: S

SubPart Title: Electrical - Design Safety Standards for Electrical Systems

Conductor Insulation

To provide adequate protection against shock and fire hazards, conductors must be insulated with approved materials. Insulating material should be the appropriate composition and thickness for the voltage and current the conductor will carry, for the temperature extremes and other environmental factors to which it will be subjected, and for the location in which it is to be placed.

Insulated conductors must also be easily identifiable, and color coding is most often used. Grounded conductors should be white or gray. Grounding conductors should be green or green with yellow stripes. Grounding conductors are permitted to be bare wires. Other types of circuit wires may be any colors except these. Figure 75 shows examples of typical insulation.

Figure 75 – Typical multiconductor insulation

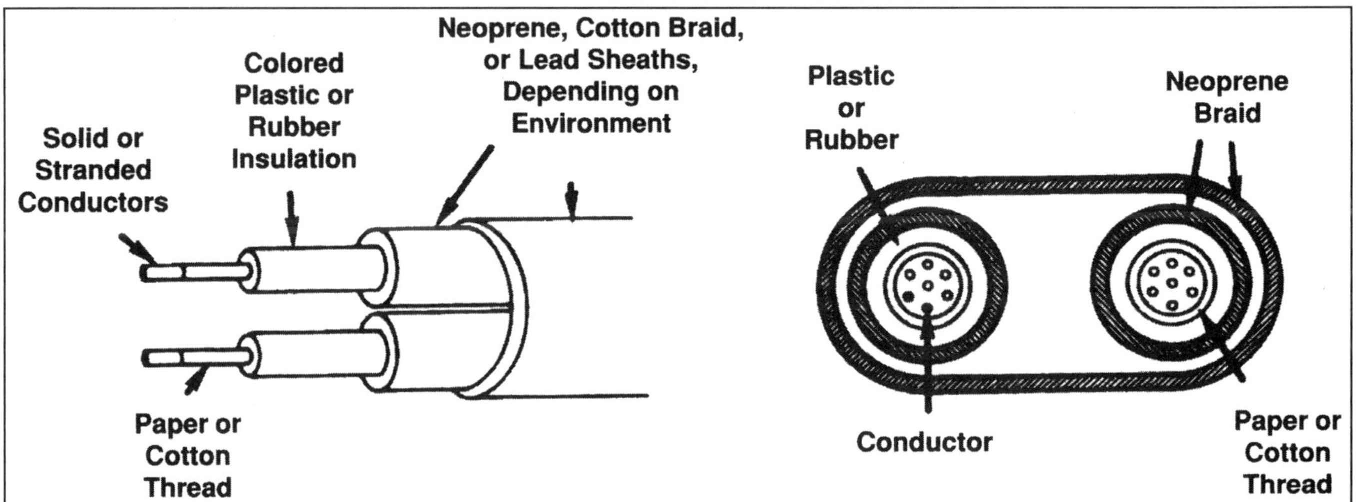

(g)	**Flexible cords and cables.**
(g)(1)	**Use of flexible cords and cables.**
(g)(1)(i)	Flexible cords and cables shall be approved for conditions of use and location.
(g)(1)(ii)	Flexible cords and cables may be used only for:
(g)(1)(ii)(A)	Pendants;
(g)(1)(ii)(B)	Wiring of fixtures;
(g)(1)(ii)(C)	Connection of portable lamps or appliances;
(g)(1)(ii)(D)	Portable and mobile signs;
(g)(1)(ii)(E)	Elevator cables;
(g)(1)(ii)(F)	Wiring of cranes and hoists;
(g)(1)(ii)(G)	Connection of stationary equipment to facilitate their frequent interchange;
(g)(1)(ii)(H)	Prevention of the transmission of noise or vibration;
(g)(1)(ii)(I)	Appliances where the fastening means and mechanical connections are designed to permit removal for maintenance and repair;

(g)(1)(ii)(J)	Data processing cables approved as a part of the data processing system;
(g)(1)(ii)(K)	Connection of moving parts; and
(g)(1)(ii)(L)	Temporary wiring as permitted in paragraph (a)(2) of this section.
(g)(1)(iii)	If used as permitted in paragraphs (g)(1)(ii)(C), (g)(1)(ii)(G), or (g)(1)(ii)(I) of this section, the flexible cord shall be equipped with an attachment plug and shall be energized from an approved receptacle outlet.
(g)(1)(iv)	Unless specifically permitted otherwise in paragraph (g)(1)(ii) of this section, flexible cords and cables may not be used:
(g)(1)(iv)(A)	As a substitute for the fixed wiring of a structure;
(g)(1)(iv)(B)	Where run through holes in walls, ceilings, or floors;
(g)(1)(iv)(C)	Where run through doorways, windows, or similar openings;
(g)(1)(iv)(D)	Where attached to building surfaces;
(g)(1)(iv)(E)	Where concealed behind building walls, ceilings, or floors; or
(g)(1)(iv)(F)	Where installed in raceways, except as otherwise permitted in this subpart.
(g)(1)(v)	Flexible cords used in show windows and showcases shall be Type S, SE, SEO, SEOO, SJ, SJE, SJEO, SJEOO, SJO, SJOO, SJT, SJTO, SJTOO, SO, SOO, ST, STO, or STOO, except for the wiring of chain-supported lighting fixtures and supply cords for portable lamps and other merchandise being displayed or exhibited.
(g)(2)	**Identification, splices, and terminations.**
(g)(2)(i)	A conductor of a flexible cord or cable that is used as a grounded conductor or an equipment grounding conductor shall be distinguishable from other conductors. Types S, SC, SCE, SCT, SE, SEO, SEOO, SJ, SJE, SJEO, SJEOO, SJO, SJT, SJTO, SJTOO, SO, SOO, ST, STO, and STOO flexible cords and Types G, G-GC, PPE, and W flexible cables shall be durably marked on the surface at intervals not exceeding 610 mm (24 in.) with the type designation, size, and number of conductors.
(g)(2)(ii)	Flexible cords may be used only in continuous lengths without splice or tap. Hard-service cord and junior hard-service cord No. 14 and larger may be repaired if spliced so that the splice retains the insulation, outer sheath properties, and usage characteristics of the cord being spliced.
(g)(2)(iii)	Flexible cords and cables shall be connected to devices and fittings so that strain relief is provided that will prevent pull from being directly transmitted to joints or terminal screws.

OSHA added several areas where flexible cords are permitted for consistency with NFPA standards. Both 1910.305(g) and Article 400 in the National Electrical Code® recognize use of flexible cords to provide the flexibility and interchange of equipment necessary in many situations. Generally the cord is a "pigtail" considered to be part of the equipment, or an extension cord set for extending the pigtail to the closest receptacle. There are a few uses of flexible cord not described here that are permitted. There may be a few instances not described above in which flexible cord would be more appropriate than fixed wiring. In making a judgment as to whether some application not anticipated by the Code® is permissible, Section 1910.305(g)(1)(iv)(A) is most helpful. If the cord is not being used as a substitution for the fixed wiring, and is being used in such a way that it does not create a hazard, then that use may be acceptable.

The types of cords identified in sections (g)(1)(v) and (g)(2)(i) can be identified on the cord itself, as such markings are required on the cord. Specific detail construction and use of each cord type can be found in Table 400.4 of the NEC®.

(h)	**Portable cables over 600 volts, nominal.** This paragraph applies to portable cables used at more than 600 volts, nominal.
(h)(1)	**Conductor construction.** Multiconductor portable cable for use in supplying power to portable or mobile equipment at over 600 volts, nominal, shall consist of No. 8 or larger conductors em-

	ploying flexible stranding. However, the minimum size of the insulated ground-check conductor of Type G-GC cables shall be No. 10.
(h)(2)	**Shielding.** Cables operated at over 2,000 volts shall be shielded for the purpose of confining the voltage stresses to the insulation.
(h)(3)	**Equipment grounding conductors.** Grounding conductors shall be provided.
(h)(4)	**Grounding shields.** All shields shall be grounded.
(h)(5)	**Minimum bending radii.** The minimum bending radii for portable cables during installation and handling in service shall be adequate to prevent damage to the cable.
(h)(6)	**Fittings.** Connectors used to connect lengths of cable in a run shall be of a type that lock firmly together. Provisions shall be made to prevent opening or closing these connectors while energized. Strain relief shall be provided at connections and terminations.
(h)(7)	**Splices.** Portable cables may not be operated with splices unless the splices are of the permanent molded, vulcanized, or other approved type.
(h)(8)	**Terminations.** Termination enclosures shall be suitably marked with a high voltage hazard warning, and terminations shall be accessible only to authorized and qualified employees.
(i)	**Fixture wires.**
(i)(1)	**General.** Fixture wires shall be approved for the voltage, temperature, and location of use. A fixture wire which is used as a grounded conductor shall be identified.
(i)(2)	**Uses permitted.** Fixture wires may be used only:
(i)(2)(i)	For installation in lighting fixtures and in similar equipment where enclosed or protected and not subject to bending or twisting in use; or
(i)(2)(ii)	For connecting lighting fixtures to the branch-circuit conductors supplying the fixtures.
(i)(3)	**Uses not permitted.** Fixture wires may not be used as branch-circuit conductors except as permitted for Class 1 power limited circuits and for fire alarm circuits.
(j)	**Equipment for general use.**
(j)(1)	**Lighting fixtures, lampholders, lamps, and receptacles.**
(j)(1)(i)	Fixtures, lampholders, lamps, rosettes, and receptacles may have no live parts normally exposed to employee contact. However, rosettes and cleat-type lampholders and receptacles located at least 2.44 m (8.0 ft) above the floor may have exposed terminals.
(j)(1)(ii)	Handlamps of the portable type supplied through flexible cords shall be equipped with a handle of molded composition or other material identified for the purpose, and a substantial guard shall be attached to the lampholder or the handle. Metal shell, paper-lined lampholders may not be used.
(j)(1)(iii)	Lampholders of the screw-shell type shall be installed for use as lampholders only. Where supplied by a circuit having a grounded conductor, the grounded conductor shall be connected to the screw shell. Lampholders installed in wet or damp locations shall be of the weatherproof type.

Lampholders

Screw-shell type lampholders must be used only as lampholders and must not be used with screwbase socket adapters. These adapters screw into the existing lamp socket and convert lampholders into receptacles. These adapters are not permissible because equipment grounding connections cannot be made through the two-blade adapters and because the fixture has been designed only for lighting. Only weatherproof lampholders may be installed in wet or damp areas (see Figure 75). Unprotected lampholders might allow moisture to enter the lampholder socket, creating an electrical shock hazard.

Figure 76 – Wet-location portable hand lamp with step-down transformer. Source: Daniel Woodhead Co., Northbrook, Ill.

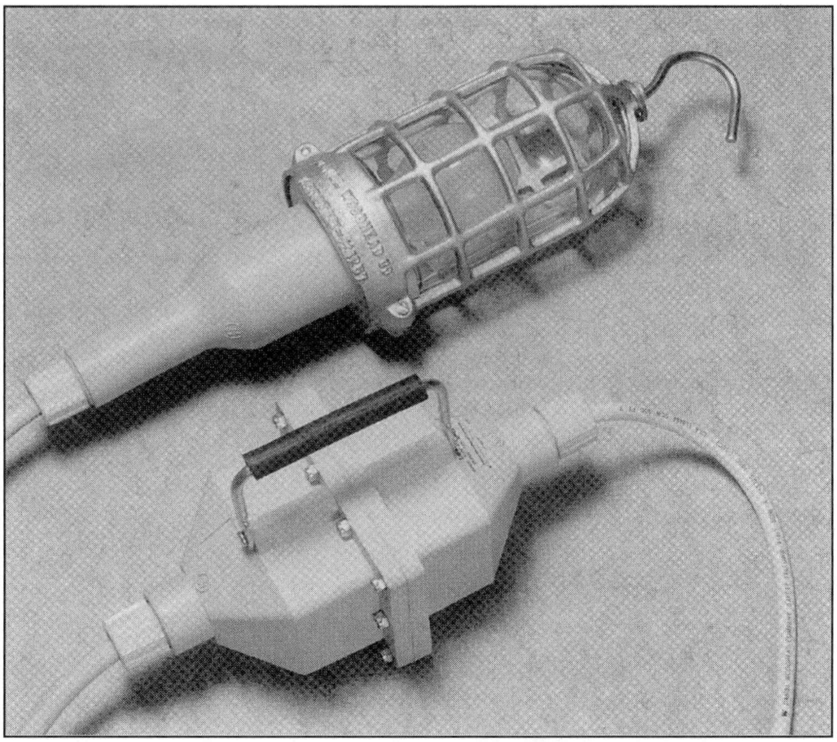

(j)(1)(iv)	Fixtures installed in wet or damp locations shall be identified for the purpose and shall be so constructed or installed that water cannot enter or accumulate in wireways, lampholders, or other electrical parts.
(j)(2)	**Receptacles, cord connectors, and attachment plugs (caps).**
(j)(2)(i)	All 15- and 20-ampere attachment plugs and connectors shall be constructed so that there are no exposed current-carrying parts except the prongs, blades, or pins. The cover for wire terminations shall be a part that is essential for the operation of an attachment plug or connector (dead-front construction). Attachment plugs shall be installed so that their prongs, blades, or pins are not energized unless inserted into an energized receptacle. No receptacles may be installed so as to require an energized attachment plug as its source of supply.
(j)(2)(ii)	Receptacles, cord connectors, and attachment plugs shall be constructed so that no receptacle or cord connector will accept an attachment plug with a different voltage or current rating than that for which the device is intended. However, a 20-ampere T-slot receptacle or cord connector may accept a 15-ampere attachment plug of the same voltage rating.
(j)(2)(iii)	Nongrounding-type receptacles and connectors may not be used for grounding-type attachment plugs.

Receptacles, Cord Connectors, and Attachment Plugs (Caps)

Cord connectors are devices that join two sections of electrical cord together. Attachment plugs are devices that are fastened onto the end of a cord so that electrical contact can be made between the conductors in the cord and the conductors in a receptacle.

Although the requirement for "dead-front" constructed attachment plugs is a new OSHA requirement, it has been previously addressed by OSHA under 1910.303(a) approval since the old wafer plugs (not dead front) are no longer listed by an OSHA Nationally Recognized Testing Lab (NRTL). This requirement is also found in the NEC®, Section 406.6(A). An example is shown in Figure 77.

Figure 77 – Non-dead-front attachment plug. This type of "wafer plug" does not meet the requirement, and is no longer listed.

Connectors, plugs, and receptacles are uniquely designed for different voltages and currents, so that only matching plugs will fit into the correct receptacle or cord connector. In this way, a piece of equipment rated for one voltage / current combination cannot be plugged into a power system that is of a different voltage or current capacity.

The only exceptions to this are 125-volt and 250-volt, 20-ampere, T-slot receptacles. One of these is illustrated in Figure 54. A 125-volt and a 250-volt, 15-ampere plug will fit into a 20-ampere T-slot receptacle or connector of the same voltage rating as well as in a 120-volt, 15-ampere grounding-type receptacle or connector of the same voltage rating. An electrical appliance that is rated for 15 amperes will not overload a 20-ampere circuit, and the 20-ampere breaker will still provide overcurrent protection for 15-ampere equipment. Note that the opposite is not necessarily true -- a 20-ampere plug will not fit into a 15-ampere receptacle or cord connector.

The National Electrical Manufacturers Association (NEMA) has standard plug and receptacle connector blade configurations. (See Appendix C.) Each has been developed to standardize the use of plugs and receptacles for different voltages, amperages, and phases from 115 volts through 600, from 10 amperes through 60, and for single- and three-phase systems.

"P" in Figure 78 identifies plugs; "R," receptacles. "W" identifies the white, neutral conductor. The terminal of the grounding conductor is marked with a "G"; in practice, it may also be marked with a green hexagonal-headed screw, a green wire connector body or similar device. The words "green" or "ground," the letters "G" or "GR," or the grounding symbol, below, may be used at terminals that are not readily visible.

Figure 78 – Receptacles and plugs with NEMA configurations.

.

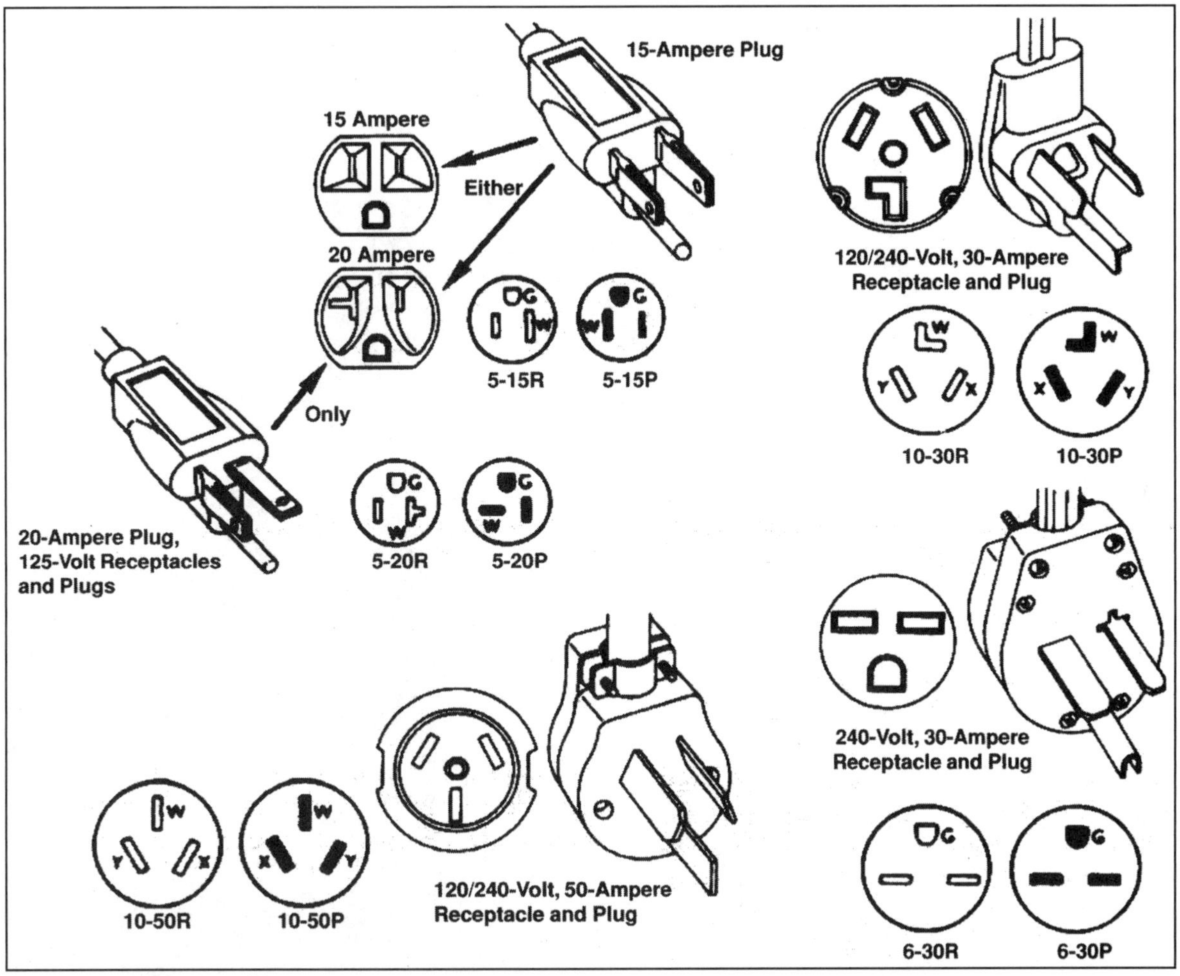

(j)(2)(iv) A receptacle installed in a wet or damp location shall be suitable for the location.

(j)(2)(v) A receptacle installed outdoors in a location protected from the weather or in other damp locations shall have an enclosure for the receptacle that is weatherproof when the receptacle is covered (attachment plug cap not inserted and receptacle covers closed).

Note to paragraph (j)(2)(v) of this section. A receptacle is considered to be in a location protected from the weather when it is located under roofed open porches, canopies, marquees, or the like and where it will not be subjected to a beating rain or water runoff.

(j)(2)(vi) A receptacle installed in a wet location where the product intended to be plugged into it is not attended while in use (for example, sprinkler system controllers, landscape lighting, and holiday lights) shall have an enclosure that is weatherproof with the attachment plug cap inserted or removed.

(j)(2)(vii) A receptacle installed in a wet location where the product intended to be plugged into it will be attended while in use (for example, portable tools) shall have an enclosure that is weatherproof when the attachment plug cap is removed.

Outlets installed in wet locations are required to have covers that protect the outlet while in use (known as "in-use covers"). If the location is considered a *damp* location, standard weather-tight snap covers are permitted, while not recommended. A number of objections were made to the term "beating rain" when the proposed rule was released, but the term is taken directly from NFPA standards and was retained for the final rule.

Figure 79 – Single outlet weatherproof cover suitable for use in wet locations (Intermatic Incorporated)

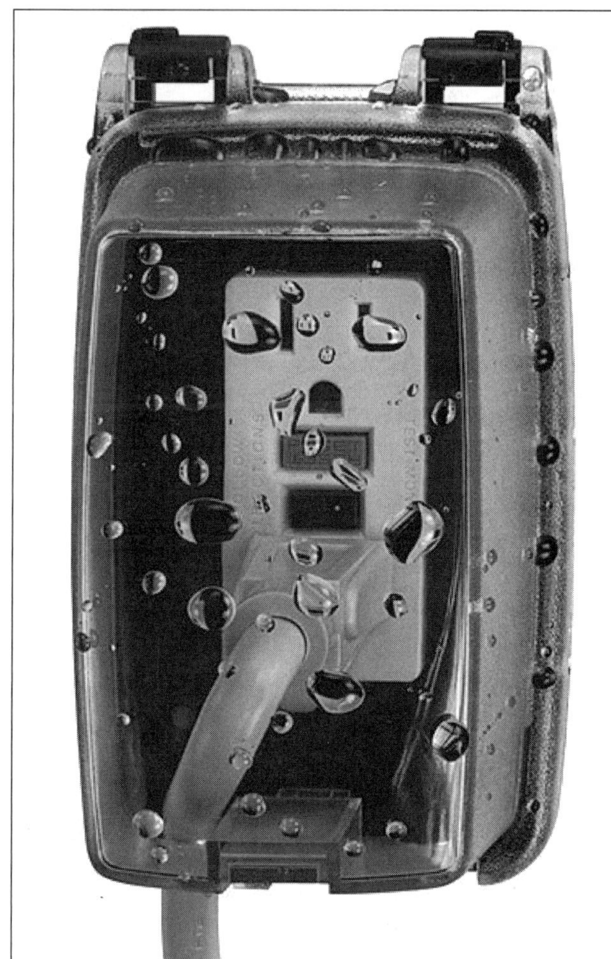

(j)(3) **Appliances.**

(j)(3)(i) Appliances may have no live parts normally exposed to contact other than parts functioning as open-resistance heating elements, such as the heating elements of a toaster, which are necessarily exposed.

Appliances

Electrical appliances such as portable air conditioning units, coffee makers, and fans must not have any exposed live wires or electrical parts that might create an electrical shock hazard.

Exceptions to this are appliances such as heaters or toasters that must have exposed current-carrying parts that operate at high temperatures to transfer heat (see Figure 80). The heat generated by these parts minimizes the possibility of direct contact and resultant electric shock. It is also recommended that space heaters be of the type equipped with a tip-over or over-temperature cutout switch.

Figure 80 – Electric space heater with exposed heating elements guarded by a protective grill

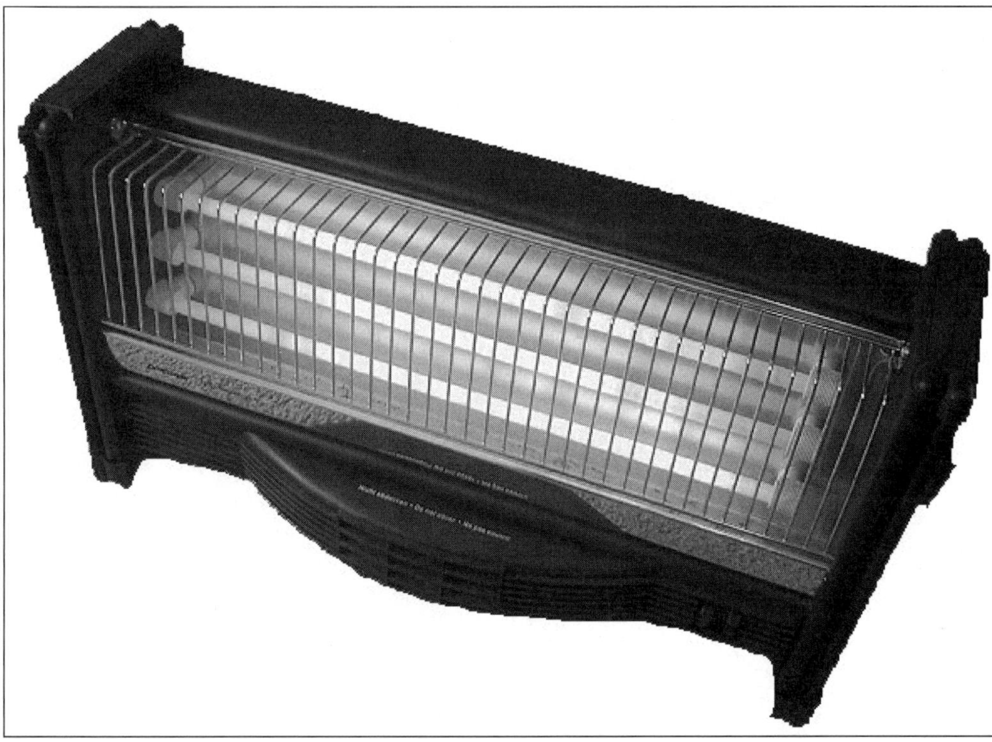

(j)(3)(ii) Each appliance shall have a means to disconnect it from all ungrounded conductors. If an appliance is supplied by more than one source, the disconnecting means shall be grouped and identified.

Appliance Disconnecting Means

A disconnecting means is a switch or plug that can open an electric circuit under load and safely stop the flow of current. All appliances must have a disconnecting means. In some cases, the disconnecting means may be a unit switch or an attachment plug. On permanently connected appliances that operate at less than 300 volt-amperes, the branch circuit overcurrent device (circuit breaker or fuse) may be used as the disconnecting means. In all cases, the disconnecting means must disconnect all ungrounded conductors.

(j)(3)(iii) Each electric appliance shall be provided with a nameplate giving the identifying name and the rating in volts and amperes, or in volts and watts. If the appliance is to be used on a specific frequency or frequencies, it shall be so marked. Where motor overload protection external to the appliance is required, the appliance shall be so marked.

(j)(3)(iv) Marking shall be located so as to be visible or easily accessible after installation.

The requirement for nameplate markings coincides with the requirements of 1910.303(e).

(j)(4)	**Motors.** This paragraph applies to motors, motor circuits, and controllers.
(j)(4)(i)	If specified in paragraph (j) (4) of this section that one piece of equipment shall be "within sight of" another piece of equipment, the piece of equipment shall be visible and not more than 15.24 m (50.0 ft) from the other.
(j)(4)(ii)	An individual disconnecting means shall be provided for each controller. A disconnecting means shall be located within sight of the controller location. However, a single disconnecting means may be located adjacent to a group of coordinated controllers mounted adjacent to each other on a multi-motor continuous process machine. The controller disconnecting means for motor branch circuits over 600 volts, nominal, may be out of sight of the controller, if the controller is marked with a warning label giving the location and identification of the disconnecting means that is to be locked in the open position.

Figure 81 – Example of a Motor Control Center (MCC)

Motor Disconnecting Means

A motor controller is a device, such as a disconnect switch or circuit breaker, that interrupts the current flow to a motor. The controller turns the power off and on and limits the current flow. A Motor Control Center (MCC) is usually made up of sections and contains multiple motor controllers (see Figure 81). A disconnecting means for the controller must be within sight from the controller (i.e., visible from the controller and located within 50 feet of the controller) (see Figure 82). If a group of controllers are located together and are used to control power to more than one mo-tor on a single continuous process machine, a single disconnect switch, located with the controllers, can be used. It should be noted that it is possible for a switch or circuit breaker to serve as both a controller and a disconnect. This depends on where the switch is located and the rating of both the motor and the switch. Detailed specifications on motor disconnecting means are given in Article 430 in the National Electrical Code®.

For larger capacity motors operating at voltages greater than 600 volts, the disconnect can be out of sight of the controller if the controller and the disconnect are labeled. Warning labels on the controller should indicate where the disconnect is located and that it can be locked out for maintenance. The disconnect must be labeled with identification, such as a number, to ensure that the correct disconnect is opened or de-energized. See Figure 83.

Figure 82 – Motor disconnecting means.

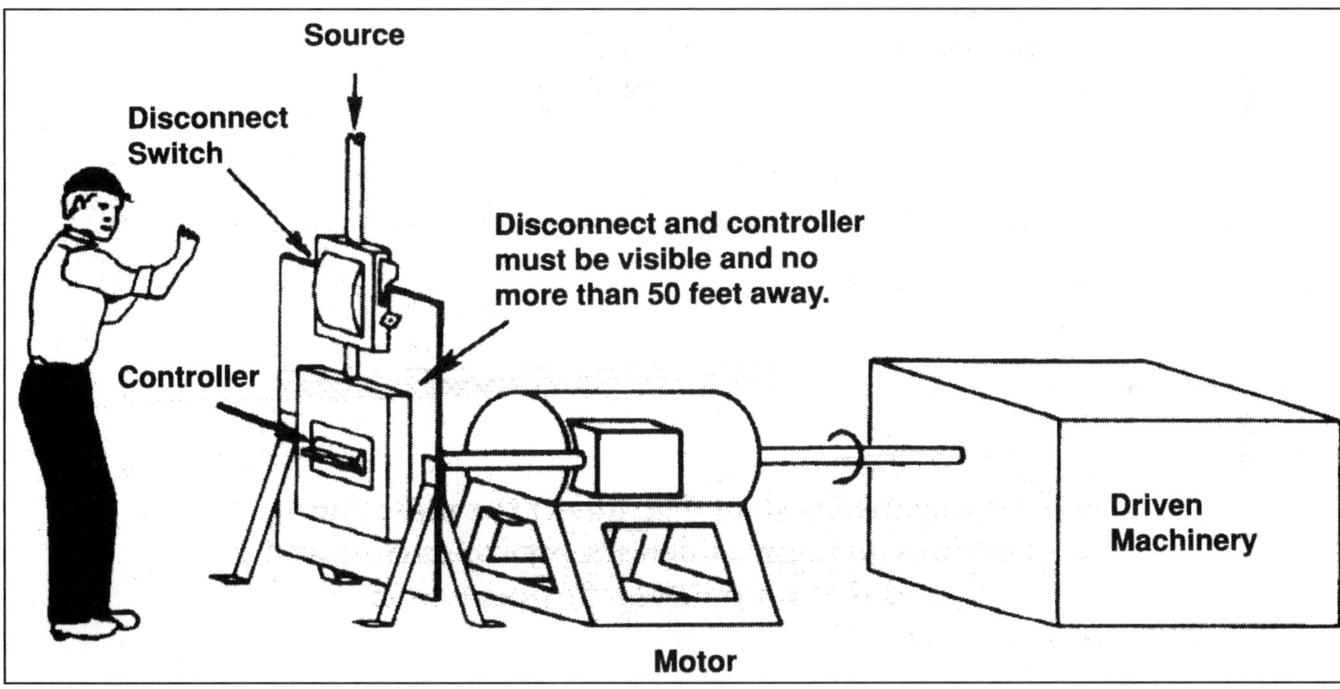

(j)(4)(iii) The disconnecting means shall disconnect the motor and the controller from all ungrounded supply conductors and shall be so designed that no pole can be operated independently.

Motor and Driven Machinery Not Within Sight of Controller

Usually, a motor and the equipment it drives should be within sight of the controller. If they are not within sight of the controller, one of two conditions must be met: (1) the controller disconnect must be designed so that it can be locked in the open, or de-energized, position to protect persons working on the motor or equipment (see Figure 84), or (2) a switch that can be manually (not magnetically) operated must be located within 50 feet of, and must be visible from, the motor (see Figure 85).

Figure 83 – Labeling required when disconnects are out of sight.

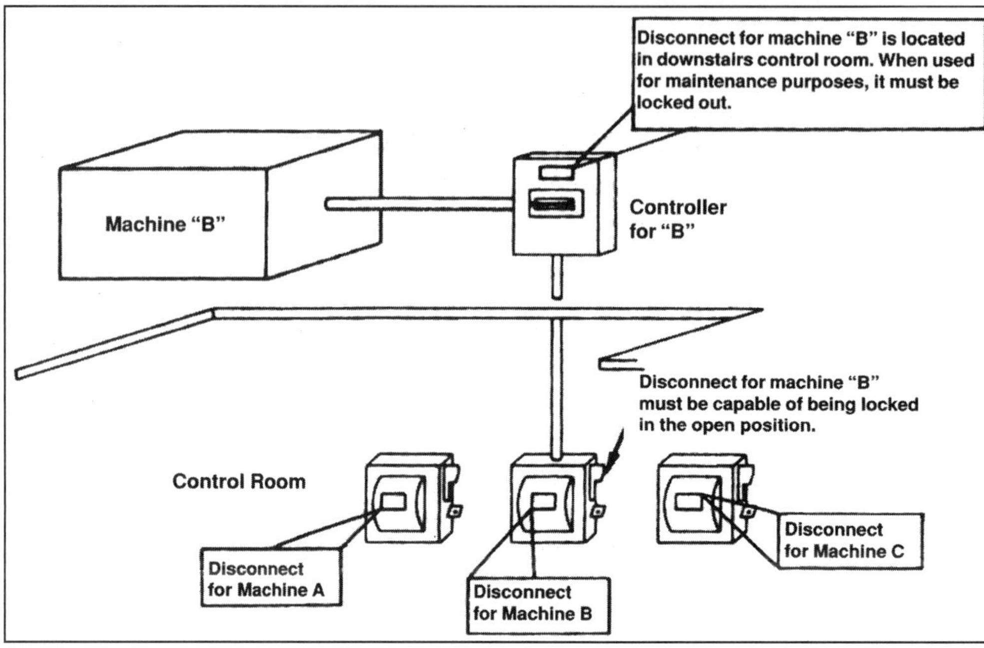

Figure 84 – Locking controller disconnecting means.

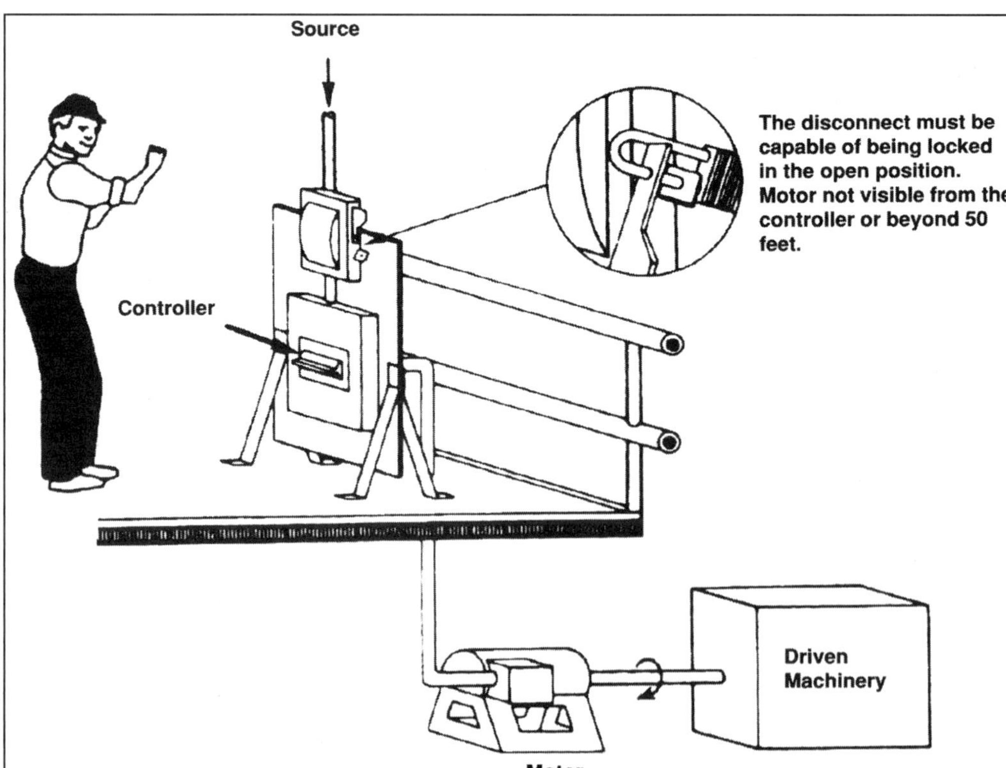

Figure 85 – Manually operable switch within sight.

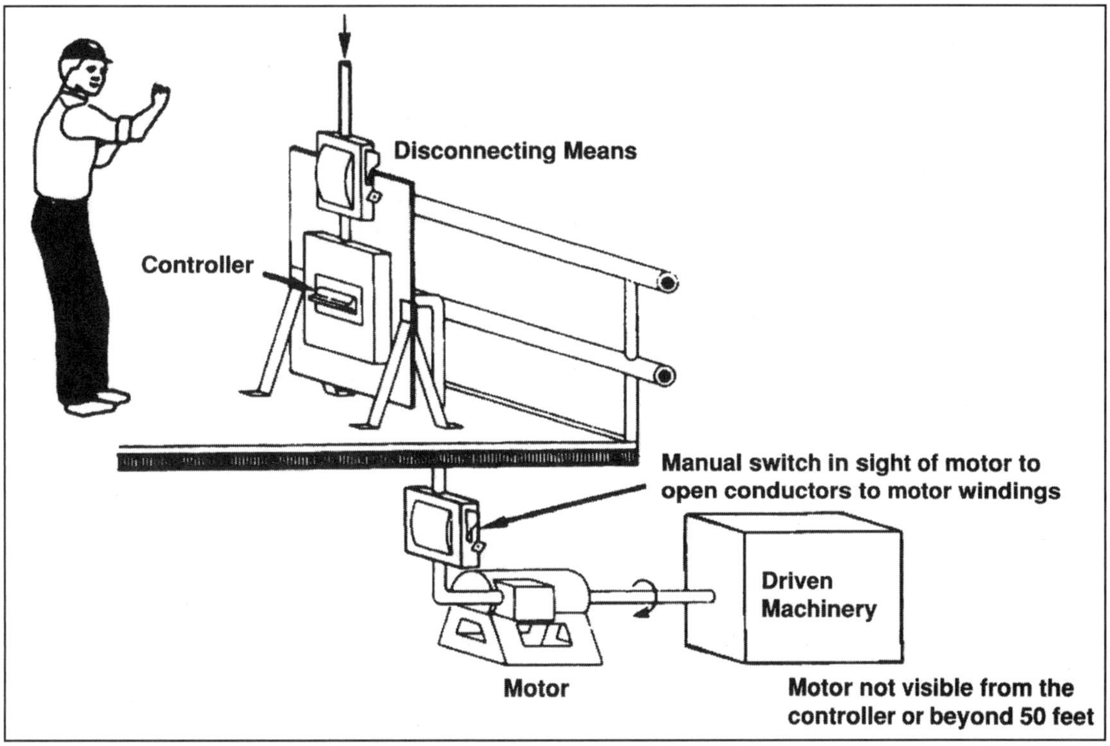

(j)(4)(iv)	The disconnecting means shall plainly indicate whether it is in the open (off) or closed (on) position.
(j)(4)(v)	The disconnecting means shall be readily accessible. If more than one disconnect is provided for the same equipment, only one need be readily accessible.
(j)(4)(vi)	An individual disconnecting means shall be provided for each motor, but a single disconnecting means may be used for a group of motors under any one of the following conditions:
(j)(4)(vi)(A)	If a number of motors drive special parts of a single machine or piece of apparatus, such as a metal or wood-working machine, crane, or hoist;
(j)(4)(vi)(B)	If a group of motors is under the protection of one set of branch-circuit protective devices; or
(j)(4)(vi)(C)	If a group of motors is in a single room within sight of the location of the disconnecting means.
(j)(4)(vii)	Motors, motor-control apparatus, and motor branch-circuit conductors shall be protected against overheating due to motor overloads or failure to start, and against short-circuits or ground faults. These provisions do not require overload protection that will stop a motor where a shutdown is likely to introduce additional or increased hazards, as in the case of fire pumps, or where continued operation of a motor is necessary for a safe shutdown of equipment or process and motor overload sensing devices are connected to a supervised alarm.

Motor Overload, Short-Circuit, and Ground-Fault Protection

All motors, motor controllers, and conductors that feed motors must be protected from overcurrent. For example, damage might occur when overheating results from an overload (excessive current flow) or from the failure of the motor to start. Damage and hazards might also result from short-circuits and ground faults. Figure 86 illustrates methods of motor protection.

Figure 86 – Motor protection methods.

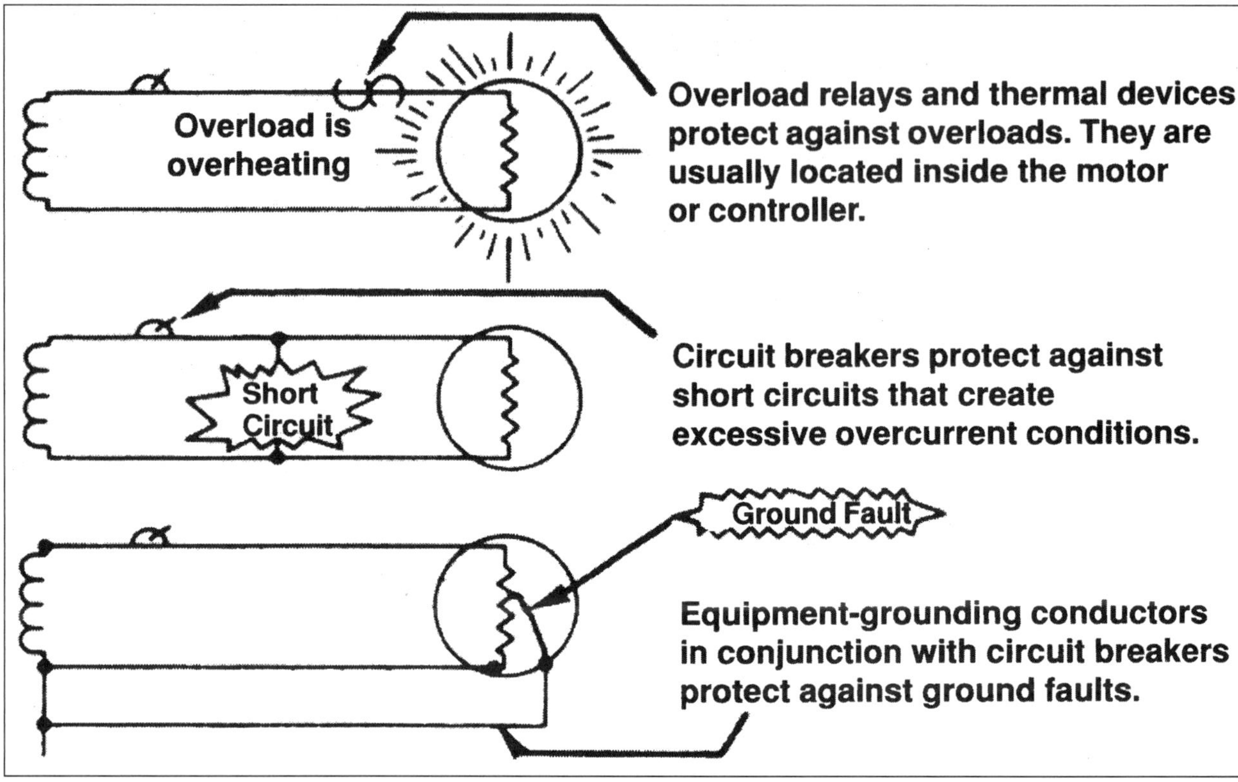

For simplicity, circuit breakers are shown in Figure 86. In most situations either a circuit breaker or a set of fuses would be appropriate. However, protection to shut down the motor is not required if it would produce a more hazardous situation. An example is a fire pump that, if overheated and shut down, is of no help in fighting a fire that might destroy a large building or other valuable property. The damage to the motor, controller, or conductors and any hazards that might result are insignificant when compared to the results of shutting off the power to this type of equipment.

In situations where a motor does not have overload protection, it must be designed so that an automatic sensing device within the motor will trigger an alarm. This alarm is usually located in a control location and must be monitored.

Protection of Live Parts, All Voltages

Motors that are fixed in place need not have their internal live parts guarded if they meet the following two conditions: (1) if the commutators (conducting members, insulated from one another, against which the brushes bear), the brush riggings (parts that support and position the brushes), and the collectors (metal rings that, through contact with the brushes, conduct current into or out of machines) are all located inside the motor end brackets, and (2) if the commutators, brush riggings, and collectors are not in electrical contact with any supply circuit that operates at more than 150 volts to ground (see Figure 87).

Figure 87 – Unguarded commutators, collectors, and brushes housed within motor

(j)(4)(viii)	Where live parts of motors or controllers operating at over 150 volts to ground are guarded against accidental contact only by location, and where adjustment or other attendance may be necessary during the operation of the apparatus, suitable insulating mats or platforms shall be provided so that the attendant cannot readily touch live parts unless standing on the mats or platforms.
(j)(5)	**Transformers.**
(j)(5)(i)	Paragraph (j)(5) of this section covers the installation of all transformers except the following:
(j)(5)(i)(A)	Current transformers;
(j)(5)(i)(B)	Dry-type transformers installed as a component part of other apparatus;
(j)(5)(i)(C)	Transformers that are an integral part of an X-ray, high frequency, or electrostatic-coating apparatus;
(j)(5)(i)(D)	Transformers used with Class 2 and Class 3 circuits, sign and outline lighting, electric discharge lighting, and power-limited fire-alarm circuits; and
(j)(5)(i)(E)	Liquid-filled or dry-type transformers used for research, development, or testing, where effective safeguard arrangements are provided.

Transformers Excluded from the Standard

The requirements of the standard for transformers apply to most transformers. Exceptions include transformers used with:

(1) Class 2 and Class 3 power-limited, signaling or remote-control circuits, which are current limited by the supply transformer so that only relatively small currents will flow. Class 2 circuits protect against electrical fires by limiting the power to 100VA for circuits that operate at 30V or less, and 0.5VA for circuits between 30V and 150V. Protection against electric shock is done by limiting the current of the circuit to 5mA or less for circuits between 30V and 150V.

Class 3 circuits are used when the power demand for circuits over 30V exceeds 0.5VA, but is not more than 100VA. Higher levels of voltage and current are permitted for Class 3 circuits (versus Class 2 circuits). In order to prevent a shock or electrocution hazard, the wiring must be rated no less than 300V. These circuits are used when the current and voltage required are not as great as those required for normal light and power systems, as with a doorbells or an oil-burner thermostat control circuit (see Figure 87);

(2) sign and outline lighting, such as neon lighting;

(3) electric discharge lighting, such as fluorescent lamps;

(4) power-limited, fire-protective signaling circuits, which have low voltage and current needs, such as smoke detectors that are tied into an alarm station in a building (see Figure 88).

Figure 88 – Typical Class 2 circuit.

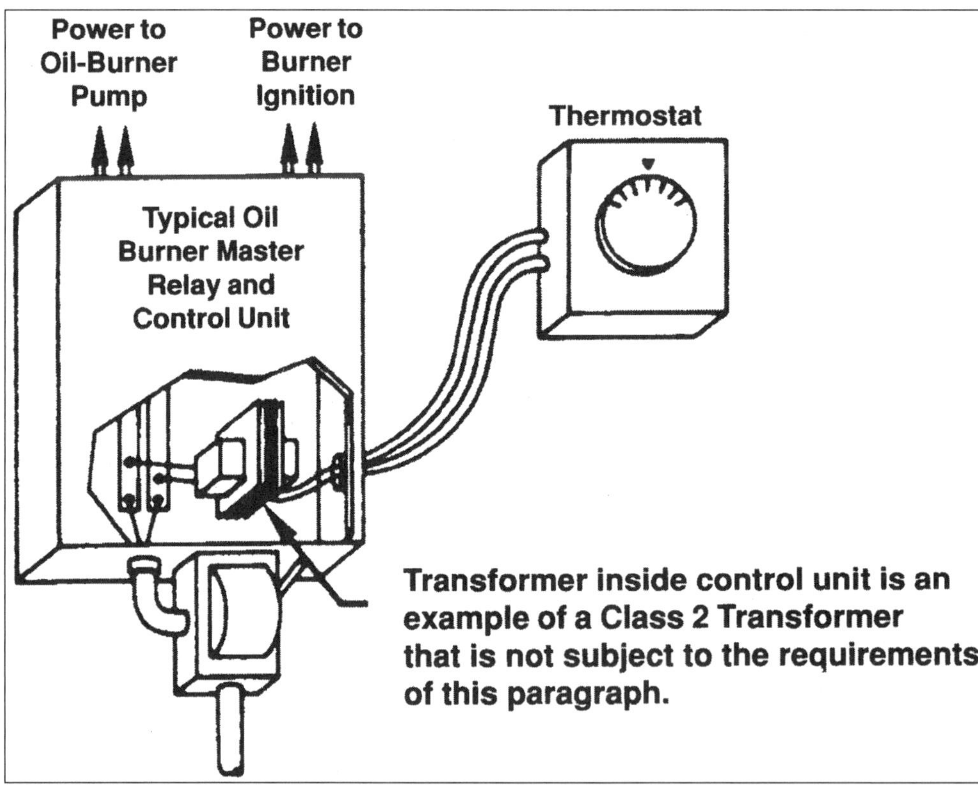

(j)(5)(ii)	The operating voltage of exposed live parts of transformer installations shall be indicated by signs or visible markings on the equipment or structure.
(j)(5)(iii)	Dry-type, high fire point liquid-insulated, and askarel-insulated transformers installed indoors and rated over 35kV shall be in a vault.
(j)(5)(iv)	Oil-insulated transformers installed indoors shall be installed in a vault.
(j)(5)(v)	Combustible material, combustible buildings and parts of buildings, fire escapes, and door and window openings shall be safeguarded from fires that may originate in oil-insulated transformers attached to or adjacent to a building or combustible material.

Transformers

Transformers that use oil as an insulator are subject to fires because of the combustible nature of oil and the heat-generating nature of electrical equipment. Therefore, if an oil-insulated transformer is installed outdoors next to or attached to a building or some other type of combustible material, the transformer must be located or otherwise protected so that the fire hazard that already exists is not increased. Fire escapes, windows, and doors, as well as combustible materials and the building (if part or all of it is constructed of combustible materials), must also be protected from the potential fire hazard. Protection can be achieved by locating the transformer away from the building (space separation); constructing fire-resistant barriers such as concrete block walls between the transformer and the building; or installing a specialized automatic sprinkler system that will provide a water spray to control an oil fire. In addition to any of these methods, a dike or curb should be installed around the transformer so that an oil fire would be contained in one area and could not spread to the building. Figure 89 illustrates two of these safeguarding methods.

Figure 89 – Transformer for power-limited, fire-protective signaling circuit

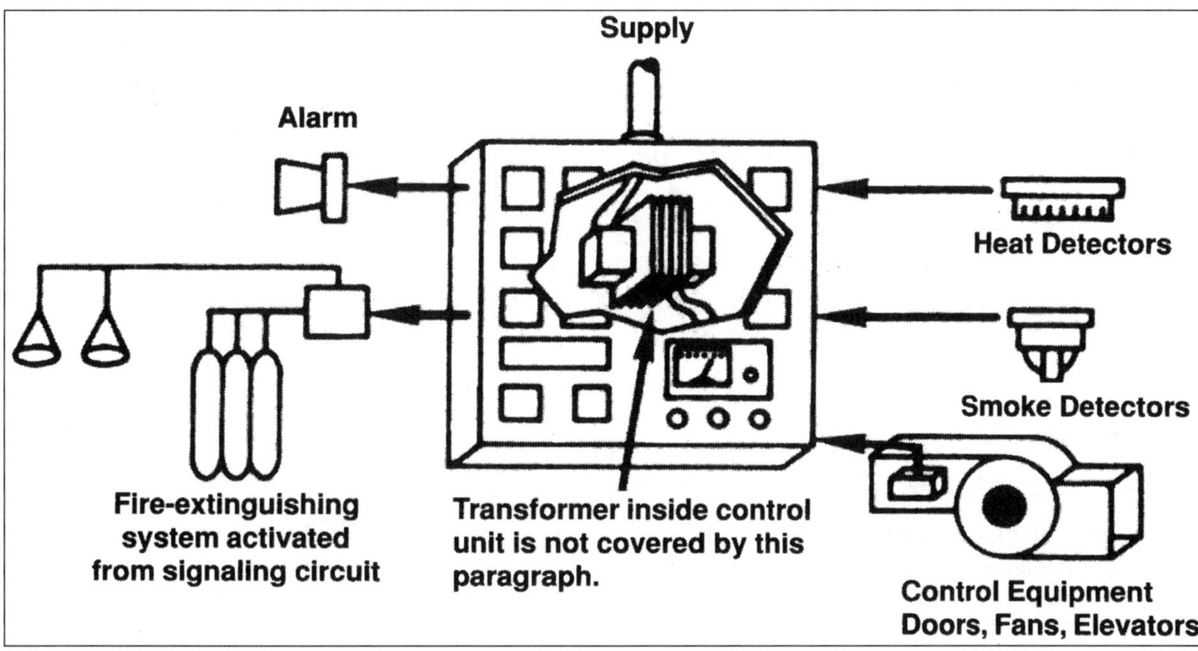

(j)(5)(vi) Transformer vaults shall be constructed so as to contain fire and combustible liquids within the vault and to prevent unauthorized access. Locks and latches shall be so arranged that a vault door can be readily opened from the inside.

(j)(5)(vii) Any pipe or duct system foreign to the vault installation may not enter or pass through a transformer vault.

Note to paragraph (j)(5)(vii) of this section. Piping or other facilities provided for vault fire protection, or for transformer cooling, are not considered foreign to the electrical installation.

(j)(5)(viii) Materials may not be stored in transformer vaults.

(j)(6) **Capacitors.**

(j)(6)(i) All capacitors, except surge capacitors or capacitors included as a component part of other apparatus, shall be provided with an automatic means of draining the stored charge after the capacitor is disconnected from its source of supply.

Figure 90 – Examples of safeguarding methods for transformers

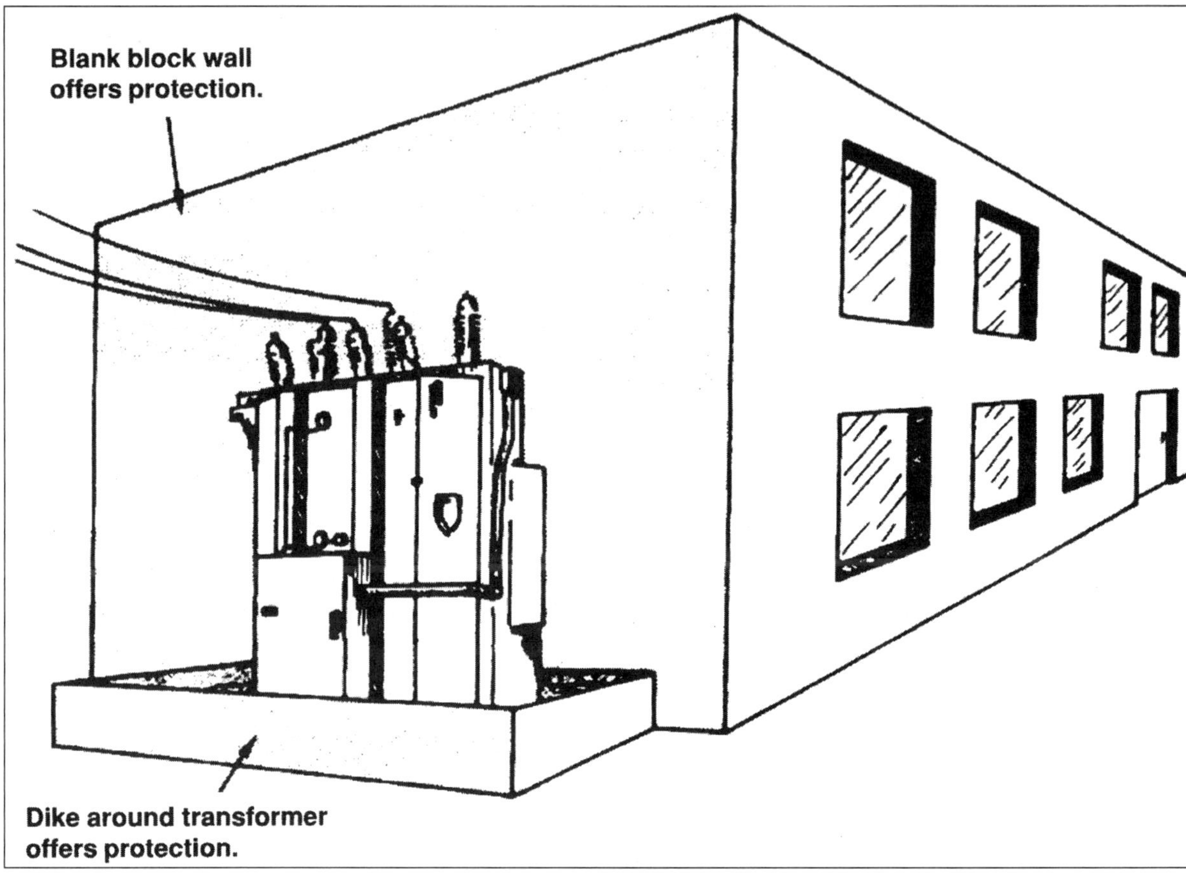

Capacitors

Capacitors store electrical charge and can be a source of severe shock unless that charge is drained when the capacitors are disconnected from the power source. Unless some type of automatic discharge is designed into a system, devices such as resistors must be permanently attached across the terminals of the capacitors to drain the charge when the circuit is open (de-energized). Most capacitors are manufactured with this type of discharge resistor already built in.

Article 340 of NFPA 70E provides some insight as to the potential hazards associated with capacitors. Section 340.5(7) states the following:

(1) A circuit of capacitance of 1 microfarad having a 10 kV capacitor charge may case ventricular fibrillation.

(2) A circuit of capacitance of 20 microfarad having a 10 kV capacitor charge may be dangerous and probably cause ventricular fibrillation.

(j)(6)(ii)	The following requirements apply to capacitors installed on circuits operating at more than 600 volts, nominal:
(j)(6)(ii)(A)	Group-operated switches shall be used for capacitor switching and shall be capable of the following:
(j)(6)(ii)(A)(1)	Carrying continuously not less than 135 percent of the rated current of the capacitor installation;
(j)(6)(ii)(A)(2)	Interrupting the maximum continuous load current of each capacitor, capacitor bank, or capacitor installation that will be switched as a unit;
(j)(6)(ii)(A)(3)	Withstanding the maximum inrush current, including contributions from adjacent capacitor installations; and
(j)(6)(ii)(A)(4)	Carrying currents due to faults on the capacitor side of the switch;
(j)(6)(ii)(B)	A means shall be installed to isolate from all sources of voltage each capacitor, capacitor bank, or capacitor installation that will be removed from service as a unit. The isolating means shall provide a visible gap in the electric circuit adequate for the operating voltage;
(j)(6)(ii)(C)	Isolating or disconnecting switches (with no interrupting rating) shall be interlocked with the load interrupting device or shall be provided with prominently displayed caution signs to prevent switching load current; and
(j)(6)(ii)(D)	For series capacitors, the proper switching shall be assured by use of at least one of the following:
(j)(6)(ii)(D)(1)	Mechanically sequenced isolating and bypass switches;
(j)(6)(ii)(D)(2)	Interlocks; or
(j)(6)(ii)(D)(3)	Switching procedure prominently displayed at the switching location.
(j)(7)	**Storage batteries.** Provisions shall be made for sufficient diffusion and ventilation of gases from storage batteries to prevent the accumulation of explosive mixtures.

Storage Batteries

Storage batteries, which are usually lead-acid or alkali, produce explosive gases, including hydrogen, if they are overcharged. These explosive gases must not accumulate in quantities that may form an explosive mixture with air. A spark or open flame could ignite the mixture and cause an explosion. Good ventilation must be provided to prevent this accumulation.

§ 1910.306 Specific Purpose Equipment and Installations

[46 FR 4056, Jan. 16, 1981; 46 FR 40185, Aug. 7, 1981; 72 FR 7205, Feb. 14, 2007]

(a)	**Electric signs and outline lighting.**
(a)(1)	**Disconnecting means.**
(a)(1)(i)	Each sign and outline lighting system, or feeder circuit or branch circuit supplying a sign or outline lighting system, shall be controlled by an externally operable switch or circuit breaker that will open all ungrounded conductors. However, a disconnecting means is not required for an exit directional sign located within a building or for cord-connected signs with an attachment plug.
(a)(1)(ii)	Signs and outline lighting systems located within fountains shall have the disconnect located at least 1.52 m (5.0 ft) from the inside walls of the fountain.
(a)(2)	**Location.**
(a)(2)(i)	The disconnecting means shall be within sight of the sign or outline lighting system that it controls. Where the disconnecting means is out of the line of sight from any section that may be energized, the disconnecting means shall be capable of being locked in the open position.
(a)(2)(ii)	Signs or outline lighting systems operated by electronic or electromechanical controllers located external to the sign or outline lighting system may have a disconnecting means located within sight of the controller or in the same enclosure with the controller. The disconnecting means shall disconnect the sign or outline lighting system and the controller from all ungrounded supply conductors. It shall be designed so no pole can be operated independently and shall be capable of being locked in the open position.
(a)(2)(iii)	Doors or covers giving access to uninsulated parts of indoor signs or outline lighting exceeding 600 volts and accessible to other than qualified persons shall either be provided with interlock switches to disconnect the primary circuit or shall be so fastened that the use of other than ordinary tools will be necessary to open them.

This section covers disconnecting means installations that are supplied directly by a branch-circuit or feeder. Disconnecting means must be within sight of the sign, or must be provided with a means to lock it in the open position. The requirements are consistent with Section 600.6 of the NEC®.

(b)	**Cranes and hoists.** This paragraph applies to the installation of electric equipment and wiring used in connection with cranes, monorail hoists, hoists, and all runways.
(b)(1)	**Disconnecting means for runway conductors**. A disconnecting means shall be provided between the runway contact conductors and the power supply. Such disconnecting means shall consist of a motor-circuit switch, circuit breaker, or molded case switch. The disconnecting means shall open all ungrounded conductors simultaneously and shall be:
(b)(1)(i)	Readily accessible and operable from the ground or floor level;
(b)(1)(ii)	Arranged to be locked in the open position; and
(b)(1)(iii)	Placed within view of the runway contact conductors.
(b)(2)	**Disconnecting means for cranes and monorail hoists.**
(b)(2)(i)	Except as provided in paragraph (b)(2)(iv) of this section, a motor-circuit switch, molded case switch, or circuit breaker shall be provided in the leads from the runway contact conductors or other power supply on all cranes and monorail hoists.
(b)(2)(ii)	The disconnecting means shall be capable of being locked in the open position.
(b)(2)(iii)	Means shall be provided at the operating station to open the power circuit to all motors of the crane or monorail hoist where the disconnecting means is not readily accessible from the crane or monorail hoist operating station.
(b)(2)(iv)	The disconnecting may be omitted where a monorail hoist or hand-propelled crane bridge installation meets all of the following conditions:
(b)(2)(iv)(A)	The unit is controlled from the ground or floor level;
(b)(2)(iv)(B)	The unit is within view of the power supply disconnecting means; and

(b)(1)(iv)(C) No fixed work platform has been provided for servicing the unit.

Figure 91 – Location of disconnect for gantry crane.

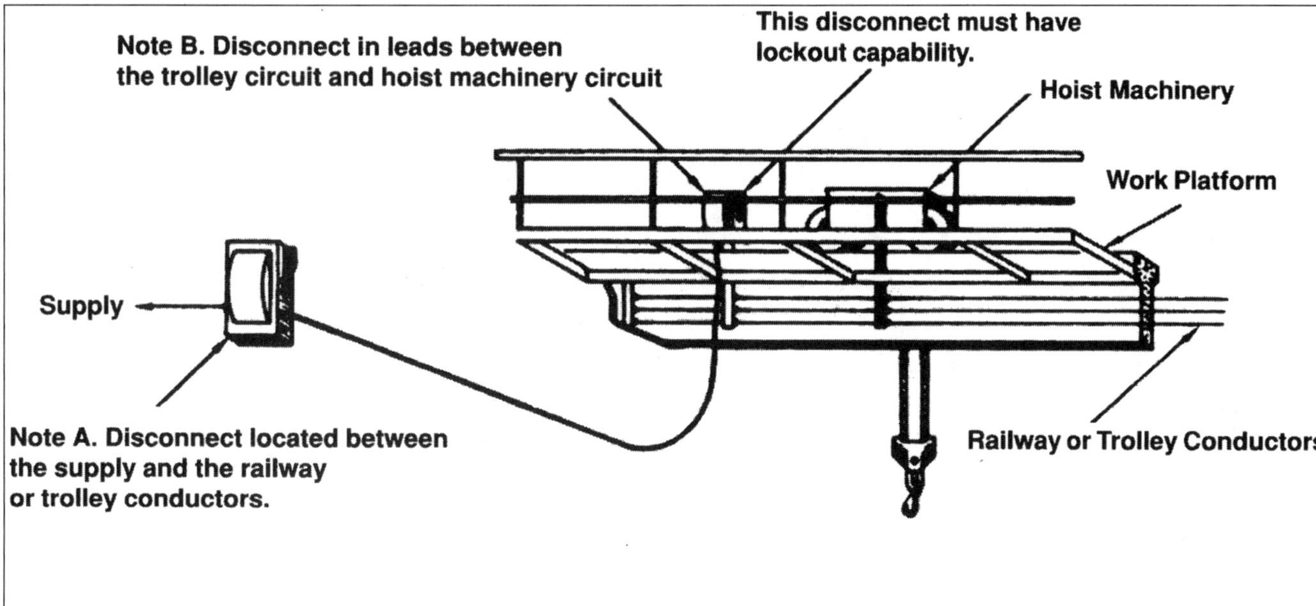

An additional control switch or a remote control switch is necessary if the second disconnecting means (B) is not accessible to the operator.

Disconnect Switches on Cranes and Hoists

Disconnect switches are required on cranes and hoists to allow maintenance and servicing operations to be performed safely. There are two key locations in a crane or hoist circuit where a disconnect is required. The first – see note A, Figure 91 – is located between the power supply and the railway or trolley contact conductors. The second – see note B, Figure 91 – is in the leads between the trolley contact conductors and the hoist machinery. This second disconnect must be capable of being locked out – i.e., locked in the open or off position – to avoid a safety hazard to workers who are servicing the hoist machinery and may not be visible from the power supply disconnect. As specified in 1910.306(b)(1)(ii), the second disconnect should be connected between the trolley conductors and the hoist machinery.

Omitting a Disconnect Switch

There are three conditions that must be met before the disconnect switch in the leads between the railway conductors and the hoist machinery can be omitted. This exception only applies to hand-propelled bridge crane installations and monorail hoists, and all three of the following conditions must be met: (1) the unit must be floor controlled, (2) the hoisting machinery unit must be within view of the power supply disconnecting means, and (3) no fixed work platform can be installed on the monorail hoist or hand-propelled bridge crane. Figure 92 illustrates a monorail hoist that meets these three conditions and may operate with only one disconnect switch located in the power supply conductor.

Figure 92 – Second disconnect not required

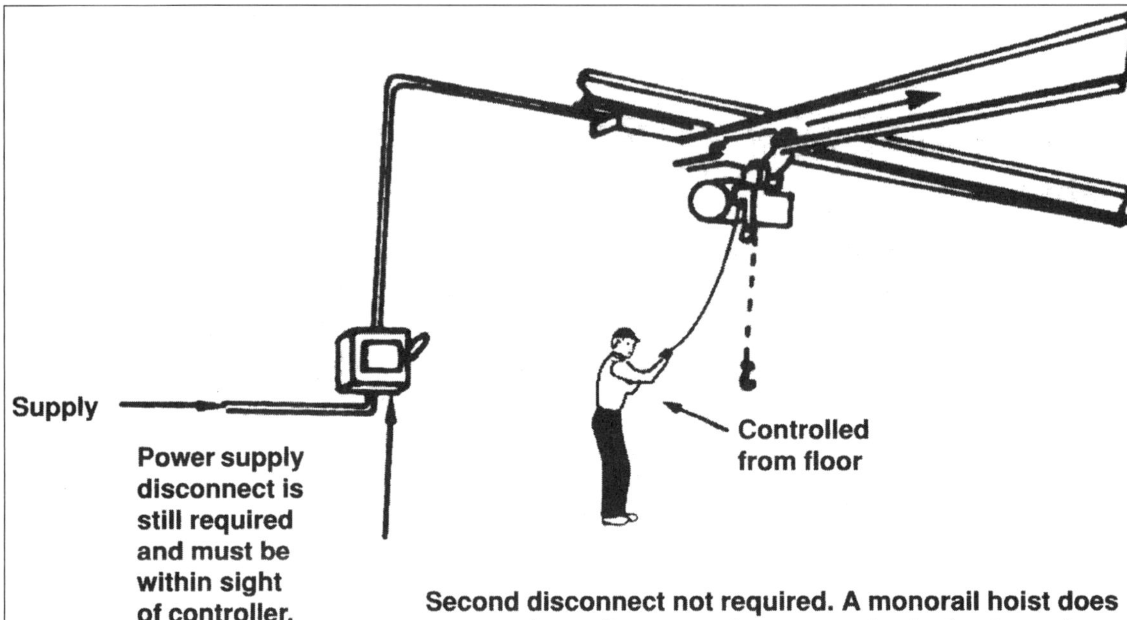

Second disconnect not required. A monorail hoist does not require a disconnecting means in the leads to the hoist machinery if it is controlled from the floor, if it is within view of the power supply disconnect, and if there is no work platform provided to service the hoist machinery.

(b)(3) **Limit switch.** A limit switch or other device shall be provided to prevent the load block from passing the safe upper limit of travel of any hoisting mechanism.

(b)(4) **Clearance.** The dimension of the working space in the direction of access to live parts that may require examination, adjustment, servicing, or maintenance while alive shall be a minimum of 762 mm (2.5 ft). Where controls are enclosed in cabinets, the doors shall either open at least 90 degrees or be removable.

(c) **Elevators, dumbwaiters, escalators, and moving walks, wheelchair lifts, and stairway chair lifts.** The following requirements apply to elevators, dumbwaiters, escalators, moving walks, wheelchair lifts, and stairway chair lifts.

(c)(1) **Disconnecting means.** Elevators, dumbwaiters, escalators, moving walks, wheelchair lifts, and stairway chair lifts shall have a single means for disconnecting all ungrounded main power supply conductors for each unit.

Disconnecting Means and Warning Signs

Each elevator, dumbwaiter, escalator, or moving walk is required to have a single disconnecting switch to shut off power to each unit. Some installations may have interconnected control panels which will remain energized even if the appropriate disconnect switch is moved to the open or off position. A common example is where elevator control panels in a high rise office building are interconnected to coordinate responses between the upper and lower floors. If a disconnect is shut off for one unit, interconnected control circuits may still be energized. When this is the case for installations made after April 16, 1981, a warning sign is required stating that a hazard still exists (see Figure 93).

Figure 93 – Interconnected control panels

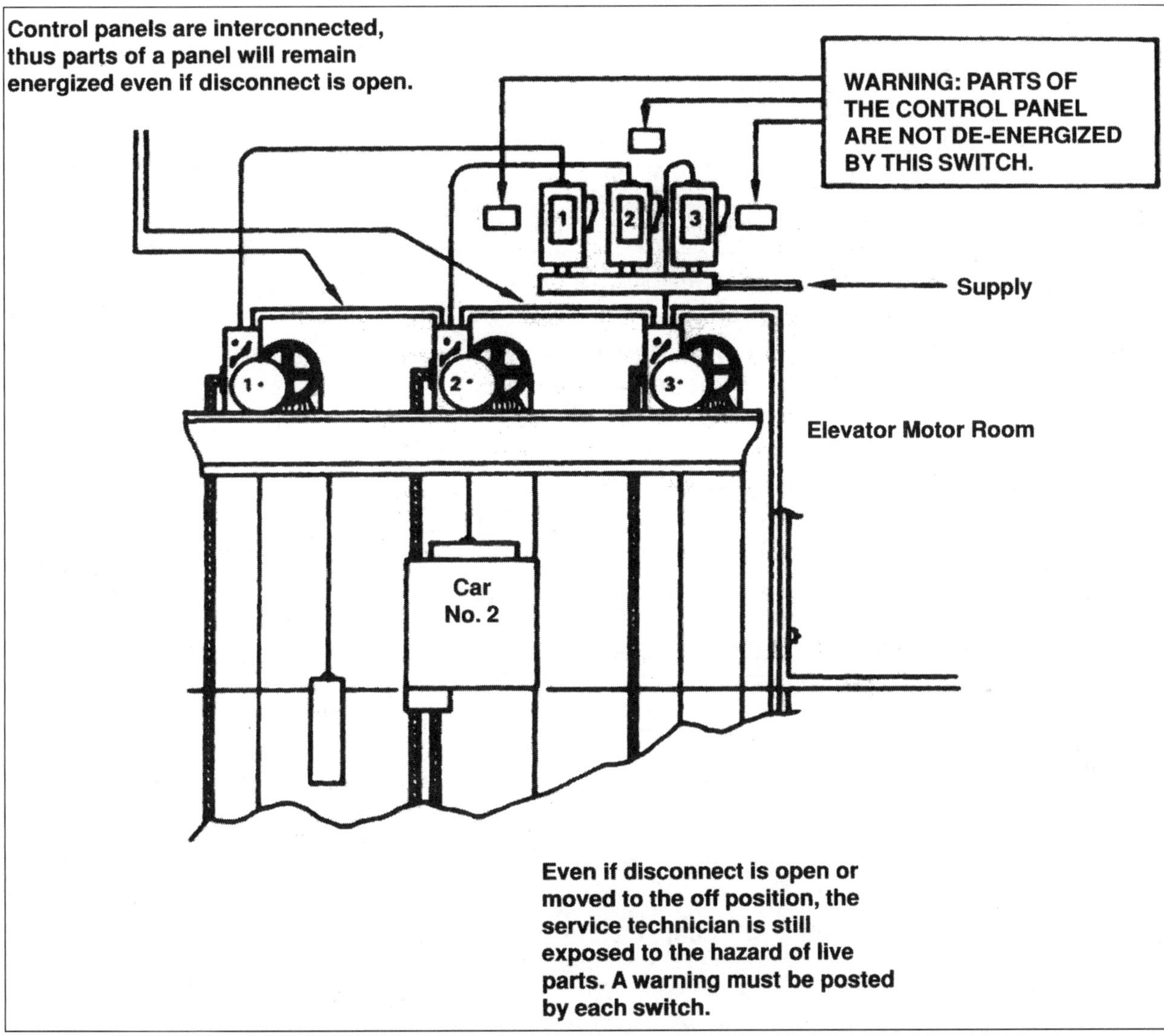

(c)(2) **Control panels.** Control panels not located in the same space as the drive machine shall be located in cabinets with doors or panels capable of being locked closed.

(c)(3) **Type**. The disconnecting means shall be an enclosed externally operable fused motor circuit switch or circuit breaker capable of being locked in the open position. The disconnecting means shall be a listed device.

(c)(4) **Operation.** No provision may be made to open or close this disconnecting means from any other part of the premises. If sprinklers are installed in hoistways, machine rooms, or machinery spaces, the disconnecting means may automatically open the power supply to the affected elevators prior to the application of water. No provision may be made to close this disconnecting means automatically (that is, power may only be restored by manual means).

(c)(5) **Location**. The disconnecting means shall be located where it is readily accessible to qualified persons.

(c)(5)(i) On elevators without generator field control, the disconnecting means shall be located within sight of the motor controller. Driving machines or motion and operation controllers not within sight of the disconnecting means shall be provided with a manually operated switch installed in the control circuit adjacent to the equipment in order to prevent starting. Where the driving machine is located in a remote machinery space, a single disconnecting means for disconnecting all ungrounded main power supply conductors shall be provided and be capable of being locked in the open position.

(c)(5)(ii) On elevators with generator field control, the disconnecting means shall be located within sight of the motor controller for the driving motor of the motor-generator set. Driving machines, motor-generator sets, or motion and operation controllers not within sight of the disconnecting means shall be provided with a manually operated switch installed in the control circuit to prevent starting. The manually operated switch shall be installed adjacent to this equipment. Where the driving machine or the motor-generator set is located in a remote machinery space, a single means for disconnecting all ungrounded main power supply conductors shall be provided and be capable of being locked in the open position.

(c)(5)(iii) On escalators and moving walks, the disconnecting means shall be installed in the space where the controller is located.

(c)(5)(iv) On wheelchair lifts and stairway chair lifts, the disconnecting means shall be located within sight of the motor controller.

(c)(6) **Identification and signs.**

(c)(6)(i) Where there is more than one driving machine in a machine room, the disconnecting means shall be numbered to correspond to the identifying number of the driving machine that they control.

(c)(6)(ii) The disconnecting means shall be provided with a sign to identify the location of the supply-side overcurrent protective device.

(c)(7) **Single-car and multicar installations.** On single-car and multicar installations, equipment receiving electrical power from more than one source shall be provided with a disconnecting means for each source of electrical power. The disconnecting means shall be within sight of the equipment served.

(c)(8) **Warning sign for multiple disconnecting means.** A warning sign shall be mounted on or next to the disconnecting means where multiple disconnecting means are used and parts of the controllers remain energized from a source other than the one disconnected. The sign shall be clearly legible and shall read "WARNING - PARTS OF THE CONTROLLER ARE NOT DEENERGIZED BY THIS SWITCH."

(c)(9) **Interconnection between multicar controllers**. A warning sign worded as required in paragraph (c)(8) of this section shall be mounted on or next to the disconnecting means where interconnections between controllers are necessary for the operation of the system on multicar installations that remain energized from a source other than the one disconnected.

(c)(10) **Motor controllers**. Motor controllers may be located outside the spaces otherwise required by paragraph (c) of this section, provided they are in enclosures with doors or removable panels capable of being locked closed and the disconnecting means is located adjacent to or is an integral part of the motor controller. Motor controller enclosures for escalators or moving walks may be located in the balustrade on the side located away from the moving steps or moving treadway. If the disconnecting means is an integral part of the motor controller, it shall be operable without opening the enclosure.

Further information on requirements for elevators and escalators can be found in American Society of Mechanical Engineers (ASME) standard A17.1 - 2010/CSA B44-10 Safety Code for Elevators and Escalators.

(d) **Electric welders - disconnecting means.**

(d)(1) **Arc welders.** A disconnecting means shall be provided in the supply circuit for each arc welder that is not equipped with a disconnect mounted as an integral part of the welder. The disconnecting means shall be a switch or circuit breaker, and its rating may not be less than that necessary to accommodate overcurrent protection.

(d)(2) **Resistance welders.** A switch or circuit breaker shall be provided by which each resistance welder and its control equipment can be disconnected from the supply circuit. The ampere rating of this disconnecting means may not be less than the supply conductor ampacity. The supply circuit switch may be used as the welder disconnecting means where the circuit supplies only one welder.

(e) **Information technology equipment.**

(e)(1) **Disconnecting means.** A means shall be provided to disconnect power to all electronic equipment in an information technology equipment room. There shall also be a similar means to disconnect the power to all dedicated heating, ventilating, and air conditioning (HVAC) systems serving the room and to cause all required fire/smoke dampers to close.

(e)(2) **Grouping**. The control for these disconnecting means shall be grouped and identified and shall be readily accessible at the principal exit doors. A single means to control both the electronic equipment and HVAC system is permitted.

(e)(3) **Exception.** Integrated electrical systems covered by § 1910.308(g) need not have the disconnecting means required by paragraph (e)(1) of this section.

Data Processing Systems - Disconnecting Means

Electrical fires are common types of fires that occur in computer rooms. This creates a need to provide positive control over two key power systems in an emergency. These are the power supply systems to the computer equipment and the ventilation system for the computer room. Should an emergency occur, shutting down the power supply to the computer, related equipment, and electronic system, wiring under the raised floors will eliminate a source of ignition that may propagate a fire. The ventilation system may contribute to the spread of the fire by fanning it or by rapidly spreading smoke through this area. A single "panic switch" device may be used to disconnect both the electronic equipment and the air-conditioning system. Locating disconnect switches near the exit door permits these systems to be shut down easily while the facility is being evacuated. These switches must be readily accessible to the operator and not blocked by equipment or a door. Lighting systems must remain separate to aid evacuation and emergency operations (see Figure 94). Figure 95 shows the operation of a disconnect for electronic equipment in a data processing room.

(f)(1)(i) A disconnecting means shall be provided in the supply circuit. The disconnecting means shall be operable from a location readily accessible from the X-ray control. For equipment connected to a 120-volt branch circuit of 30 amperes or less, a grounding-type attachment plug cap and receptacle of proper rating may serve as a disconnecting means.

(f)(1)(ii) If more than one piece of equipment is operated from the same high-voltage circuit, each piece or each group of equipment as a unit shall be provided with a high-voltage switch or equivalent disconnecting means. The disconnecting means shall be constructed, enclosed, or located so as to avoid contact by employees with its live parts.

(f)(2) **Control.** The following requirements apply to industrial and commercial laboratory equipment.

(f)(2)(i) Radiographic and fluoroscopic-type equipment shall be effectively enclosed or shall have interlocks that deenergize the equipment automatically to prevent ready access to live current-carrying parts.

Figure 94 – Location of disconnect in data processing center

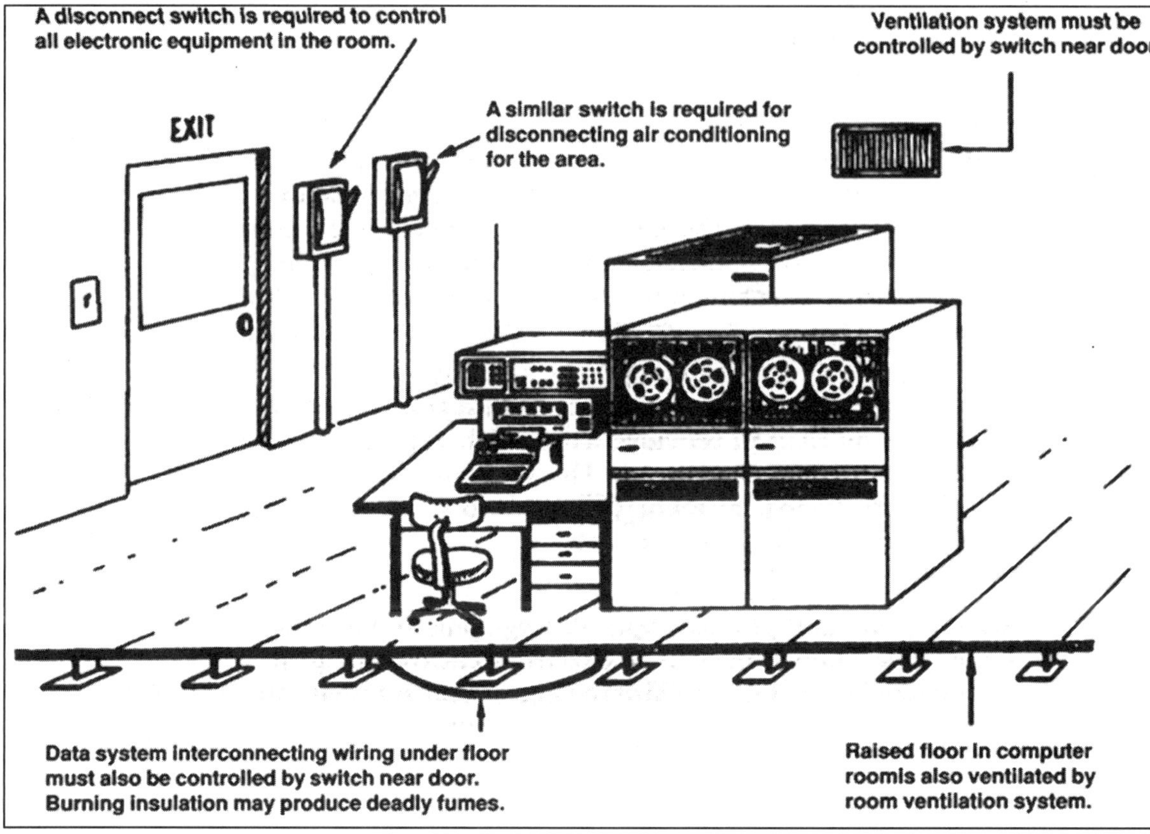

Figure 95 – Typical operation of disconnecting means.

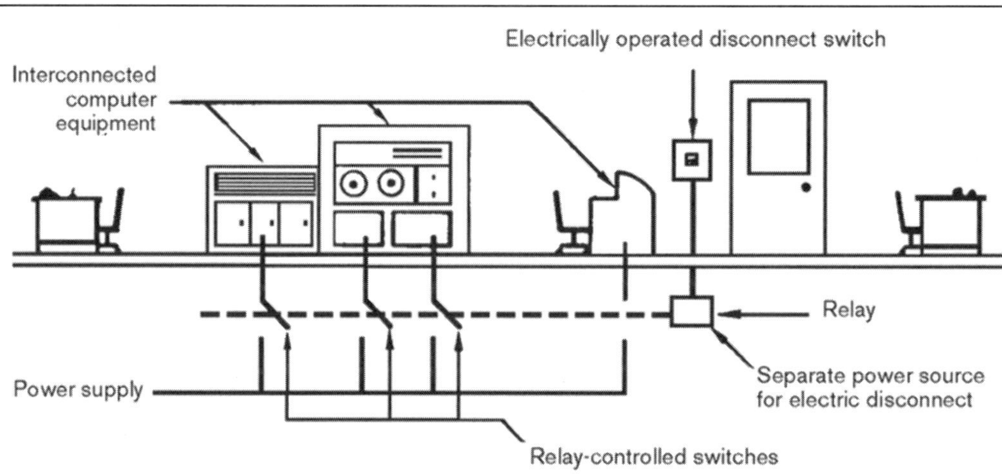

An electrically operated disconnect switch is commonly used in this application. This type of switch serves as a remote control switch that operates a relay to open the power supply conductors to the computer equipment.

Radiographic and Fluoroscopic Equipment

Radiographic (X-Ray) and fluoroscopic equipment operates at high energy levels and high voltages that would present a serious hazard if a service technician were to open access panels while the equipment is energized. To prevent exposure to this hazard, radiographic and fluoroscopic-type X-Ray equipment must be located in metal enclosures constructed to prevent access by unqualified persons. Should access be required (e.g., to make an adjustment), then the access panel or door must be equipped with interlocks that will shut off power to live equipment inside when the door is opened (See Figure 96).

(f) **X-ray equipment.** This paragraph applies to X-ray equipment.

(f)(1) **Disconnecting means.**

(f)(1)(i) A disconnecting means shall be provided in the supply circuit. The disconnecting means shall be operable from a location readily accessible from the X-ray control. For equipment connected to a 120-volt branch circuit of 30 amperes or less, a grounding-type attachment plug cap and receptacle of proper rating may serve as a disconnecting means.

Figure 96.- Illustration showing the operation of an interlock.

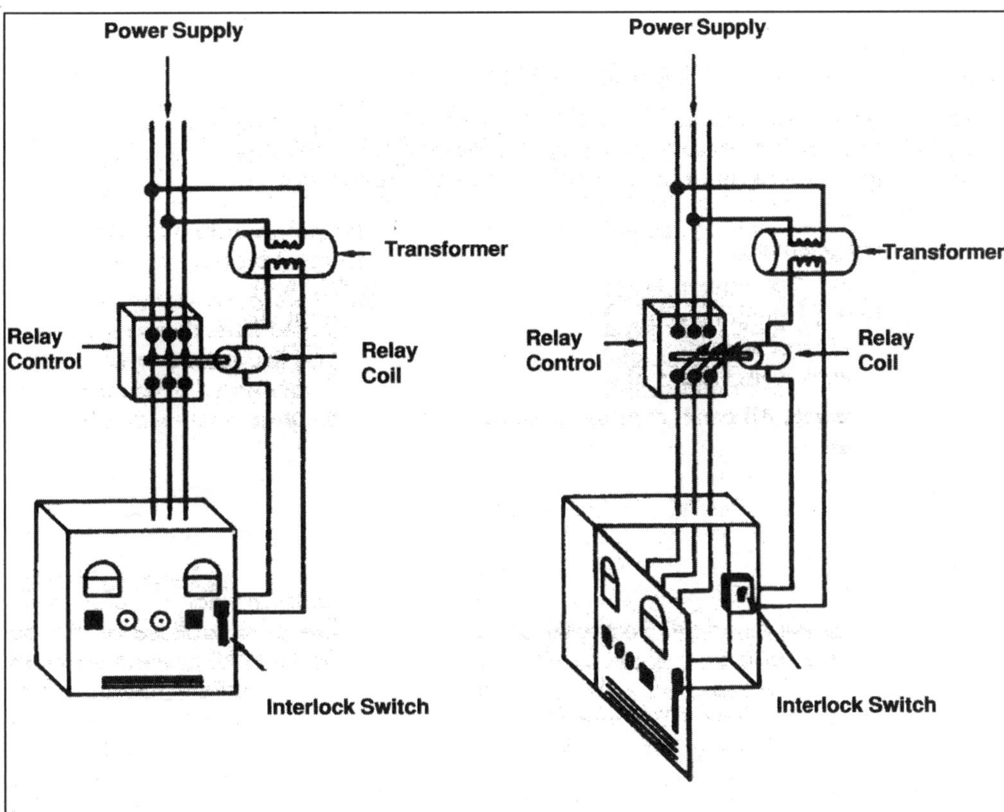

A. Power is supplied through a relay control that is connected to an interlock circuit.

B. When the panel door opens, a micro switch operates a relay coil that opens the power supply conductors and de-energizes the equipment inside the enclosure.

(f)(1)(ii)	If more than one piece of equipment is operated from the same high-voltage circuit, each piece or each group of equipment as a unit shall be provided with a high-voltage switch or equivalent disconnecting means. The disconnecting means shall be constructed, enclosed, or located so as to avoid contact by employees with its live parts.
(f)(2)	**Control.** The following requirements apply to industrial and commercial laboratory equipment.
(f)(2)(i)	Radiographic and fluoroscopic-type equipment shall be effectively enclosed or shall have interlocks that deenergize the equipment automatically to prevent ready access to live current-carrying parts.
(f)(2)(ii)	Diffraction- and irradiation-type equipment shall have a pilot light, readable meter deflection, or equivalent means to indicate when the equipment is energized, unless the equipment or installation is effectively enclosed or is provided with interlocks to prevent access to live current-carrying parts during operation.
(g)	**Induction and dielectric heating equipment.** This paragraph applies to induction and dielectric heating equipment and accessories for industrial and scientific applications, but not for medical or dental applications or for appliances.
(g)(1)	**Guarding and grounding.**
(g)(1)(i)	The converting apparatus (including the dc line) and high-frequency electric circuits (excluding the output circuits and remote-control circuits) shall be completely contained within enclosures of noncombustible material.
(g)(1)(ii)	All panel controls shall be of dead-front construction

Panel Controls

Induction heating equipment uses high frequency alternating currents. Control panels for induction and dielectric heating equipment are required to be of dead-front construction; that is, constructed so that there are no live parts exposed on the operating side of the equipment. All connections and terminations must be behind the front panel.

(g)(1)(iii)	Doors or detachable panels shall be employed for internal access. Where doors are used giving access to voltages from 500 to 1000 volts ac or dc, either door locks shall be provided or interlocks shall be installed. Where doors are used giving access to voltages of over 1000 volts ac or dc, either mechanical lockouts with a disconnecting means to prevent access until circuit parts within the cubicle are deenergized, or both door interlocking and mechanical door locks, shall be provided. Detachable panels not normally used for access to such parts shall be fastened in a manner that will make them difficult to remove (for example, by requiring the use of tools).
(g)(1)(iv)	Warning labels or signs that read “DANGER - HIGH VOLTAGE - KEEP OUT” shall be attached to the equipment and shall be plainly visible where persons might contact energized parts when doors are opened or closed or when panels are removed from compartments containing over 250 volts ac or dc.
(g)(1)(v)	Induction and dielectric heating equipment shall be protected as follows:
(g)(1)(v)(A)	Protective cages or adequate shielding shall be used to guard work applicators other than induction heating coils.
(g)(1)(v)(B)	Induction heating coils shall be protected by insulation or refractory materials or both.
(g)(1)(v)(C)	Interlock switches shall be used on all hinged access doors, sliding panels, or other such means of access to the applicator, unless the applicator is an induction heating coil at dc ground potential or operating at less than 150 volts ac.
(g)(1)(v)(D)	Interlock switches shall be connected in such a manner as to remove all power from the applicator when any one of the access doors or panels is open.

(g)(1)(vi) A readily accessible disconnecting means shall be provided by which each heating equipment can be isolated from its supply circuit. The ampere rating of this disconnecting means may not be less than the nameplate current rating of the equipment. The supply circuit disconnecting means is permitted as a heating equipment disconnecting means where the circuit supplies only one piece of equipment.

(g)(2) **Remote control.**

(g)(2)(i) If remote controls are used for applying power, a selector switch shall be provided and interlocked to provide power from only one control point at a time.

(g)(2)(ii) Switches operated by foot pressure shall be provided with a shield over the contact button to avoid accidental closing of the switch.

Remote Controls

Remote controls must have an interlocked selector switch so only one control point can operate the power to a particular heating process at a time. For example, induction heating is well suited for surface hardening metal parts to protect against wear. A common production application is a set-up in which gear teeth are surface-hardened by induction heating. This set-up may have two or more control points to facilitate production and allow the operator to observe the process from a remote location. An interlocked selector switch must be provided to allow control from only one point at a time, thus protecting against accidental operation at unattended control locations (See Figure 97).

Figure 97 – Induction heat-treating process.

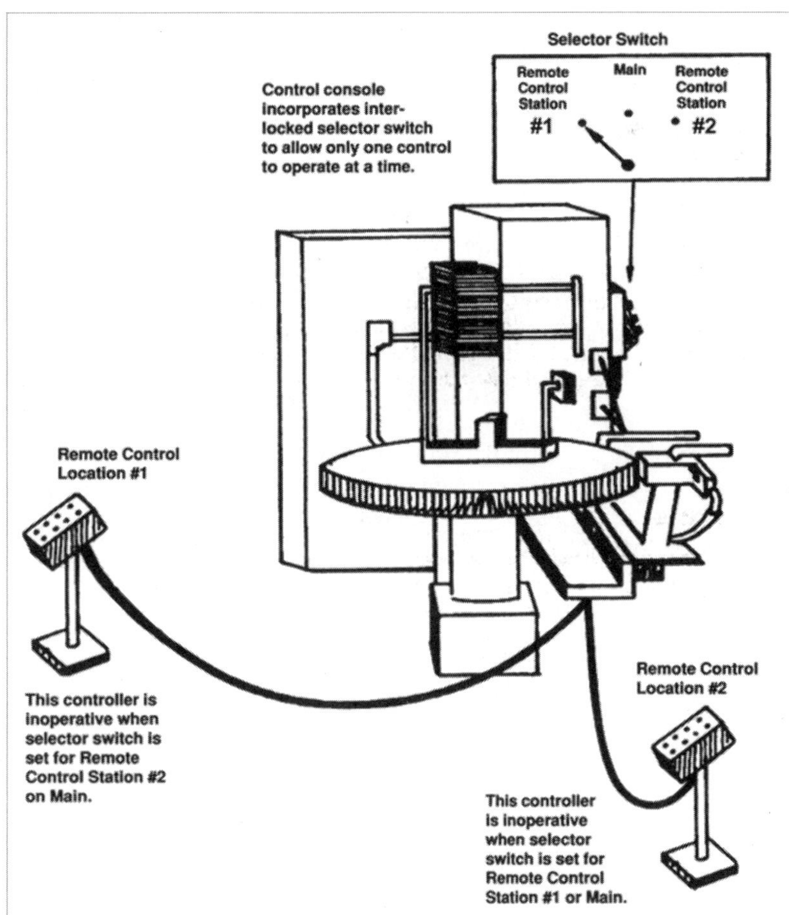

(h) **Electrolytic cells.** This paragraph applies to the installation of the electrical components and accessory equipment of electrolytic cells, electrolytic cell lines, and process power supply for the production of aluminum, cadmium, chlorine, copper, fluorine, hydrogen peroxide, magnesium, sodium, sodium chlorate, and zinc. Cells used as a source of electric energy and for electroplating processes and cells used for production of hydrogen are not covered by this paragraph.

Cell Line Working Zone

Paragraph 1910.303(g)(2) requires live parts of equipment operating at 50 volts or more to be guarded. Recognizing a special case, paragraph 1910.306(h) permits electrolytic cells to operate with live parts exposed, provided that specific alternate safety measures are employed.Electrolytic cells differ in certain respects from standard electrical installations. In a standard electrical circuit, all live parts are required to be insulated or guarded. One side of the circuit is connected to ground to provide a common reference and to stabilize voltages, and thus a potential exists between any live conductor and any grounded surface. A person coming in contact with both a live conductor and a grounded surface will provide a path for electric current to flow and will receive a shock varying in severity with the amount of current that flows through the body. The purpose of the guarding provisions of the NEC® is to prevent contact with live parts, since contact with grounded surfaces is very common.In an electrolytic cell line, however, the actual working parts and some working surfaces are, by necessity, the bare live parts of an electric circuit. Employee contact with these live parts is necessary in normal electrolytic cell working procedures. Therefore, whereas a standard circuit necessitates guarding of live parts, employees working on electrolytic cells must be protected against contact with ground. The provisions of paragraph 1910.306(h) recognize the different working conditions of electrolytic cell lines and outline the procedures to be used as an alternative to guarding of live parts.

Figure 98 illustrates a typical aluminum reduction cell line arrangement showing the location of the cell line working zone boundaries.

An electrolytic cell line and its DC power supply circuits, all within the cell line working zone, are considered a single unit just as a machine tool is considered a single unit supplied from a single power source. Cell line installations may cover several acres of floor space and may have loads in excess of 400,000 amperes DC at voltages in excess of 1000 volts DC.

The heavy bus bars and connections carrying this current are usually bare; therefore, it is necessary to minimize the possibility of workers touching the bars while also touching other conductive surfaces at a different potential. To do this, the NEC® defines boundaries to the cell line unit.

The boundaries of the cell line working zone are specified in Section 668.10 of the NEC® and are keyed to vertical and horizontal measurements from energized surfaces of the cell line or their attachments.

Figure 98 – The boundaries of the cell line as a single functional unit

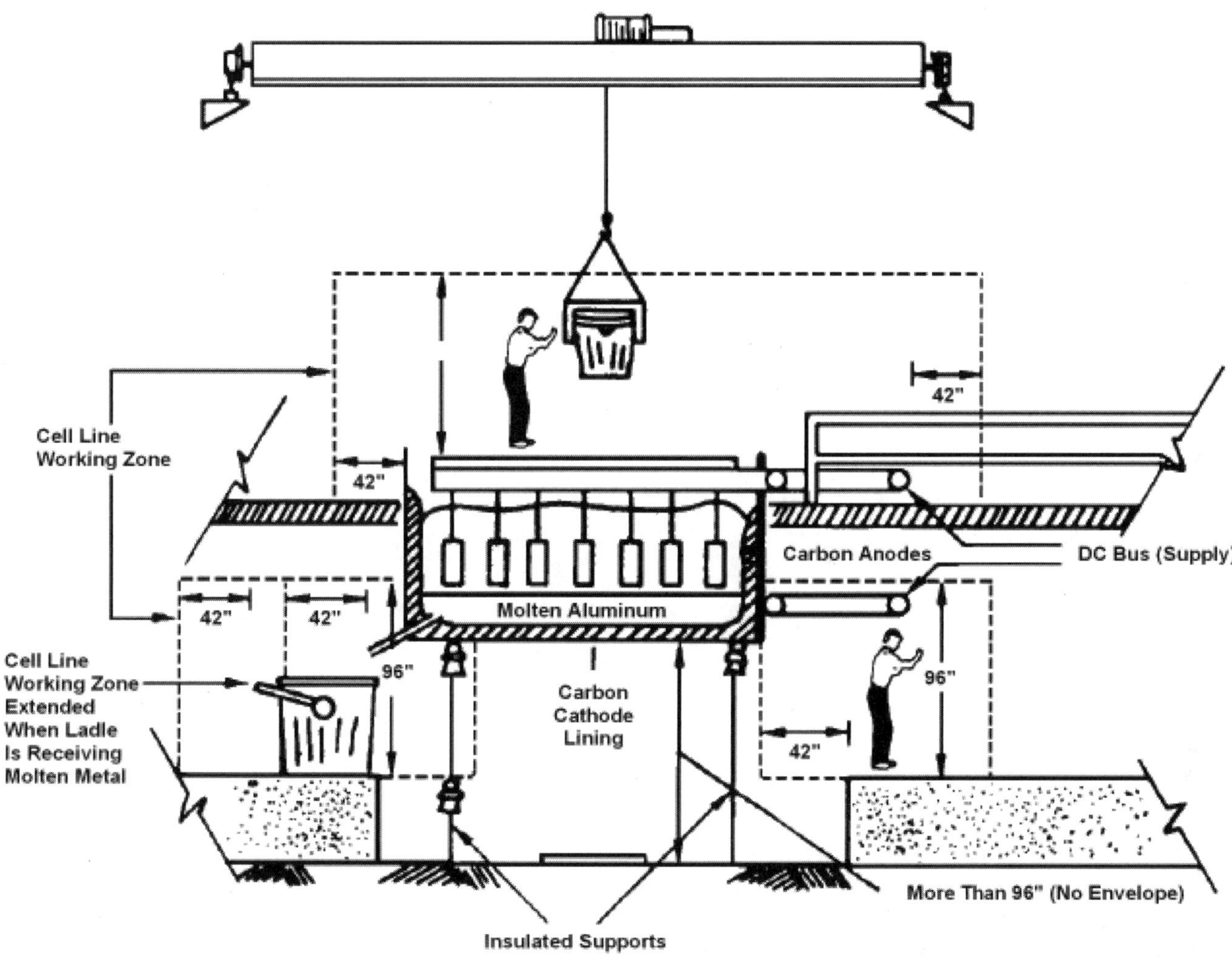

These measurements include: the space within 2.5m (96 inches) above any energized part, the space below energized surfaces provided the headroom is less than 2.5m (96 inches), and the space within 1.0m (42 inches) horizontally from energized surfaces or from the 2.5m (96-inches) vertical envelope. These space requirements do not extend beyond walls or other fixed barriers.

(h)(1)	**Application.** Installations covered by paragraph (h) of this section shall comply with all applicable provisions of this subpart, except as follows:
(h)(1)(i)	Overcurrent protection of electrolytic cell dc process power circuits need not comply with the requirements of § 1910.304(f);
(h)(1)(ii)	Equipment located or used within the cell line working zone or associated with the cell line dc power circuits need not comply with the provisions of § 1910.304(g); and
(h)(1)(iii)	Electrolytic cells, cell line conductors, cell line attachments, and the wiring of auxiliary equipment and devices within the cell line working zone need not comply with the provisions of § 1910.303, and § 1910.304 (b) and (c).

(h)(2)	**Disconnecting means.** If more than one dc cell line process power supply serves the same cell line, a disconnecting means shall be provided on the cell line circuit side of each power supply to disconnect it from the cell line circuit. Removable links or removable conductors may be used as the disconnecting means.
(h)(3)	**Portable electric equipment.**
(h)(3)(i)	The frames and enclosures of portable electric equipment used within the cell line working zone may not be grounded, unless the cell line circuit voltage does not exceed 200 volts DC or the frames are guarded.
(h)(3)(ii)	Ungrounded portable electric equipment shall be distinctively marked and shall employ plugs and receptacles of a configuration that prevents connection of this equipment to grounding receptacles and that prevents inadvertent interchange of ungrounded and grounded portable electric equipment.
(h)(4)	**Power supply circuits and receptacles for portable electric equipment.**
(h)(4)(i)	Circuits supplying power to ungrounded receptacles for hand-held, cord- and plug-connected equipment shall meet the following requirements:
(h)(4)(i)(A)	The circuits shall be electrically isolated from any distribution system supplying areas other than the cell line working zone and shall be ungrounded;
(h)(4)(i)(B)	The circuits shall be supplied through isolating transformers with primaries operating at not more than 600 volts between conductors and protected with proper overcurrent protection;
(h)(4)(i)(C)	The secondary voltage of the isolating transformers may not exceed 300 volts between conductors; and
(h)(4)(i)(D)	All circuits supplied from the secondaries shall be ungrounded and shall have an approved overcurrent device of proper rating in each conductor.
(h)(4)(ii)	Receptacles and their mating plugs for ungrounded equipment may not have provision for a grounding conductor and shall be of a configuration that prevents their use for equipment required to be grounded.
(h)(4)(iii)	Receptacles on circuits supplied by an isolating transformer with an ungrounded secondary:
(h)(4)(iii)(A)	Shall have a distinctive configuration;
(h)(4)(iii)(B)	Shall be distinctively marked; and
(h)(4)(iii)(C)	May not be used in any other location in the facility.

Figure 99 – Isolating transformer supplying hand-held double-insulated tool for use in cell line working zone.

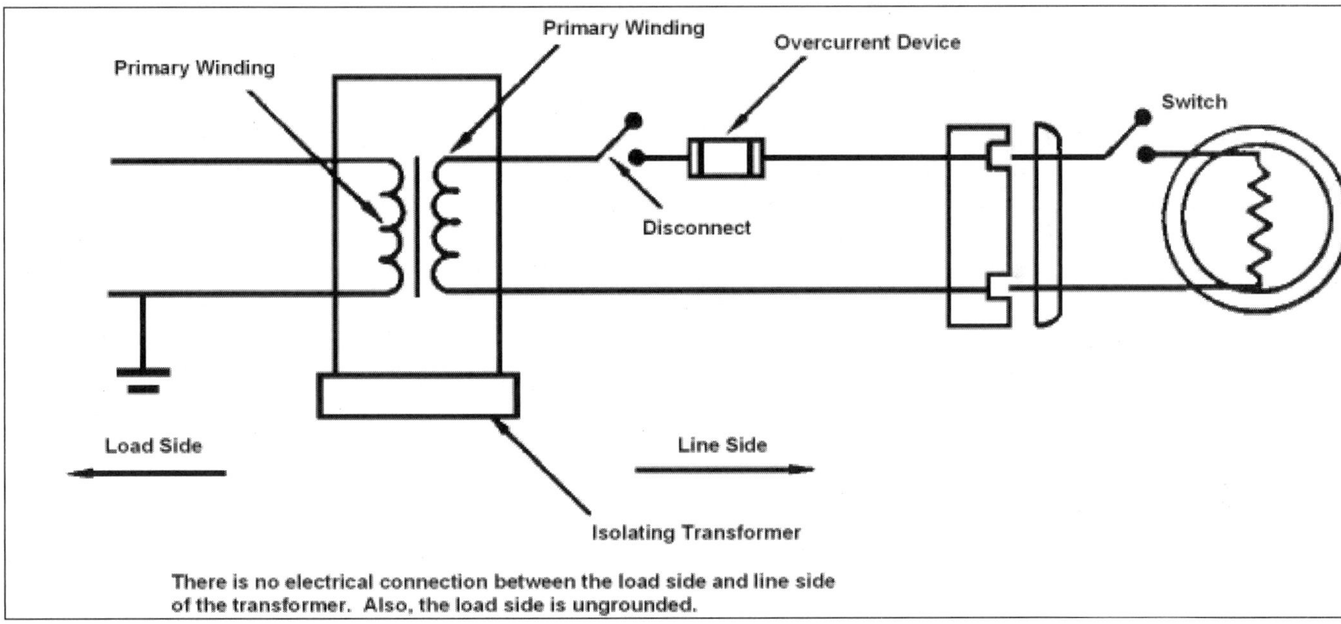

Grounding Portable Electric Equipment

The cell line DC circuits operate at high voltage and power ratings. Aluminum reduction, for instance, can be rated at 600 volts DC and 155,000 amperes DC.

Voltage to ground from these current-carrying parts is generally unknown. Therefore, it is necessary to minimize the possibility of workers accidentally providing a path to ground at any point. To prevent the introduction of inadvertent grounds, tool frames must be insulated from ground rather than grounded, and even the circuits supplying such tools cannot have a grounded conductor. Grounding is permitted, however, if the cell line voltage is 200 volts or less.

Isolating transformers are used to supply power to hand-held, cord-connected equipment used in the cell line working zone. The secondary windings of isolating transformers have no electrical connection to the primary windings. Therefore, the load side of the circuit is not influenced by the grounding of the distribution system supplying the primary windings. These isolated circuits must not be grounded. Figure 99 illustrates a schematic drawing of an isolated circuit used to supply hand-held power tools.

Plugs and receptacles used for equipment connected to a circuit supplied by an isolating transformer must have a configuration different from any other used in the plant.

(h)(5)	**Fixed and portable electric equipment.**
(h)(5)(i)	The following need not be grounded:
(h)(5)(i)(A)	AC systems supplying fixed and portable electric equipment within the cell line working zone; and
(h)(5)(i)(B)	Exposed conductive surfaces, such as electric equipment housings, cabinets, boxes, motors, raceways and the like that are within the cell line working zone.
(h)(5)(ii)	Auxiliary electric equipment, such as motors, transducers, sensors, control devices, and alarms, mounted on an electrolytic cell or other energized surface shall be connected to the premises wiring systems by any of the following means:

(h)(5)(ii)(A)	Multiconductor hard usage or extra hard usage flexible cord;
(h)(5)(ii)(B)	Wire or cable in suitable nonmetallic raceways or cable trays; or
(h)(5)(ii)(C)	Wire or cable in suitable metal raceways or metal cable trays installed with insulating breaks such that they will not cause a potentially hazardous electrical condition.
(h)(5)(iii)	Fixed electric equipment may be bonded to the energized conductive surfaces of the cell line, its attachments, or auxiliaries. If fixed electric equipment is mounted on an energized conductive surface, it shall be bonded to that surface.
(h)(6)	**Auxiliary nonelectric connections.** Auxiliary nonelectric connections, such as air hoses, water hoses, and the like, to an electrolytic cell, its attachments, or auxiliary equipment may not have continuous conductive reinforcing wire, armor, braids, and the like. Hoses shall be of a nonconductive material.
(h)(7)	**Cranes and hoists.**
(h)(7)(i)	The conductive surfaces of cranes and hoists that enter the cell line working zone need not be grounded. The portion of an overhead crane or hoist that contacts an energized electrolytic cell or energized attachments shall be insulated from ground.
(h)(7)(ii)	Remote crane or hoist controls that may introduce hazardous electrical conditions into the cell line working zone shall employ one or more of the following systems:
(h)(7)(ii)(A)	Isolated and ungrounded control circuit;
(h)(7)(ii)(B)	Nonconductive rope operator;
(h)(7)(ii)(C)	Pendant pushbutton with nonconductive supporting means and with nonconductive surfaces or ungrounded exposed conductive surfaces; or
(h)(7)(ii)(D)	Radio.
(i)	**Electrically driven or controlled irrigation machines.**
(i)(1)	**Lightning protection.** If an irrigation machine has a stationary point, a grounding electrode system shall be connected to the machine at the stationary point for lightning protection.
(i)(2)	**Disconnecting means.**
(i)(2)(i)	The main disconnecting means for a center pivot irrigation machine shall be located at the point of connection of electrical power to the machine or shall be visible and not more than 15.2 m (50 ft) from the machine.
(i)(2)(ii)	The disconnecting means shall be readily accessible and capable of being locked in the open position.
(i)(2)(iii)	A disconnecting means shall be provided for each motor and controller.
(j)	**Swimming pools, fountains, and similar installations.** This paragraph applies to electric wiring for and equipment in or adjacent to all swimming, wading, therapeutic, and decorative pools and fountains; hydromassage bathtubs, whether permanently installed or storable; and metallic auxiliary equipment, such as pumps, filters, and similar equipment. Therapeutic pools in health care facilities are exempt from these provisions.
(j)(1)	**Receptacles.**
(j)(1)(i)	A single receptacle of the locking and grounding type that provides power for a permanently installed swimming pool recirculating pump motor may be located not less than 1.52 m (5 ft) from the inside walls of a pool. All other receptacles on the property shall be located at least 3.05 m (10 ft) from the inside walls of a pool.
(j)(1)(ii)	Receptacles that are located within 4.57 m (15 ft), or 6.08 m (20 ft) if the installation was built after August 13, 2007, of the inside walls of the pool shall be protected by ground-fault circuit interrupters.
(j)(1)(iii)	Where a pool is installed permanently at a dwelling unit, at least one 125-volt, 15- or 20-ampere receptacle on a general-purpose branch circuit shall be located a minimum of 3.05 m (10 ft) and

not more than 6.08 m (20 ft) from the inside wall of the pool. This receptacle shall be located not more than 1.98 m (6.5 ft) above the floor, platform, or grade level serving the pool.

Note to paragraph (j)(1) of this section: In determining these dimensions, the distance to be measured is the shortest path the supply cord of an appliance connected to the receptacle would follow without piercing a floor, wall, or ceiling of a building or other effective permanent barrier.

Figure 100 – NEMA configurations for 2-pole 3-wire grounding plugs and receptacles

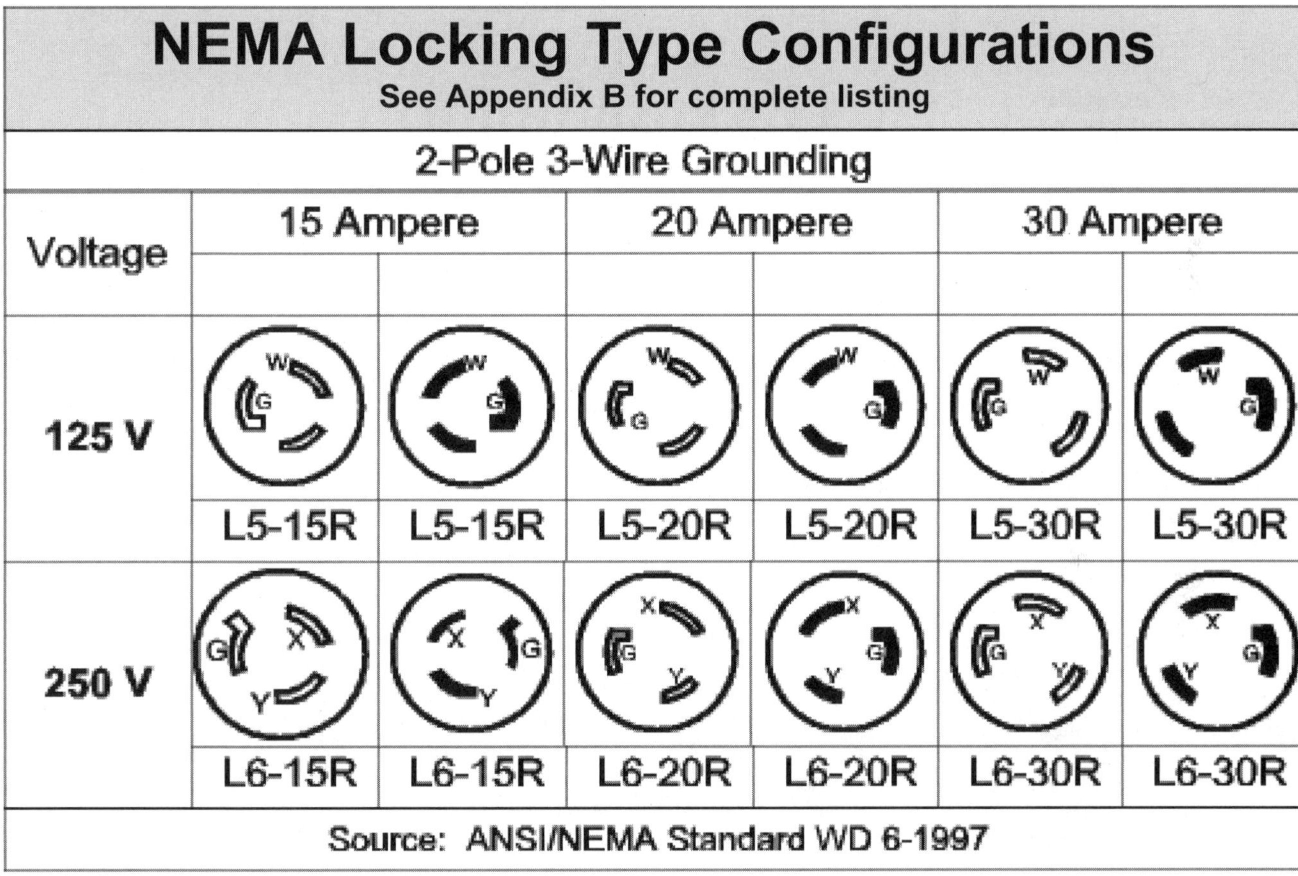

Receptacles

A single receptacle for the pool recirculating pump may be located as close as 1.52m (5 feet) from the inside wall of the pool. This allows for the use of a cord-connected pump motor and provides a means to easily remove the motor during the winter to prevent freezing. The single receptacle and plug must be the locking and grounding type. Figure 100 illustrates the National Electrical Manufacturers Association (NEMA) configuration for locking plugs and receptacles. It is very important to note that the National Electrical Code® specifies 1.83m (6 feet) as the minimum distance, and additionally specifies that this receptacle be ground-fault circuit interrupter (GFCI) protected.

All other receptacles must be at least 3.05m (10 feet) from the inside wall of the pool. The NEC® currently allows for these receptacles be a minimum of 1.83m (6 feet) from the inside wall of the pool. Any receptacle within 4.57m (15 feet) or 6.0m (20 feet) after August 13, 2007, of the pool wall must be equipped with a GFCI. Figure 101 illustrates these key distances where receptacles are permitted. The 6.0m (20 feet) requirement mirrors the current requirements of the NEC®.

Figure 101 – Distance of receptacles to permanently installed swimming pool

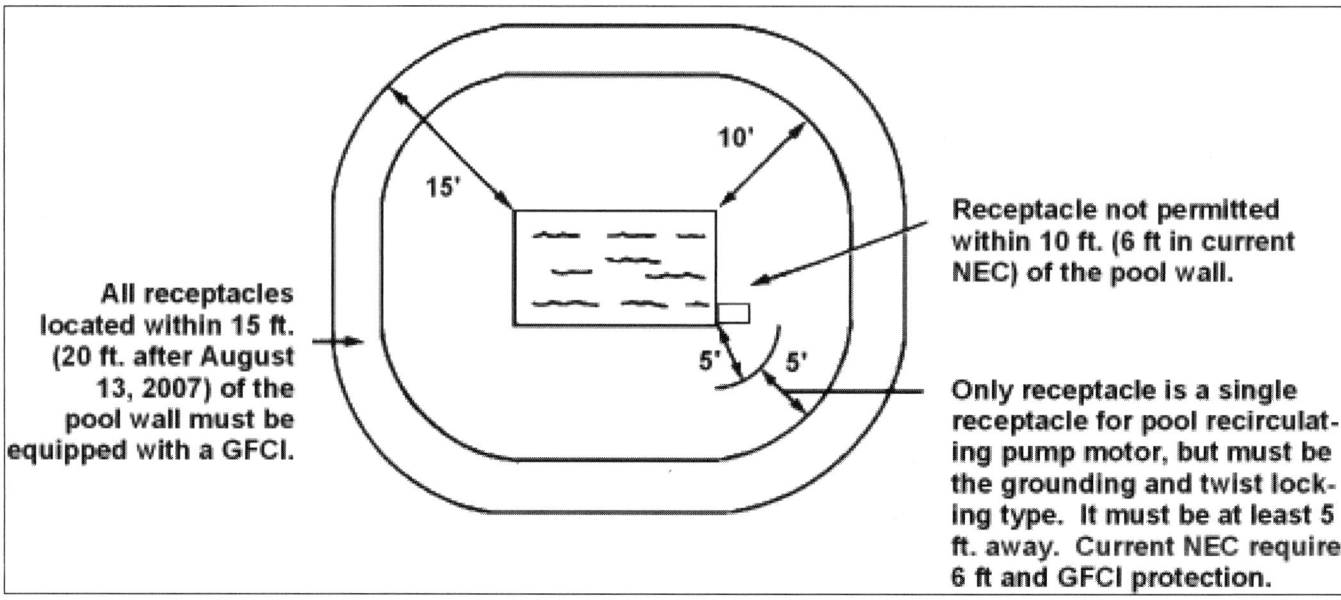

This requirement for a GFCI is directed toward the hazards of shock and electrocution. It is generally accepted that appliances and other electrical equipment (such as filters and pumps) will be used in and around swimming pools. In the event that one of these pieces of electrical equipment failed (faulted), a person in contact with the water, or wet surfaces around the pool, could provide a path from that energized circuit to ground, resulting in electrocution.

The GFCI is a fast-acting device which senses small current leakage to ground and, in a fraction of a second, opens the circuit and "interrupts" its faulty flow to ground.

Placed between the electrical service and the tool or appliance it serves, the GFCI continually matches the amount of current going to and from the tool along the normal path of the circuit conductors.

Whenever the amount "going" differs from the amount "returning" by a set trip level (5mA ±1mA on currently approved GFCIs), the GFCI opens the circuit within 1/40th of a second. This difference in current is called leakage current to ground, and the path it takes to ground could be through a person – in which case, the rapid response of the GFCI is fast enough to prevent electrocution. This protection provided by the GFCI is independent of the condition of the equipment grounding conductor. Thus, the GFCI can provide protection even if the equipment grounding conductor becomes inoperative.

It should be noted that a fuse or circuit breaker cannot possibly provide this kind of personnel protection, because it monitors only the amount of current flowing, not the path of current, and cannot detect leakage current through a fault. For example, a 20-ampere circuit breaker will not trip out even if all 20 amperes are flowing through a ground fault, but the GFCI will trip if from .004 to .006 amperes begins to flow through a ground fault. This small amount of current flowing for the extremely short time required to trip the GFCI would not electrocute a person.

The paragraphs describe ideal operation for the GFCI. However, circumstances and defects can change this. Variable circumstances and possible defects can change this. Therefore, one should never assume that this device will absolutely prevent all electrocutions.

Figure 102A illustrates a schematic of a GFCI. Figure 102B shows a GFCI incorporated into a receptacle.

Figure 102A – Ground-fault circuit interrupter

Different transformer continuously monitors the circuit to ensure that all current that flows Out to a motor or appliance returns to the source via the circuit conductors. If any current leaks to a fault, the sensing circuit opens the circuit breaker and stops all current flow.

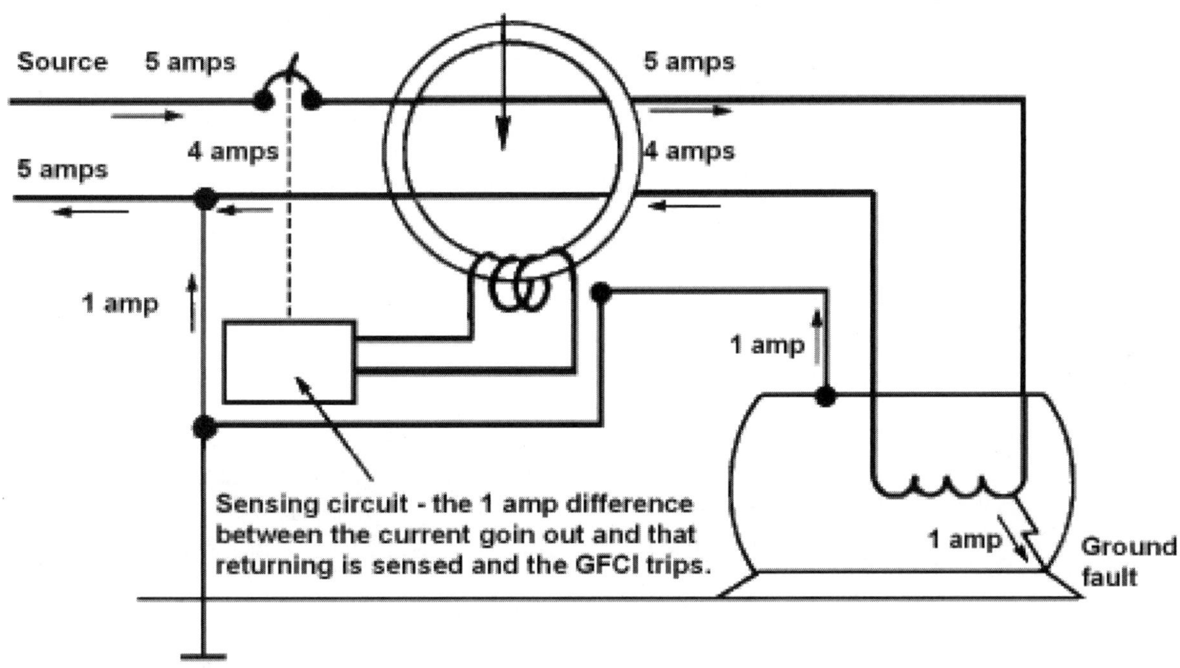

Figure 102B – Receptacle-type ground-fault circuit interrupter (GFCI).

(j)(2)	**Lighting fixtures, lighting outlets, and ceiling suspended (paddle) fans.**
(j)(2)(i)	In outdoor pool areas, lighting fixtures, lighting outlets, and ceiling-suspended (paddle) fans may not be installed over the pool or over the area extending 1.52 m (5 ft) horizontally from the inside walls of a pool unless no part of the lighting fixture of a ceiling-suspended (paddle) fan is less than 3.66 m (12 ft) above the maximum water level. However, a lighting fixture or lighting outlet that was installed before April 16, 1981, may be located less than 1.52 m (5 ft) measured horizontally from the inside walls of a pool if it is at least 1.52 m (5 ft) above the surface of the maximum water level and is rigidly attached to the existing structure. It shall also be protected by a ground-fault circuit interrupter installed in the branch circuit supplying the fixture.
(j)(2)(ii)	Lighting fixtures and lighting outlets installed in the area extending between 1.52 m (5 ft) and 3.05 m (10 ft) horizontally from the inside walls of a pool shall be protected by a ground-fault circuit interrupter unless installed 1.52 m (5 ft) above the maximum water level and rigidly attached to the structure adjacent to or enclosing the pool.
(j)(3)	**Cord- and plug-connected equipment.** Flexible cords used with the following equipment may not exceed 0.9 m (3 ft) in length and shall have a copper equipment grounding conductor with a grounding-type attachment plug:
(j)(3)(i)	Cord- and plug-connected lighting fixtures installed within 4.88 m (16 ft) of the water surface of permanently installed pools; and
(j)(3)(ii)	Other cord- and plug-connected, fixed or stationary equipment used with permanently installed pools.
(j)(4)	**Underwater equipment.**

(j)(4)(i) A ground-fault circuit interrupter shall be installed in the branch circuit supplying underwater fixtures operating at more than 15 volts. Equipment installed underwater shall be identified for the purpose.

(j)(4)(ii) No underwater lighting fixtures may be installed for operation at over 150 volts between conductors.

(j)(4)(iii) A lighting fixture facing upward shall have the lens adequately guarded to prevent contact by any person.

(j)(5) **Fountains.** All electric equipment, including power supply cords, operating at more than 15 volts and used with fountains shall be protected by ground-fault circuit interrupters.

OSHA did not originally define the term "fountains". Although 1910.306(j) is entitled, *Swimming pools, fountains, and similar installations,* it was often argued that the term referred to "drinking fountains". "Fountain" was defined in Section 680 of the NEC®, and the definition now appears in 1910.399; "Fountains, ornamental pools, display pools, and reflection pools." A note to the definition further clarifies the meaning by stating, "This definition does not include drinking fountains."

Figure 103 – GFCI requirement for fountains applies to ornamental fountains of this type.

(k) **Carnivals, circuses, fairs, and similar events.** This paragraph covers the installation of portable wiring and equipment, including wiring in or on all structures, for carnivals, circuses, exhibitions, fairs, traveling attractions, and similar events.

(k)(1) **Protection of electric equipment**. Electric equipment and wiring methods in or on rides, concessions, or other units shall be provided with mechanical protection where such equipment or wiring methods are subject to physical damage.

(k)(2) **Installation.**

(k)(2)(i) Services shall be installed in accordance with applicable requirements of this subpart, and, in addition, shall comply with the following:

(k)(2)(i)(A) Service equipment may not be installed in a location that is accessible to unqualified persons, unless the equipment is lockable; and

(k)(2)(i)(B) Service equipment shall be mounted on solid backing and installed so as to be protected from the weather, unless the equipment is of weatherproof construction.

(k)(2)(ii) Amusement rides and amusement attractions shall be maintained not less than 4.57 m (15 ft) in any direction from overhead conductors operating at 600 volts or less, except for the conductors supplying the amusement ride or attraction. Amusement rides or attractions may not be located under or within 4.57 m (15 ft) horizontally of conductors operating in excess of 600 volts.

(k)(2)(iii) Flexible cords and cables shall be listed for extra-hard usage. When used outdoors, flexible cords and cables shall also be listed for wet locations and shall be sunlight resistant.

(k)(2)(iv) Single conductor cable shall be size No. 2 or larger.

(k)(2)(v) Open conductors are prohibited except as part of a listed assembly or festoon lighting installed in accordance with § 1910.304(c).

(k)(2)(vi) Flexible cords and cables shall be continuous without splice or tap between boxes or fittings. Cord connectors may not be laid on the ground unless listed for wet locations. Connectors and cable connections may not be placed in audience traffic paths or within areas accessible to the public unless guarded.

(k)(2)(vii) Wiring for an amusement ride, attraction, tent, or similar structure may not be supported by another ride or structure unless specifically identified for the purpose.

(k)(2)(viii) Flexible cords and cables run on the ground, where accessible to the public, shall be covered with approved nonconductive mats. Cables and mats shall be arranged so as not to present a tripping hazard.

(k)(2)(ix) A box or fitting shall be installed at each connection point, outlet, switch point, or junction point.

(k)(3) **Inside tents and concessions**. Electrical wiring for temporary lighting, where installed inside of tents and concessions, shall be securely installed, and, where subject to physical damage, shall be provided with mechanical protection. All temporary lamps for general illumination shall be protected from accidental breakage by a suitable fixture or lampholder with a guard.

(k)(4) **Portable distribution and termination boxes**. Employers may only use portable distribution and termination boxes that meet the following requirements:

(k)(4)(i) Boxes shall be designed so that no live parts are exposed to accidental contact. Where installed outdoors, the box shall be of weatherproof construction and mounted so that the bottom of the enclosure is not less than 152 mm (6 in.) above the ground;

(k)(4)(ii) Busbars shall have an ampere rating not less than the overcurrent device supplying the feeder supplying the box. Busbar connectors shall be provided where conductors terminate directly on busbars;

(k)(4)(iii) Receptacles shall have overcurrent protection installed within the box. The overcurrent protection may not exceed the ampere rating of the receptacle, except as permitted in § 1910.305(j)(4) for motor loads;

(k)(4)(iv) Where single-pole connectors are used, they shall comply with the following:

(k)(4)(iv)(A) Where ac single-pole portable cable connectors are used, they shall be listed and of the locking type. Where paralleled sets of current-carrying single-pole separable connectors are provided as input devices, they shall be prominently labeled with a warning indicating the presence of internal parallel connections. The use of single-pole separable connectors shall comply with at least one of the following conditions:

(k)(4)(iv)(A)(1) Connection and disconnection of connectors are only possible where the supply connectors are interlocked to the source and it is not possible to connect or disconnect connectors when the supply is energized; or

(k)(4)(iv)(A)(2) Line connectors are of the listed sequential-interlocking type so that load connectors are connected in the following sequence:

(k)(4)(iv)(A)(2)(i) Equipment grounding conductor connection;

(k)(4)(iv)(A)(2)(ii) Grounded circuit-conductor connection, if provided; and

(k)(4)(iv)(A)(2)(iii) Ungrounded conductor connection; and so that disconnection is in the reverse order; or

(k)(4)(iv)(A)(3) A caution notice is provided adjacent to the line connectors indicating that plug connection must be in the following sequence:

(k)(4)(iv)(A)(3)(i) Equipment grounding conductor connection;

(k)(4)(iv)(A)(3)(ii) Grounded circuit-conductor connection, if provided; and

(k)(4)(iv)(A)(3)(iii) Ungrounded conductor connection; and indicating that disconnection is in the reverse order; and

(k)(4)(iv)(B) Single-pole separable connectors used in portable professional motion picture and television equipment may be interchangeable for ac or dc use or for different current ratings on the same premises only if they are listed for ac/dc use and marked to identify the system to which they are connected;

(k)(4)(v) Overcurrent *protection* of equipment and conductors shall be provided; and

(k)(4)(vi) The following equipment connected to the same source shall be bonded:

(k)(4)(vi)(A) Metal raceways and metal sheathed cable;

(k)(4)(vi)(B) Metal enclosures of electrical equipment; and

(k)(4)(vi)(C) Metal frames and metal parts of rides, concessions, trailers, trucks, or other equipment that contain or support electrical equipment.

(k)(5) **Disconnecting means.**

(k)(5)(i) Each ride and concession shall be provided with a fused disconnect switch or circuit breaker located within sight and within 1.83 m (6 ft) of the operator's station.

(k)(5)(ii) The disconnecting means shall be readily accessible to the operator, including when the ride is in operation.

(k)(5)(iii) Where accessible to unqualified persons, the enclosure for the switch or circuit breaker shall be of the lockable type.

(k)(5)(iv) A shunt trip device that opens the fused disconnect or circuit breaker when a switch located in the ride operator's console is closed is a permissible method of opening the circuit.

From 1991 to 2002, OSHA received reports of 29 serious accidents associated with carnivals, circuses, exhibitions, fairs, and similar events. Eleven of these accidents, resulting in 10 fatalities and 5 injuries, involved electric shock. Eight of those 11 cases (8 fatalities and 1 injury) involved electric wiring and equipment covered by the installation requirements in Subpart S. OSHA believes that the new electrical requirements for these events will prevent similar accidents in the future. As for the NEC®, Article 525 already covers these requirements; the biggest issue continues to be enforcement of the requirements by local authorities.

1910.306(k) requires mechanical protection of electric equipment (paragraph (k)(1)) and of wiring methods in and around rides, concessions, or other units subject to physical damage (paragraph (k)(2)). Inside tents and concession stands, the electrical wiring for temporary lighting must be secured and protected from physical damage (paragraph (k)(3)). Paragraph (k)(4) sets requirements for portable distribution and termination boxes. These new provisions will provide more electrical safety for employees working in and around this equipment.

1910.306(k)(5) requires the disconnecting means to be readily accessible to the operator; that is, the fused disconnect switch or circuit breaker must be located within sight and within 1.83 meters (6 feet) of the operator for concession stands and rides. This provision provides protection by enabling the operator to stop the equipment in an emergency. The disconnecting means must also be lockable if it is exposed to unqualified persons, to prevent such persons from operating it.

§ 1910.307 Hazardous (Classified) Locations

[46 FR 4056, Jan. 16, 1981; 46 FR 40185, Aug. 7, 1981; 72 FR 7210, Feb. 14, 2007]

(a)	**Scope.**
(a)(1)	**Applicability.** This section covers the requirements for electric equipment and wiring in locations that are classified depending on the properties of the flammable vapors, liquids or gases, or combustible dusts or fibers that may be present therein and the likelihood that a flammable or combustible concentration or quantity is present. Hazardous (classified) locations may be found in occupancies such as, but not limited to, the following: aircraft hangars, gasoline dispensing and service stations, bulk storage plants for gasoline or other volatile flammable liquids, paint-finishing process plants, health care facilities, agricultural or other facilities where excessive combustible dusts may be present, marinas, boat yards, and petroleum and chemical processing plants. Each room, section or area shall be considered individually in determining its classification.
(a)(2)	**Classifications.**
(a)(2)(i)	These hazardous (classified) locations are assigned the following designations:
(a)(2)(i)(A)	Class I, Division 1
(a)(2)(i)(B)	Class I, Division 2
(a)(2)(i)(C)	Class I, Zone 0
(a)(2)(i)(D)	Class I, Zone 1
(a)(2)(i)(E)	Class I, Zone 2
(a)(2)(i)(F)	Class II, Division 1
(a)(2)(i)(G)	Class II, Division 2
(a)(2)(i)(H)	Class III, Division 1
(a)(2)(i)(I)	Class III, Division 2
(a)(2)(ii)	For definitions of these locations, see § 1910.399.
(a)(3)	**Other sections of this subpart.** All applicable requirements in this subpart apply to hazardous (classified) locations unless modified by provisions of this section.
(a)(4)	**Division and zone classification**. In Class I locations, an installation must be classified as using the division classification system meeting paragraphs (c), (d), (e), and (f) of this section or using the zone classification system meeting paragraph (g) of this section. In Class II and Class III locations, an installation must be classified using the division classification system meeting paragraphs (c), (d), (e), and (f) of this section.
(b)	**Documentation.** All areas designated as hazardous (classified) locations under the Class and Zone system and areas designated under the Class and Division system established after August 13, 2007 shall be properly documented. This documentation shall be available to those authorized to design, install, inspect, maintain, or operate electric equipment at the location.

Hazardous Locations

The discussion on hazardous locations on this and the following pages provides a general overview of the guidelines contained in NEC® Chapter 5, as an aid to understanding design concepts and equipment selection. Application of the rules for design and installation purposes, however, should not be done without referring to the actual text in the Code®.

OSHA will now also require employers to document the designation of hazardous locations within their facilities in 1910.307(b). The documentation must denote the boundaries of each division or zone so that employees who install, inspect, maintain, or operate equipment in these areas will be able to determine whether the equipment is safe for the location. OSHA is requiring documentation for the dividivision system only for new installations that use that system. The document requirement does apply, however, to all installations made under the zone system.

1910.307

Title: Hazardous (classified) locations.

Subpart: S

SubPart Title: Electrical - Design Safety Standards for Electrical Systems

Traditional Classifications

Hazardous locations are areas where flammable liquids, gases, or vapors, combustible dusts, or other easily ignitable materials exist in sufficient quantities to produce an explosion or fire. In hazardous locations, specially designed equipment and special installation techniques must be used to protect against the explosive and flammable potential of these substances.

Hazardous locations are classified as Class I, Class II, or Class III, depending on what type of hazardous substance is or may be present. In general, Class I locations are those in which flammable vapors and gases may be present. Class II locations are those in which combustible dusts may be found. Class III locations are those in which there are ignitable fibers and flyings.

Each of these classes is divided into two hazard categories, Division 1 and Division 2, depending on the likelihood of the presence of a flammable or ignitable concentration of a substance. Division 1 locations are designated as such because a flammable gas, vapor, dust or easily ignitable material is normally present in hazardous quantities. In Division 2 locations, the existence of hazardous quantities of these materials is not normal, but they may occasionally exist either accidentally or when material in storage is handled. In general, the installation requirements for Division 1 locations are more stringent than for Division 2 locations. Additionally, Class I and Class II locations are also subdivided into groups of gases, vapors, and dusts having similar properties.

Table 307-1 summarizes the various hazardous (classified) locations. The definitions of the locations are given in Paragraph 1910.399(a) of the Standard.

Table 307-1 – Summary of Class I, II, III Hazardous Locations

Classes	Groups	Divisions	
		1	2
I. Gases, Vapors and Liquids (Article 501)	A. Acetylene B. Hydrogen, etc. C. Ethyl ether, etc. D. Acetone, ammonia, ethylene oxide, etc.	Normally explosive	Not normally present in an explosive concentration (but may accidentally exist)
II. Dusts (Article 502)	E. Combustible metal dust, AL, Mg, etc. F. Combustible carbonaceous dusts G. Flour, wood, grain, plastic or chemical dusts	Ignitable quantities of dust normally in or may be in suspension	Dust not normally suspended in an ignitable concentration (but may accidentally exist). Dust layers are present.
III. Fibers and Flyings (Article 503)	Textiles, woodworking, etc. (easily ignitable)	Handled or used in manufacturing	Stored or handled in storage (exclusive of manufacturing)

1910.307

Title: Hazardous (classified) locations.

Subpart: S
SubPart Title: Electrical - Design Safety Standards for Electrical Systems

---

Table 307-2 – Permitted Techniques in Class I and Zoned Locations

Division 1: — Explosion-proof. — Purged and pressurized (Type X or Y). — Intrinsically safe.	**Zone 0:** — Intrinsically safe "ia". — Class I, Division 1 intrinsically safe. **Division 1:** — Explosion-proof. — Purged and pressurized (Type X or Y). — Intrinsically safe. **Zone 1:** — Flameproof "d". — Purged and pressurized. — Intrinsically safe "ib". — Oil immersion "o". — Increased safety "e". — Encapsulation "m". — Powder filling "q". — Any Class I, Division 1 method. — Any Class I, Zone 0 method.
Division 2: — Purged and pressurized (Type Z). — Intrinsically safe. — Nonincendive. — Oil immersion. — Hermetically sealed. — Any Class I, Zone 0 or 1 method. — Any Class I, Zone 0, Zone 1, or Zone 2 method.	**Zone 2:** — Non-sparking "nA". — Protected sparking "nC". — Restricted breathing "nR". — Any Class I, Division 1 or 2 method. — Any Class I, Zone 0 or 1 method.

Additional Classification Method for Class I Locations

NFPA 70E and the NEC® incorporate an alternative system (in addition to the division classification system) for installing electric equipment in Class I locations. (Class II locations continue under the division system.) This system is called the "zone classification system," or the "zone system." (See Table 307-2.) The zone system designates three classifications: Class I, Zone 0; Class I, Zone 1; and Class I, Zone 2 (see NEC® Article 505). The zone system is based on various European standards that were developed by the International Electrotechnical Commission (IEC). The IEC prepares and publishes international standards for all electrical, electronic and related technologies. This global organization is made up of members from more than 60 participating countries, including the U.S...

A modified version of this system was first adopted into the NEC® in the 1996 edition. Although the zone and division classification systems differ in concept, individual equipment can be approved for use under both systems when the equipment incorporates protective techniques for both systems (as determined by the NRTL that lists or labels the equipment). Based on the successful use of the zone system in European countries for many years and the acceptance of the zone system by the NEC® and international standards, OSHA believes that an installation conforming to requirements for this system is as safe as one conforming to requirements for the division system. The zone system is an alternative method to the division system; employers may use either system for installations of electric equipment in Class I hazardous locations. OSHA will recognize the use of the zone system under 1910.307 and any other OSHA standard that references 1910.307.

Instead of using Division 1 and Division 2 to determine the extent of the hazardous conditions that exist in a particular Class I location, this alternate classification method uses the designations Zone 0, Zone 1 and Zone 2 to identify locations where hazardous conditions always exist, normally exist or do not usually exist, respectively. See NEC® Article 505 for the complete definitions.

For the purpose of Article 505, Group D flammable liquids and gases are identified as Group IIA; Group C liquids and gases as Group IIB, and Groups A and B as Group IIC.

The Article 505 classification method appears to allow some "fine tuning" of hazardous locations and permit some economies in the equipment and wiring installations for Class I hazardous locations. For the persons who do not routinely work with the various designations for hazardous locations, including the enforcing authorities, the situation has become much more complex and confusing. The nearly duplicating classifications, presented in reverse order ("0," normally thought of to be without hazard, is the most hazardous location, and the grouping of materials has been reversed in order), call for close attention to correctly designating and correctly identifying hazardous locations in the field.

General Requirements

General-purpose electrical equipment can cause explosions and fires in areas where flammable vapors, liquids, and gases, and combustible dusts or fibers are present. These areas require special electrical equipment which is designed for the specific hazard involved. This includes explosion-proof equipment for flammable vapor, liquid and gas hazards and dust-ignition-proof equipment for combustible dust. Other equipment used includes nonsparking equipment, intrinsically safe equipment, and purged and pressurized equipment.

Many pieces of electrical equipment include certain parts that arc, spark, or produce heat under normal operating conditions. For example, circuit controls, switches, and contacts may arc or spark when operated. Motors and lighting fixtures are examples of equipment that may heat up. These energy sources can produce temperatures high enough to cause ignition (see Figure 104). Electrical equipment should not be installed in known or potentially hazardous locations unless absolutely necessary. However, when electrical equipment must be installed in these areas, the sparking, arcing, and heating nature of the equipment must be controlled.

Installations in hazardous locations must be: (1) intrinsically safe, (2) approved for the hazardous location, or (3) of a type and design that provides protection from the hazards arising from the combustibility and flammability of the vapors, liquids, gases, dusts, or fibers that will be present.

Installations can be one or any combination of these options.

Figure 104 – Explosion occurring in general-purpose equipment

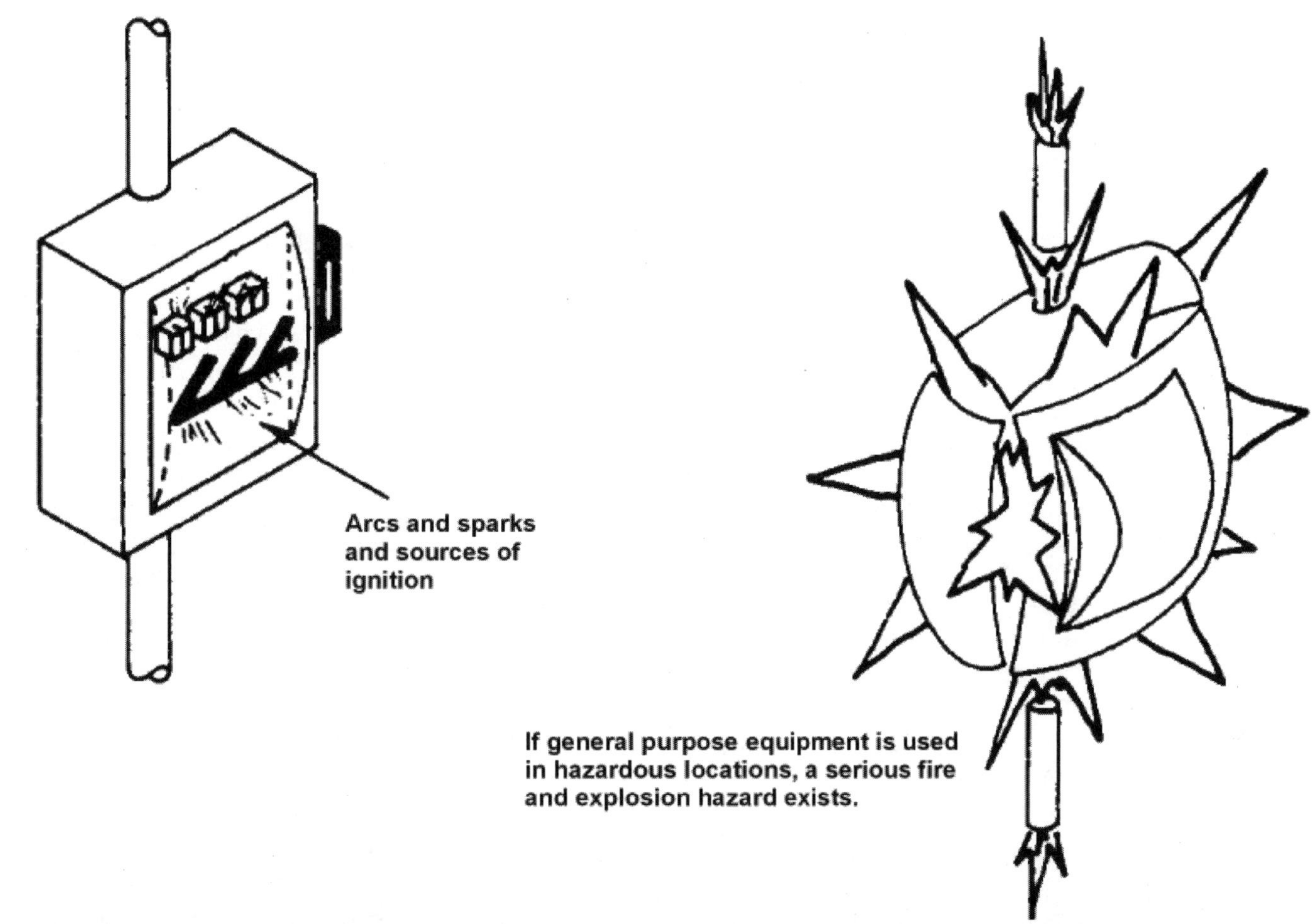

(c) **Electrical installations.** Equipment, wiring methods, and installations of equipment in hazardous (classified) locations shall be intrinsically safe, approved for the hazardous (classified) location, or safe for the hazardous (classified) location. Requirements for each of these options are as follows:

(c)(1) **Intrinsically safe.** Equipment and associated wiring approved as intrinsically safe shall be permitted in any hazardous (classified) location for which it is approved.

The only equipment obtainable that has been designed, manufactured, tested, and listed as intrinsically safe is instrumentation. For additional information see Article 504, Intrinsically Safe Systems, which first appeared in the 1990 National Electrical Code® and has been revised in subsequent editions. See also ANSI/ISA RP 12.6-1995.

Intrinsically Safe

Equipment and wiring approved as intrinsically safe is acceptable in any hazardous (classified) location for which it is designed. Intrinsically safe equipment is not capable of releasing sufficient electrical or thermal energy under normal or abnormal conditions to cause ignition of a specific flammable or combustible atmospheric mixture in its most easily ignitable concentration.

To avoid contaminating nonhazardous locations, the passage of flammable gases and vapors through the equipment must be prevented. Additionally, all interconnections between circuits must be evaluated to be sure that an unexpected source of ignition is not introduced through other nonintrinsically safe equipment. Separation of intrinsically safe and non-intrinsically safe wiring may be necessary to ensure that the circuits in hazardous (classified) locations remain safe.

(c)(2) **Approved for the hazardous (classified) location.**

(c)(2)(i) Equipment shall be approved not only for the class of location but also for the ignitable or combustible properties of the specific gas, vapor, dust, or fiber that will be present.

Note to paragraph (c)(2)(i) of this section: NFPA 70, the National Electrical Code, lists or defines hazardous gases, vapors, and dusts by "Groups" characterized by their ignitable or combustible properties.

(c)(2)(ii) Equipment shall be marked to show the class, group, and operating temperature or temperature range, based on operation in a 40 degrees C ambient, for which it is approved. The temperature marking may not exceed the ignition temperature of the specific gas or vapor to be encountered. However, the following provisions modify this marking requirement for specific equipment:

(c)(2)(ii)(A) Equipment of the nonheat-producing type, such as junction boxes, conduit, and fittings, and equipment of the heat-producing type having a maximum temperature not more than 100° C (212° F) need not have a marked operating temperature or temperature range;

(c)(2)(ii)(B) Fixed lighting fixtures marked for use in Class I, Division 2 or Class II, Division 2 locations only need not be marked to indicate the group;

(c)(2)(ii)(C) Fixed general-purpose equipment in Class I locations, other than lighting fixtures, that is acceptable for use in Class I, Division 2 locations need not be marked with the class, group, division, or operating temperature;

(c)(2)(ii)(D) Fixed dust-tight equipment, other than lighting fixtures, that is acceptable for use in Class II, Division 2 and Class III locations need not be marked with the class, group, division, or operating temperature; and

(c)(2)(ii)(E) Electric equipment suitable for ambient temperatures exceeding 40°C (104°F) shall be marked with both the maximum ambient temperature and the operating temperature or temperature range at that ambient temperature; and

Approved for the Hazardous (Classified) Location

Only OSHA, with jurisdictional authority, approves qualified laboratories. Qualified laboratories such as Underwriters Laboratories Inc. and Factory Mutual Inc. can only test, list and label equipment as meeting certain requirements. See definitions in back of book for "acceptable," "approved," "listed," and "labeled."

Under this option, equipment must be approved for the class, division, and group of location. There are two types of equipment specifically designed for hazardous (classified) locations — explosion-proof and dust-ignition-proof. Explosion-proof apparatus is intended for Class I locations, while dust-ignition-proof equipment is primarily intended for Class II and III locations. Equipment approved specifically for hazardous locations carries an Underwriters' Laboratories Inc. (UL) label and indicates in what class, division, and location group it may be installed. Equipment used in locations having deposits of readily combustible paint residue (such as inside of most spray booths) must have specific listings for such locations. Equipment (such as the fluorescent hand lamp shown in Figure 105) approved for use in a Division 1 location may be installed in a Division 2 location of the same class and group.

Figure 105 – A fluorescent hazardous-duty hand lamp for Class I, Division 1, Groups C and D, and Class II, Division 1, Groups F and G locations. *Source: Daniel Woodhead Co., Northbrook, Ill.*

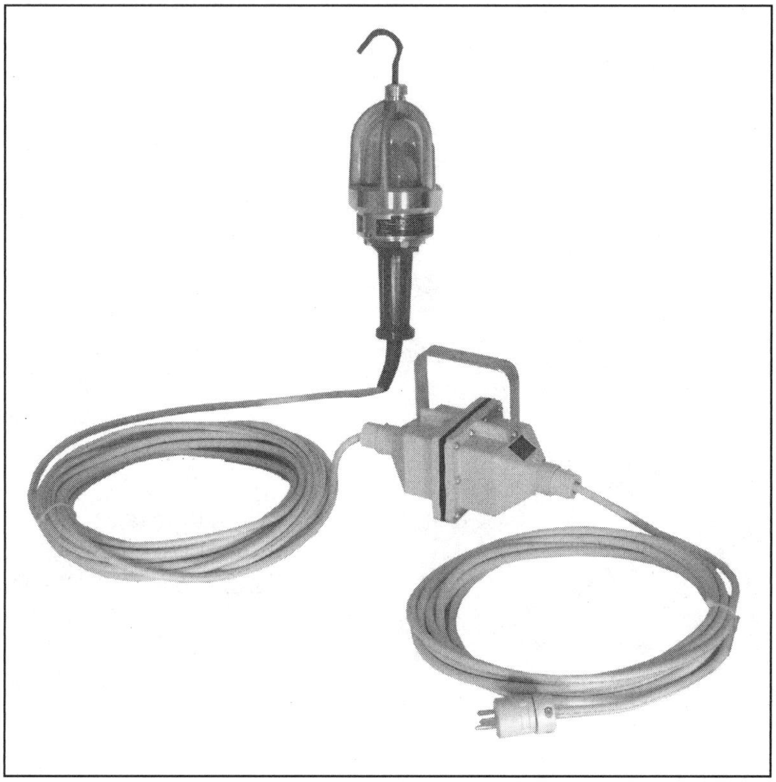

Explosion-Proof Equipment

Generally, equipment installed in Class I locations must be approved as explosion-proof. Figure 106 shows an internal view of a totally enclosed, fan-cooled explosion-proof motor. The rotor and its windings and fan are completely enclosed. An internal seal, a wide flange on the starter cover, and a close-fitting journal prevent the escape of hot gases or flames from the enclosure. A nonferrous internal fan circulates air inside the enclosure, transferring the heat from the windings to the enclosure. The frame or enclosure is explosion-proof and may have an external fan that forces air over its outside surfaces. The forced external circulation will provide more effective motor cooling than natural air circulation. However, none of this external air comes in contact with the windings. Since it is impractical to keep flammable gases outside of enclosures, arcing equipment must be installed in enclosures that are designed to withstand an explosion. This minimizes the risk of having an external explosion occur when a flammable gas enters the enclosure and is ignited by the arcs (see Figure 107). Not only must the equipment be strong enough to withstand an internal explosion, but the enclosures must be designed to vent the resulting explosive gases. This venting must ensure that the gases are cooled to a temperature below the ignition temperature of the hazardous substance involved before being released into the hazardous atmosphere.

When an internal explosion occurs, it causes the shape of the enclosure to change from rectangular to elliptical (see Figure 108). Adequate strength is one requirement for the design of an explosion-proof enclosure; a safety factor of 4 is generally used. To prevent failure of the enclosure, openings are designed to relieve the pressure of the expanding gases. All joints and flanges are held to narrow tolerances—the accurately machined joints act to cool the hot gases resulting from an internal explosion so that by the time they reach the outside hazardous atmosphere, they are too cool to cause ignition.

There are two common enclosure designs: threaded-joint enclosures (see Figure 109) and ground-joint enclosures (see Figure 110). When hot gases travel through the very small openings in either of these joints, they are cooled before reaching the surrounding hazardous atmosphere.

Figure 106 – Design features of a totally enclosed, fan-cooled, explosion-proof motor. ***(Courtesy of General Electric Co.)***

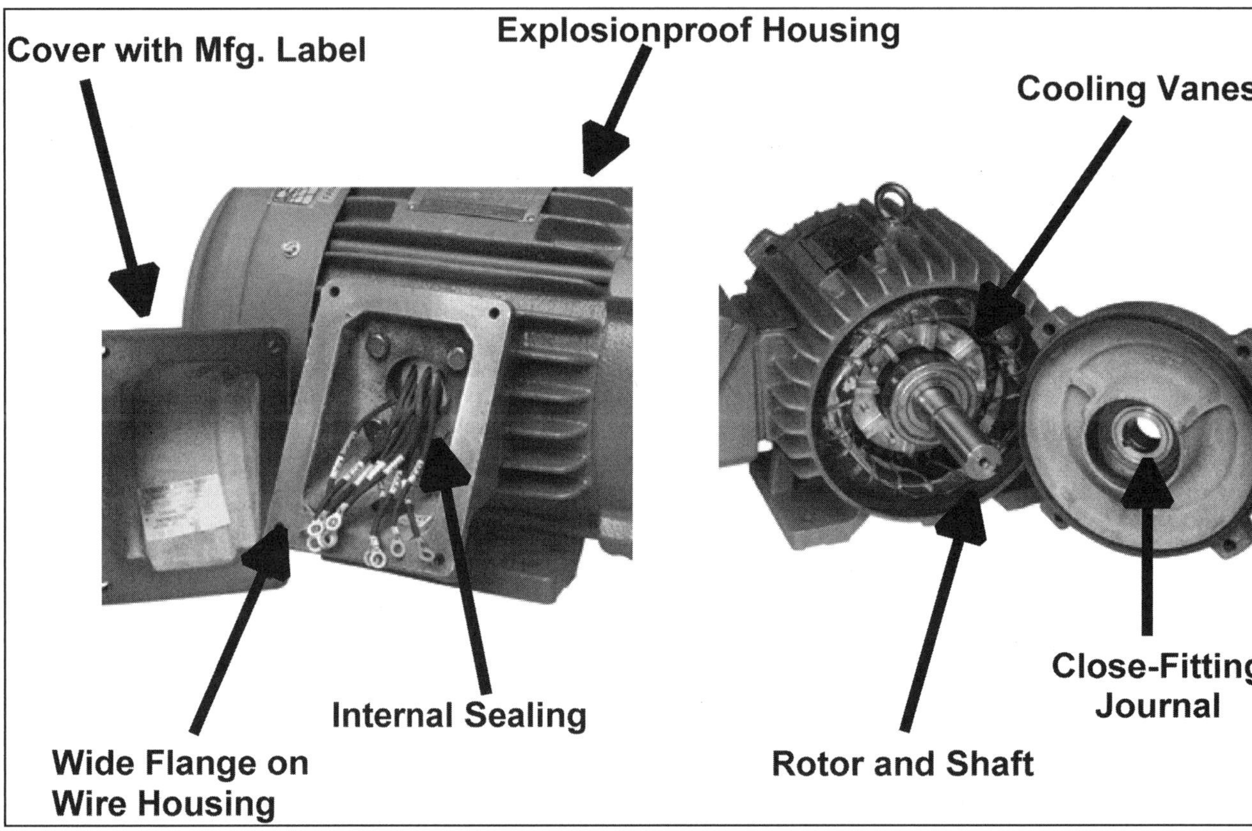

Other design requirements, such as sealing, prevent the passage of gases, vapors or fumes from one portion of an electrical system to another. Motors, which typically contain sparking brushes or commutators and tend to heat up, must also be designed to provide for the control of internal explosions.

Because the explosion characteristics of hazardous substances vary with the specific material involved, each group requires special design considerations. For Class I hazardous locations, there are four groups — A, B, C, and D. (See Table 307-3A.) Design characteristics for these four groups require the containment of maximum explosion pressure, maximum safe clearance between parts of enclosures, including threaded joints or joints that are ground to narrow tolerances, and operation at a temperature below the ignition temperature of the atmospheric mixture involved.

The classification of chemical into groups was greatly expanded in the 1981 NEC®. The specific naming of individual chemicals has since been significantly reduced. We have included Table 307-3A for information only, as an expanded listing of flammable liquids and gases that have like hazardous properties for the purpose of the Class I groups. A more complete guide to classifying these materials for electrical equipment is contained in NFPA 497M. See also Table 307-1.

Figure 107 – Explosion occurring in approved equipment

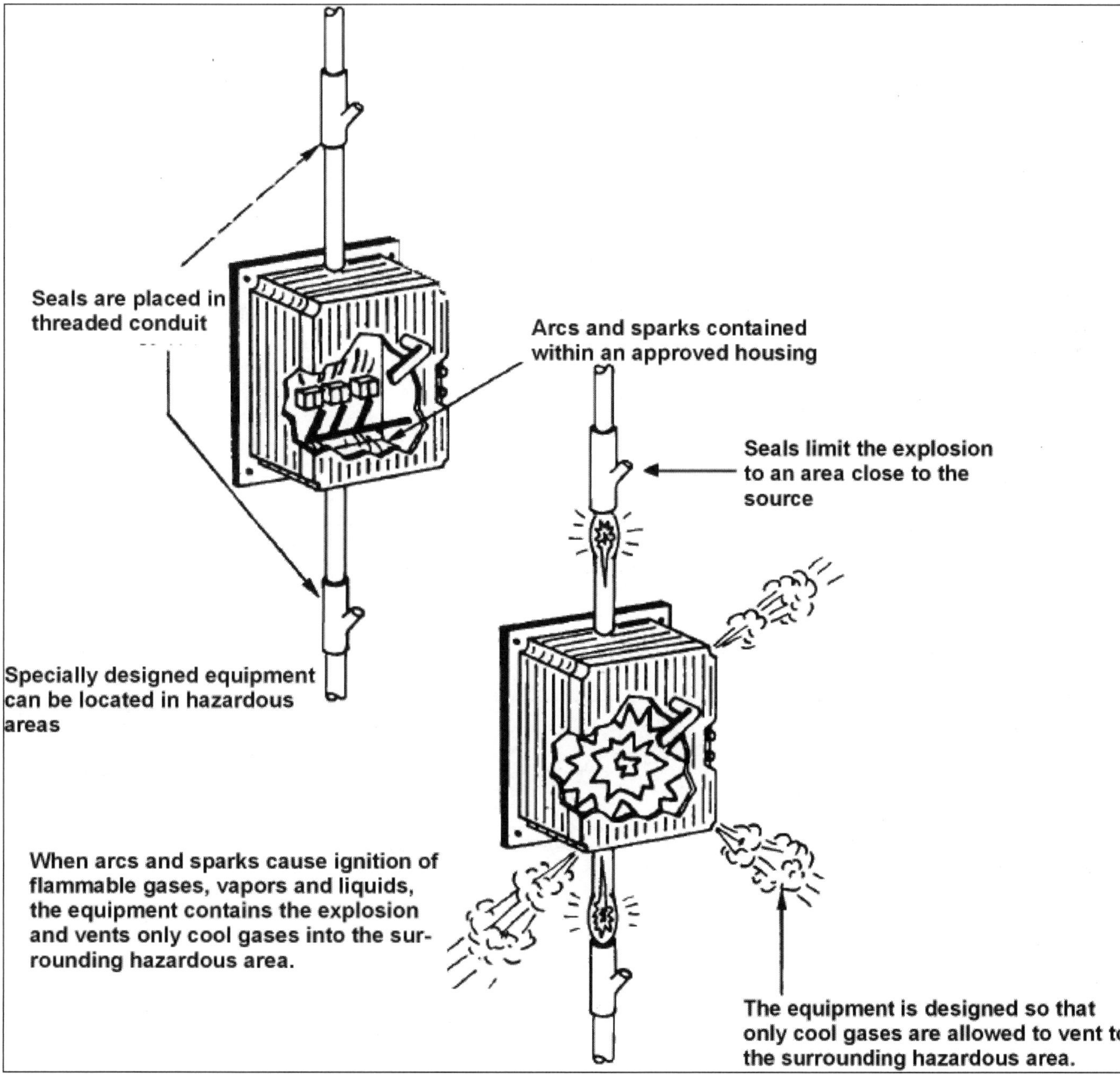

Table 307-3A – Chemicals by Group – Class I

Group A Atmospheres:	Acetylene
Group B Atmospheres:	Acrolein (inhibited), Arsine, Butadiene, Ethylene Oxide, Hydrogen, manufactured gases containing more than 30% Hydrogen (by volume), Propylene Oxide, Propylnitrate.
Group C Atmospheres:	Acetaldehyde, Allyl Alcohol, N-Butyraldehyde, Carbon Monoxide, Crotonalclehyde, Cyclopropane, Diethyl Ether, Diethylamine, Epichlorohydrine, Ethylene, Ethylenimine, Ethyl Mercaptan, Ethyl Sulfide, Hydrogen Cyanide, Hydrogen Sulfide, Morpholine, 2 Nitropropane, Tetrahydrofuran, Unsymmetrical Oimethyl Hydrazine (UDMH 1,1 Dimetbyl Hydrazine)
Group D Atmospheres:	Acetic Acid (Glacial), Acetone, Acrylonitrile, Ammonia, Benzene, Butane, IButanol (Butyl Alcohol), 2-Butanol (secondary Butyl Alcohol), N-Butyl Acetate, Isobutyl Acetate, Di-Isobutylene, Ethane, Ethanol (Ethyl Alcohol), Ethyl Acetate, Ethyl Acrylate (inhibited), Ethylene Diamine (Anhydrous), Ethylene Dichloride, Ethylene Gycol Monomethyl Ether, Gasoline, Heptanes, Hexanes, Isoprene, Isopropryl Ether, Mesityll Oxide, Methane (natural gas), Methanol (Methyl Alcohol), 3-Methyl-I-Butanol (Isoamyl Alcohol), Methyl Ethyl Ketone, Methyl Isobutyl Ketone, 2-Methyl-I Propanol (Isobutyl Alcohol), 2-Methyl-2-Propanol (Tertiary Butyl Alcohol), Petroleum Naphtha, Pyridine, Octanes, Pentanes, I-Pentanol (Amyl Alcohol), Propane, I-Propanol (Propyl Alcohol), 2-Propanol (Isopropyl Alcohol), Propylene, Styrene, Toluene, Vinyl Acetate, Vinyl Chloride, Xylenes.

Figure 108 – Internal explosive pressure

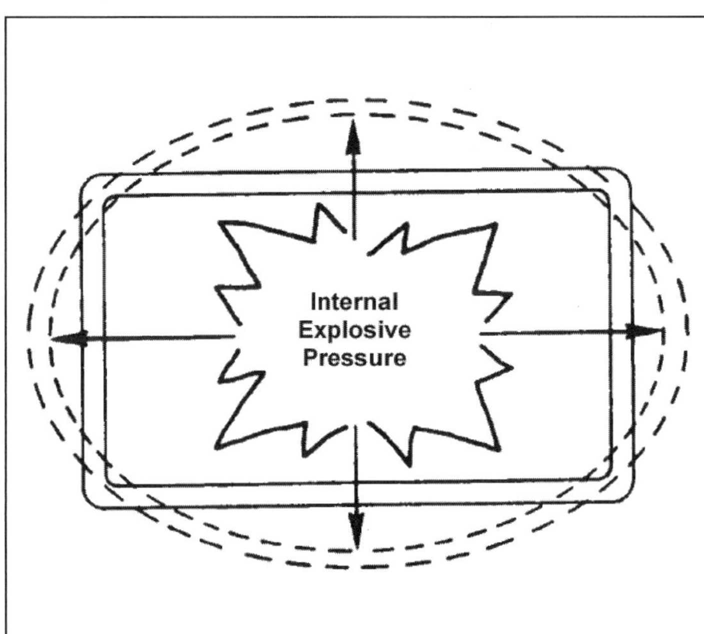

1910.307

Title: Hazardous (classified) locations.

Subpart: S

SubPart Title: Electrical - Design Safety Standards for Electrical Systems

---

Figure 109 – Threaded joint design

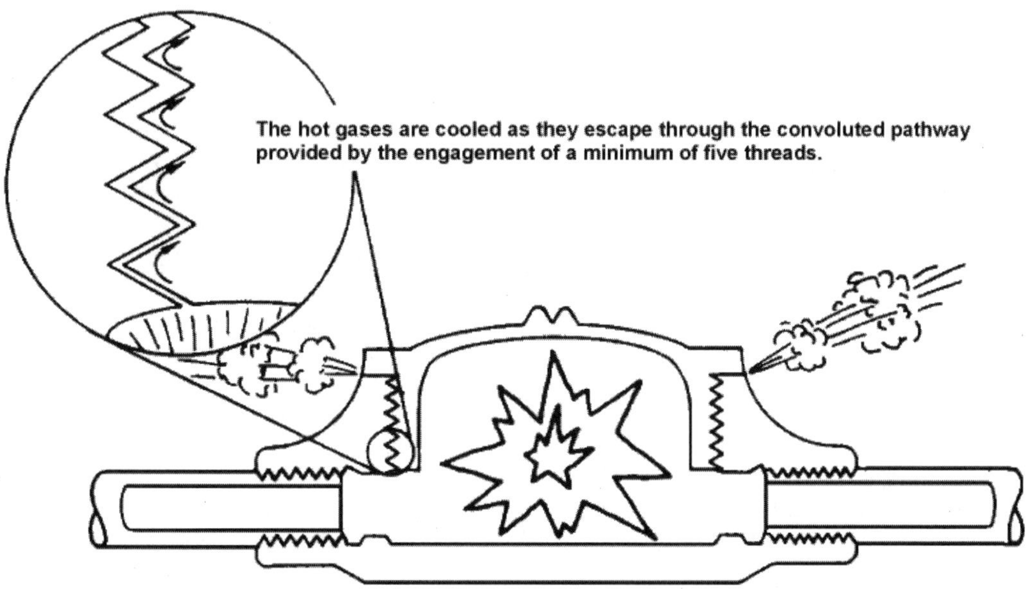

The case of the enclosure is made of cast metal, strong enough to withstand the maximum explosion pressure of a specific group of hazardous gases or vapors.

Figure 110 – Openings designed into ground joint

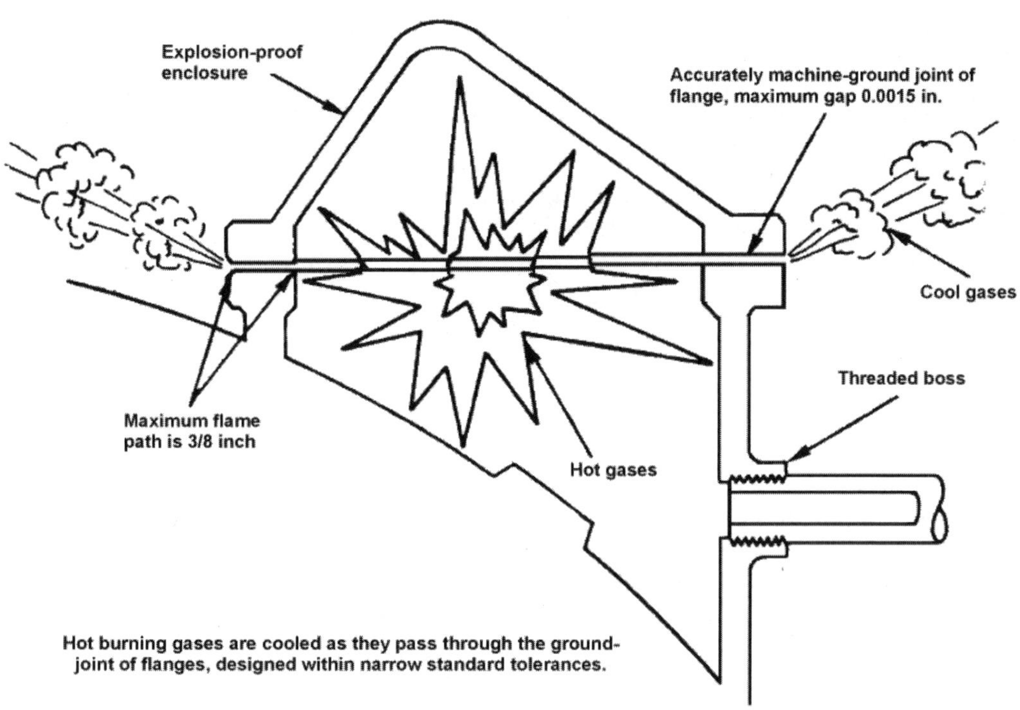

Dust-Ignition-Proof Equipment

In Class II, Division 1 locations, equipment must generally be dust-ignition-proof. Section 500.2 of the NEC® defines dust-ignition-proof as equipment "enclosed in a manner that excludes dusts and does not permit arcs, sparks, or heat otherwise generated or liberated inside of the enclosure to cause ignition of exterior accumulations or atmospheric suspensions of a specified dust on or in the vicinity of the enclosure."

Dust-ignition-proof equipment is designed to keep ignitable amounts of dust from entering the enclosure. In addition, dust may accumulate on electrical equipment, causing overheating of the equipment, as well as the dehydration or gradual carbonization of organic dust deposits. Overheated equipment may malfunction and cause a fire. Dust that has carbonized is susceptible to spontaneous ignition or smoldering. Therefore, equipment must also be designed to operate below the ignition temperature of the specific dust involved even when blanketed. The shape of the enclosure must be designed to minimize dust accumulation when fixtures are out of reach of normal housekeeping activities, e.g., lighting fixture canopies.

In Class II hazardous locations there are three groups—E, F, and G. (See Table 307-3B.) Special designs are required to prevent dust from entering into the electrical equipment enclosure. Assembly joints and motor shaft openings must be tight enough to prevent dust from entering the enclosure. In addition, the design must take into account the insulating effects of dust layers on equipment and must ensure that the equipment will operate below the ignition temperature of the dust involved. If conductive combustible dusts are present, the design of equipment must take the special nature of these dusts into account.

In general, equipment which is approved explosion-proof is not designed for, and is not acceptable for use in, Class II locations, unless specifically approved for use in such locations. For example, since grain dust has a lower ignition temperature than many flammable vapors, equipment approved for Class I locations may operate at a temperature that is too high for Class II locations. On the other hand, equipment that is dust-ignition-proof is generally acceptable for use in Class III locations, since the same design considerations are involved.

Table 307-3B – Chemicals by Group – Class II

Group E Atmospheres	Combustible metal dusts, including aluminum, magnesium, and their commercial alloys, and other combustible dusts of similarly hazardous characteristics.
Group F Atmospheres	Combustible carbonaceous dusts, including carbon black, charcoal, coal, or dusts that have been sensitized by other materials.
Group G Atmospheres	Combustible dusts not included in groups E or F, including flour, grain, wood, plastic, or chemicals.
Source: National Electrical Code® Section 500.6(B). For more detailed information see NFPA 497M.	

Marking

Approved equipment must be marked to indicate the class, group, and operating temperature range (based on a 40°C ambient temperature) in which it is designed to be used. Furthermore, the temperature marked on the equipment must not be greater than the ignition temperature of the specific gases or vapors in the area.

There are, however, four exceptions to this marking requirement. First, equipment that does not produce heat (for example, junction boxes or conduits) and equipment that does produce heat but that has a maximum surface temperature of less than 100°C (or 212°F) are not required to be marked with the operating temperature range. The heat normally released from this equipment cannot ignite gases, liquids, vapors, or dusts.

Second, any permanent lighting fixtures that are approved and marked for use in Class I, Division 2 locations do not need to be marked to show a specific group. This is because these fixtures are acceptable for use with all of the chemical groups for Class I (that is, groups A, B, C, and D).

Third, general-purpose equipment, other than lighting fixtures, that is acceptable for use in Division 2 locations does not have to be labeled according to class, group, division, or operating temperature. This type of equipment does not contain any devices that might produce arcs or sparks and, therefore, is not a potential ignition source. For example, squirrel-cage induction motors without brushes, switching mechanisms or similar arc-producing devices are permitted in Class I, Division 2 locations (See NEC® Section 501.8(B)); therefore, they need no marking.

Fourth, for Class II, Division 2 and Class III locations, dust-tight equipment (other than lighting fixtures) is not required to be marked. In these locations, dust-tight equipment does not present a hazard, so it need not be identified.

Equipment which meets the requirements for Class III locations is permitted even though it may not be labeled as “dust-tight” or “suitable for Class III locations.”

(c)(3) **Safe for the hazardous (classified) location.** Equipment that is safe for the location shall be of a type and design that the employer demonstrates will provide protection from the hazards arising from the combustibility and flammability of vapors, liquids, gases, dusts, or fibers involved.

Note to paragraph (c)(3) of this section: The National Electrical Code, NFPA 70, contains guidelines for determining the type and design of equipment and installations that will meet this requirement. Those guidelines address electric wiring, equipment, and systems installed in hazardous (classified) locations and contain specific provisions for the following: wiring methods, wiring connections; conductor insulation, flexible cords, sealing and drainage, transformers, capacitors, switches, circuit breakers, fuses, motor controllers, receptacles, attachment plugs, meters, relays, instruments, resistors, generators, motors, lighting fixtures, storage battery charging equipment, electric cranes, electric hoists and similar equipment, utilization equipment, signaling systems, alarm systems, remote control systems, local loud speaker and communication systems, ventilation piping, live parts, lightning surge protection, and grounding.

Safe for the Hazardous (Classified) Location

Under this option, equipment installed in hazardous (classified) locations must be of a type and design that provides protection from the hazards arising from the combustibility and flammability of vapors, liquids, gases, dusts, or fibers. The employer has the responsibility of demonstrating that the installation meets this requirement. Guidelines for installing equipment under this option are contained in the National Electrical Code® in effect at the time of installation of that equipment. Compliance with these guidelines is not the only means of complying with this option; however, the employer must demonstrate that his installation is safe for the hazardous (classified) location.

The following paragraphs summarize installation practices given in the NEC®. These practices would be an acceptable means of complying with this third option given for equipment in hazardous locations.

Class I, Divisions 1 and 2.

Article 501 of the National Electrical Code® contains installation requirements for electrical wiring and equipment used in Class I hazardous areas. The requirements as they pertain to Class I, Division 1 hazardous locations are summarized in Figure 111 and Table 307-4.

The requirements for Class I, Division 2 locations are summarized in Figure 115 and Table 307-5.

The following sections further explain requirements for Type MI cable, sealing, motors, grounding, and bonding.

Table 307-4 – Summary of Equipment Requirements for Class I, Division 1 Hazardous Locations

A	Meters, relays, and instruments, such as voltage or current meters and pressure or temperature sensors, must be in enclosures identified for Class I, Division 1 locations. Such enclosures include explosion proof and purged and pressurized enclosures. See NEC® Section 501.105.
B	Wiring methods acceptable for use in Class I Division 1 locations include threaded rigid metal or steel intermediate metal conduit and type MI cable. Flexible fittings, such as motor terminations, must be approved for Class I locations. All boxes and enclosures must be explosion-proof and threaded for conduit or cable terminations. All joints must be wrench tight with a minimum of five threads engaged. See NEC® 501.10(A)(1)(a).
C	Sealing is required for conduit and cable systems to prevent the passage of gases, vapors, and flame from one part of the electrical installation to another through the conduit. Type MI cable inherently prevents this from happening by its construction; however, it must be sealed to keep moisture and other fluids from entering the cable at terminations. See Figure 112 and NEC® Section 501.15. **(1)** Seals are required where conduit passes from Division 1 to Division 2 or non-hazardous locations. **(2)** Seals are required within 18 inches from enclosures containing arcing devices. **(3)** Seals are required if conduit is 2 inches in diameter or larger entering an enclosure containing terminations, splices, or taps. See Figure 113 for a description of seals.
D	Drainage is required where liquid or condensed vapor may be trapped within an enclosure or raceway. An approved system of preventing accumulations or to permit periodic drainage are two methods to control condensation of vapors and liquid accumulation. See NEC® Section 501.15(F).
E	Arcing devices, such as switches, circuit breakers, motor controllers, and fuses, must be identified for Class I locations. See NEC® Section 501.115(A).
F	Motors shall be **(1)** Identified for use in Class I, Division 1 locations; **(2)** Totally enclosed with positive pressure ventilation; **(3)** Totally enclosed inert-gas-filled with a positive pressure within the enclosure; or **(4)** Submerged in a flammable liquid or gas. The last kind of installation is permissible, however, only when there is pressure on the enclosure that is greater than atmospheric pressure and the liquid or gas is only flammable in air. This type of motor is not permitted to be energized until it has been purged of all air. The latter three types of motors must be arranged to be de-energized should the pressure fail or the supply of liquid or gas fail—as with the submerged type. Types (2) and (3) may not operate at a surface temperature above 80 percent of the ignition temperature of the gas or vapor involved. See NEC® 501.125(A).
G	Lighting fixtures, both fixed and portable, must be explosion-proof and guarded against physical damage. See NEC® Section 501.130(A).
H	Flexible cords must be designed for extra hard usage, contain an equipment grounding conductor (see Figure 111), be supported so that there will be no tension on the terminal connections, and be provided with seals where they enter explosion-proof enclosures. See NEC® Section 501.140.
I	Receptacles and attachment plugs for use with portable equipment must be identified explosion-proof and provided with an equipment grounding connection. See NEC® Section 501.145.
J	Signaling, alarm, remote control and communications systems are required to be identified for Class I, Division 1 locations regardless of voltage. See NEC® Section 501.150(A).
K	Equipment grounding is required of all non-current-carrying metal parts of the electrical system. In addition, locknuts and bushings must not be relied upon for electrical connection between raceways and equipment. If locknuts and bushings are used, bonding jumpers are required. See NEC® Section 501.30(A).

Figure 111 – Class 1, Division 1 hazardous location. (See Table 307-4.)

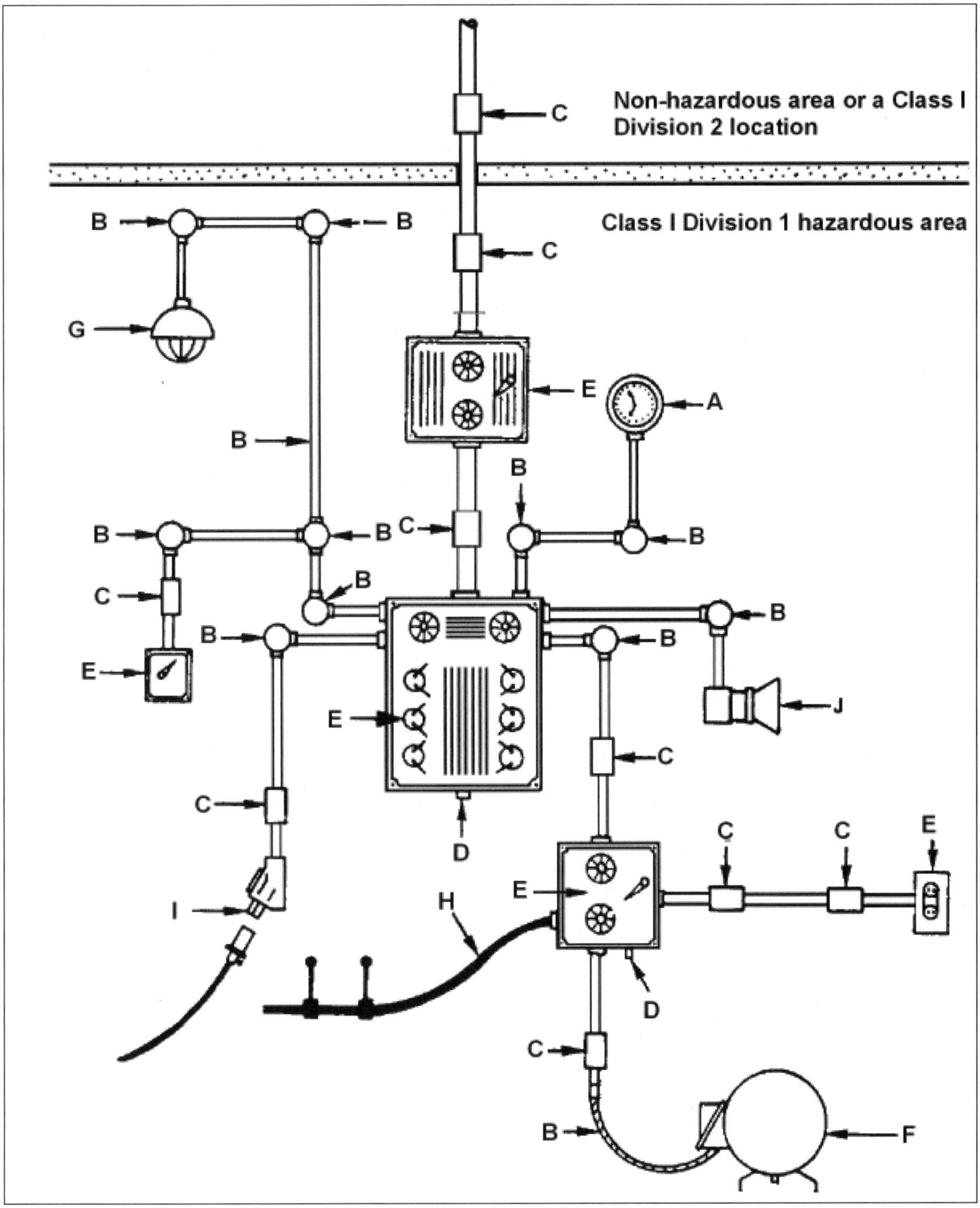

Figure 112 – Construction of Type MI (mineral insulated) cable.

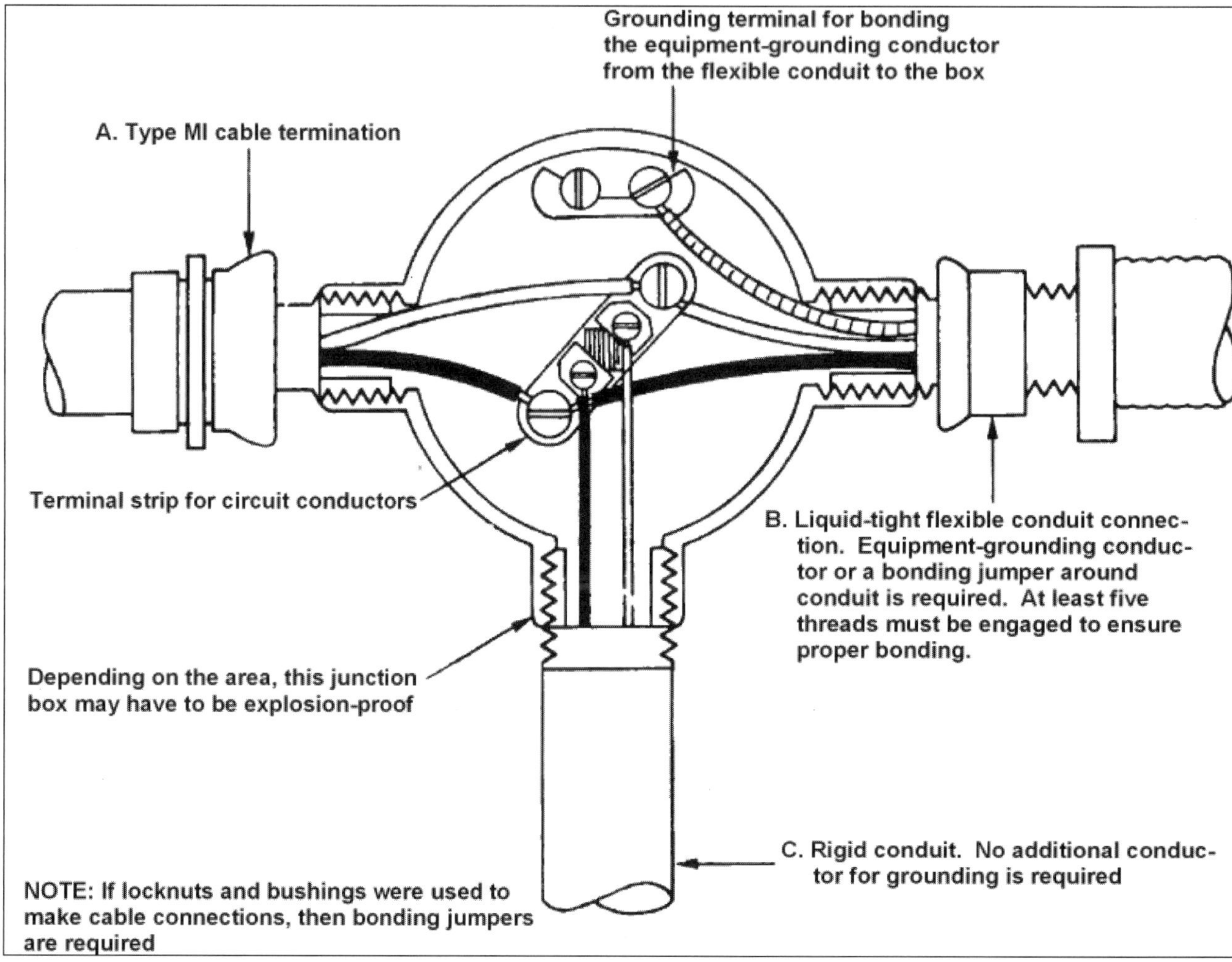

Figure 113 – Sealing.

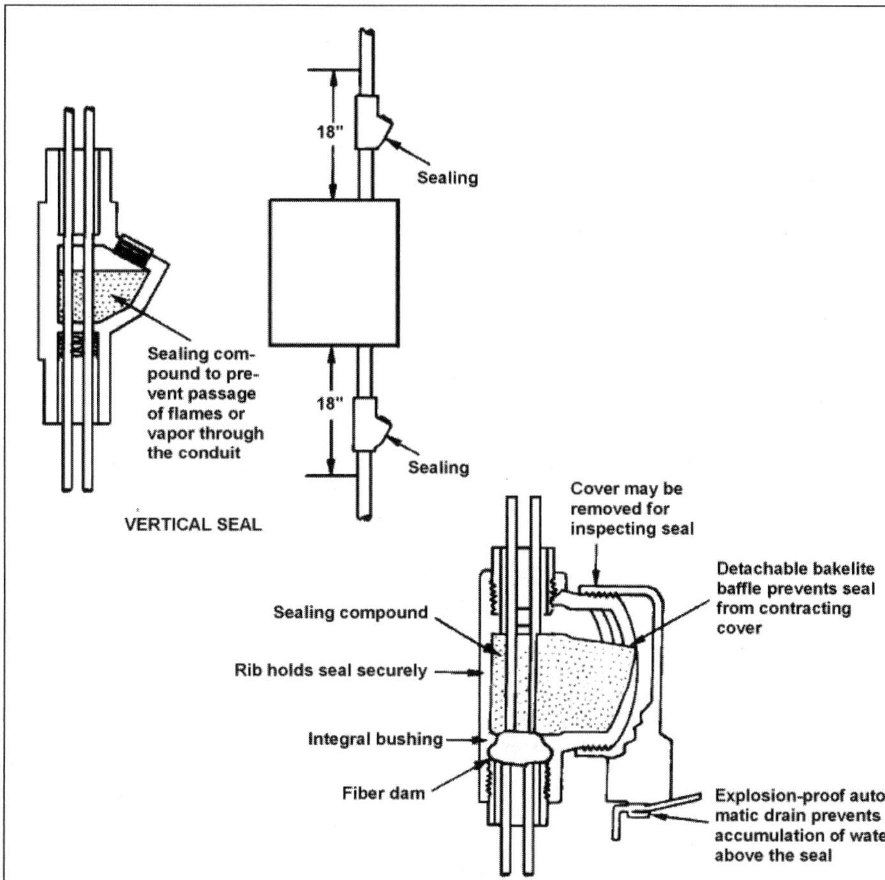

Figure 114 – Bonding in Class 1 hazardous (classified) locations.

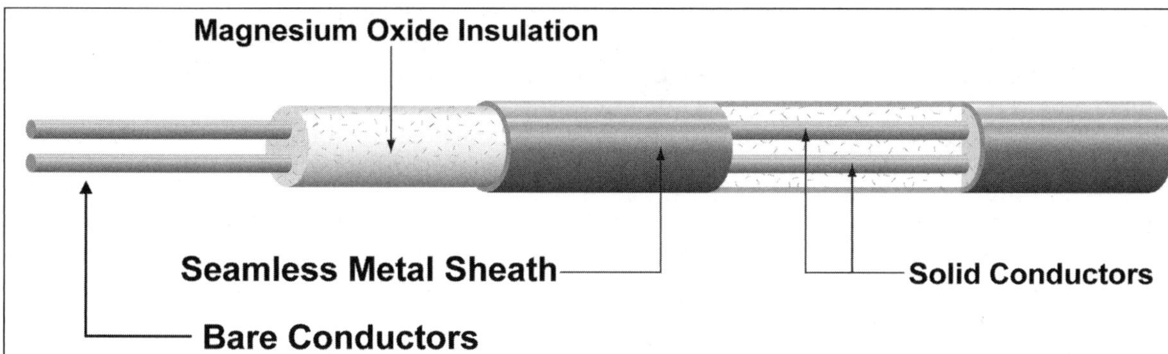

Type MI cable is a mineral-insulated cable of copper conductors in tightly compressed magnesium oxide that is enclosed in a liquidtight and gas tight copper sheath. Since Type MI cable fittings suitable for nonhazardous locations may not be explosionproof, fittings used with the cable must be specially designed for Class I locations. Boxes, fittings, and joints used with the cable must also be explosionproof.

In Class I, Division 1 and 2 hazardous locations, exposed non-current-carrying metal parts of equipment, such as frames and cabinets, must be grounded. The ground must provide a path back to the source if an accidental fault occurs.

Bonding is also required to provide a permanent ground for exposed metal parts. To be considered effective, bonding must prevent the occurrence of arcs or sparks caused by poor connections. Figure 114 shows a typical grounding and bonding technique. See NEC® Article 250 and Section 501.30(A).

Special care must be taken to make proper bonding connections not only to assure that there is a continuous equipment grounding path but to be positive that no arcing or sparking will take place between connections. Locknut bushings and double-locknut connectors cannot be relied upon for bonding purposes. Figure 114 illustrates bonding methods and three typical arrangements of conduit and cable connections to an explosion-proof enclosure. These are: a) Type MI cable termination, b) flexible conduit connection, and c) rigid conduit connection.

The requirements for Class I, Division 2 hazardous locations are summarized in Figure 115 and Table 307-5.

Table 307-5 – Summary of Class I, Division 2 Hazardous Locations (see Figure 115)

A	Meters, instruments and relays in Class I, Division 2 locations must be in identified explosion-proof enclosures. However, general-purpose equipment may be used if circuit-interrupting contacts are immersed in oil or enclosed in a hermetically sealed chamber or in circuits that do not release enough energy to ignite the hazardous atmosphere. See NEC® Section 501.105(B).
B	Wiring methods: generally, threaded rigid or intermediate conduit or types PLTC, MI, MC, MV, TC, or ITC cable systems must be used. Boxes and fittings are not required to be explosion-proof unless they enclose arcing or sparking devices. Numerous restrictions apply to cable use; see NEC® Section 501.10(B).
C	Seals are required for all conduit systems connected to explosion-proof enclosures. Seals are also required where conduit passes from hazardous to non-hazardous areas. See NEC® Section 501.15(B).
D	Drainage is required where liquid or condensed vapor may be trapped within an enclosure or along a raceway. See NEC® Section 501.15(F).
E	Most arcing devices are required to be in explosion-proof enclosures. These include items such as switches, circuit breakers, motor controllers and fuses. However, general purpose enclosures may be used for Class I, Division 2 locations, if the arcing and sparking parts are contained in a hermetically sealed chamber or are oil immersed or if solid-state devices without switching contacts are used. See NEC® Section 501.115(B).
F	Motors, generators and other rotating electrical machinery suitable for use in Class I, Division 1 locations are also acceptable in Class 1, Division 2 locations. Other motors must have their contacts, switching devices, and resistance devices in enclosures suitable for Class I, Division 2 locations (see note E, above). Motors without brushes, switching mechanisms, or similar arc-producing devices are also acceptable. See NEC® Section 501.125(B).
G	Lighting fixtures in Class I Division 2 locations must be totally enclosed and protected from physical damage. If normal operating surface temperatures exceed 80 percent of the ignition temperature of the gas, liquid or vapor involved, then explosion-proof fixtures must be installed. See NEC® Section 501.130.
H	Flexible cords in Divisions 1 and 2 are required to: 1) be suitable for extra hard usage, 2) contain

	an equipment grounding conductor, 3) be connected to terminals in an approved manner, 4) be properly supported, and 5) be provided with suitable seals where necessary.
I	Receptacles and attachment plugs must be approved for Class 1 locations.
J	Signaling systems and other similar systems: See NEC® Section 501.150(B).
K	Equipment grounding is required of all non-current-carrying metal parts of the electrical system. In addition, lock nuts and bushings must not be relied upon for electrical connection between raceways and equipment. If locknuts and bushings are used, bonding jumpers are required. A conduit threaded into a box or enclosure must be engaged wrenchtight to ensure electrical continuity and be acceptable as a grounding connection. See NEC® Section 501.30(A).

Figure 115 – Class I, Division 2 hazardous locations. (See Table 307-5.)

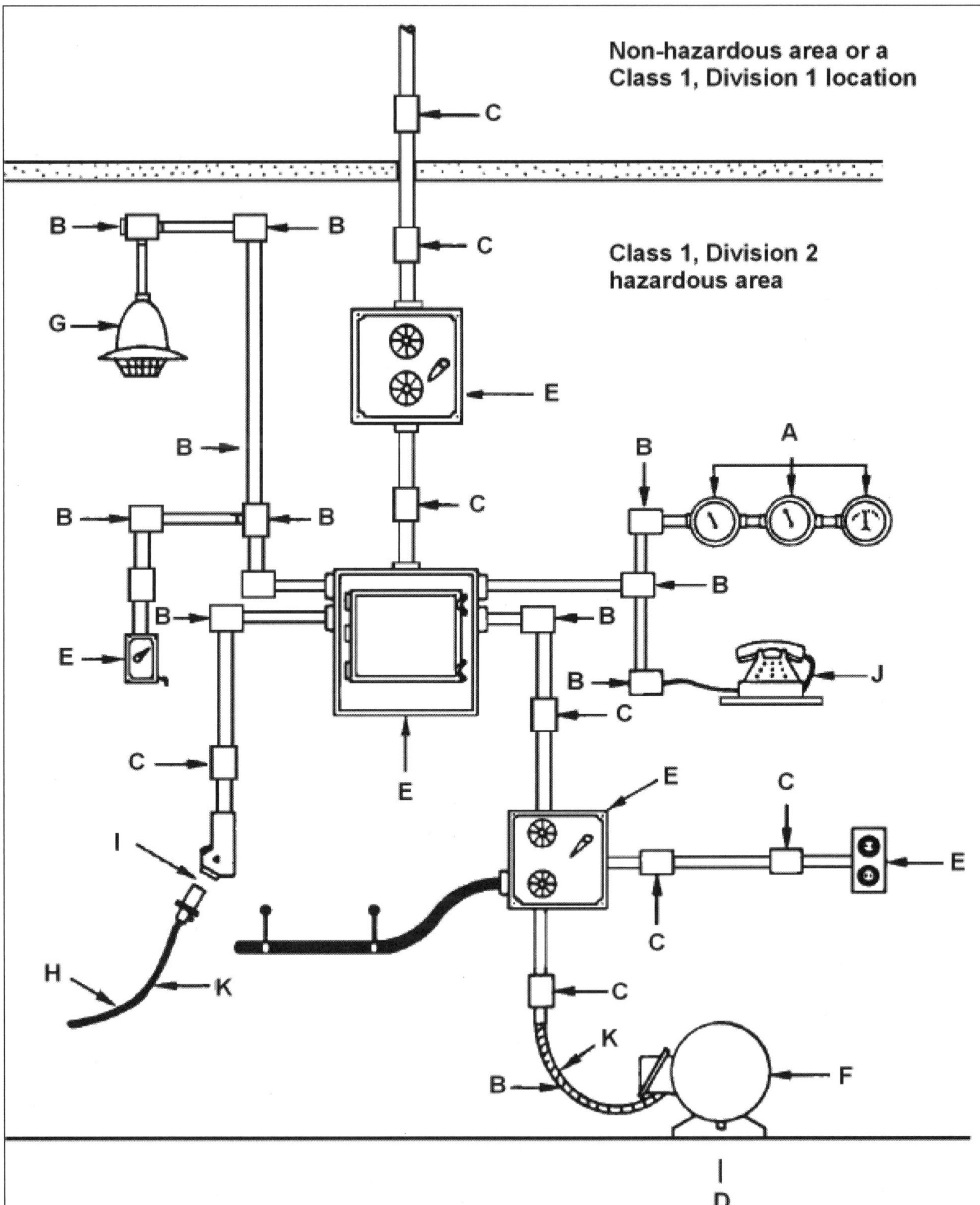

Class II Hazardous Locations

Article 502 of the National Electrical Code® is concerned with the installation requirements for electrical wiring and equipment used in Class II hazardous areas. The requirements as they pertain to Class II, Division 1 and Division 2 locations are summarized in Figure 116 and Table 307-6. The requirements for transformers and capacitors, sealing, pipe ventilation, and grounding and bonding in Class II Hazardous Locations are then explained.

Class II locations are hazardous because of the presence of combustible dust. As discussed previously, these dusts are broken down into three groups — E, F, and G.

Figure 116 – Class II hazardous locations. (See Table 307-6.)

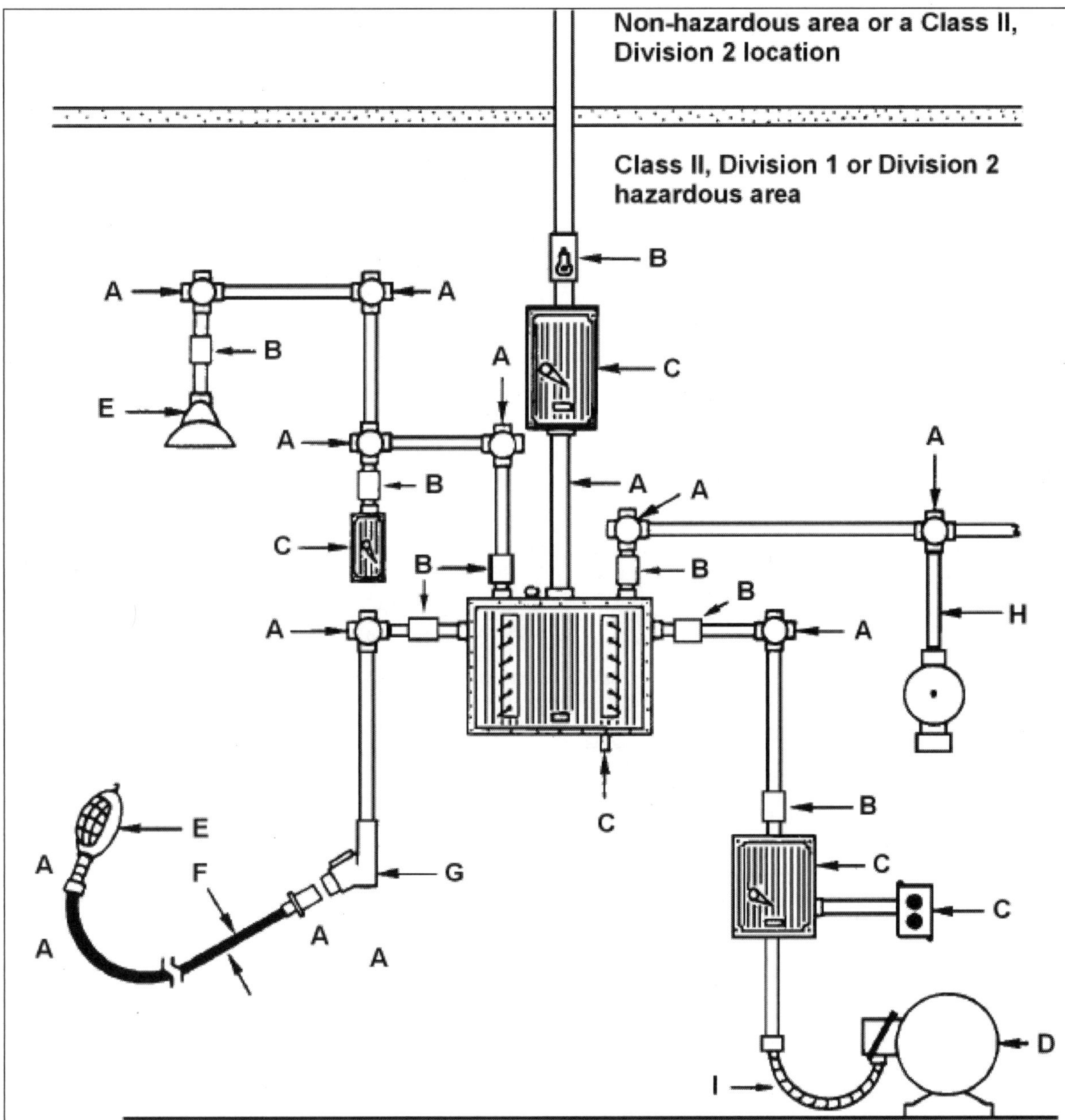

Table 307-6 – Summary of Class II Hazardous Locations (see Figure 116)

A	Wiring methods for Class II, Division 1 locations: boxes and fittings containing arcing and sparking parts are required to be in dust-ignition-proof enclosures. Threaded metal conduit or type MI cable with listed terminations is required for Class II, Division 1 locations; some use of Type MC cable is permitted. See NEC® Section 502.10(A)(1). In Class II Division 2 locations, boxes and fittings are not required to be dust-ignition proof but must be designed to minimize the entrance of dust and prevent the escape of sparks or burning material. In addition to the wiring systems suitable for Division 1 locations, the following systems are suitable for Division 2 locations: electrical metallic tubing, dust-tight wireways and the following types of cables with many restrictions: MC, MT, PLTC, ITC, TC. See NEC® Section 502.10(B)(1).
B	Suitable means of preventing the entrance of dust into a dust-ignition-proof enclosure must be provided where a raceway provides a path to the dust-ignition-proof enclosure from another enclosure that could allow the entrance of dust. See Figures 117-119. Also see NEC® Section 502.15.
C	Switches, circuit breakers, motor controllers, and fuses installed in Class II, Division 1 locations must be identified dust-ignition proof. Also see NEC® Section 502.115(A). In Class II, Division 2 areas, enclosures for fuses, switches, circuit breakers, and motor controllers must be dust-tight. Also see NEC® Section 502.115(B).
D	In Class II, Division 1 locations, motors, generators, and other rotating electrical machinery must be identified dust-ignition proof or totally enclosed pipe ventilated (see Figure 120). See NEC® Section 502.125(A). In Class II, Division 2 areas, rotating equipment must be one of the following types: 1) dust-ignition-proof, 2) totally enclosed, pipe ventilated, 3) totally enclosed, nonventilated, 4) totally enclosed, fan cooled, or 5) totally enclosed, water/air cooled. Under certain conditions, standard open-type machines and self-cleaning squirrel-cage motors may be used. See NEC® Section 502.125(B).
E	In Class II, Division 1 locations, lighting fixtures must be dust-ignition-proof. See NEC® Section 502.130(A). Lighting fixtures in Class II Division 2 locations must be designed to minimize accumulation of dust and must be enclosed to prevent the release of sparks or hot metal. In both divisions, each fixture must be clearly marked for the maximum wattage of the lamp, so that the maximum permissible surface temperature for the fixture is not exceeded. Additionally, fixtures must be protected from damage. See NEC® Section 502.130(B).
F	Flexible cords in Divisions 1 and 2 are required to: 1) be suitable for extra hard usage, 2) contain an equipment grounding conductor, 3) be connected to terminals in an approved manner, 4) be properly supported, and 5) be provided with suitable seals where necessary. See NEC® Section 502.140.
G	Receptacles and attachment plugs used in Class II, Division 1 areas are required to be identified for Class II locations and provided with a connection for an equipment grounding conductor. See NEC® Section 502.145(A). In Division 2 areas, the receptacle must be designed so the connection to the supply circuit cannot be made or broken while the parts are exposed. This is commonly done with an interlocking arrangement between a circuit breaker and the receptacle. The plug cannot be removed until the circuit breaker is in the off position, and the breaker cannot be switched to the on position unless the plug is inserted in the receptacle. See NEC® Section 502.145(B).

Table 307-6 – Summary of Class II Hazardous Locations (Continued)

H	Signaling systems and other similar systems: See NEC® Section 502.150.
I	Equipment grounding is required of all non-current-carrying metal parts of the electrical system. Lock nuts and bushings must not be relied upon for electrical connection between raceways and equipment enclosures. If locknuts or bushings are used, bonding jumpers are required. A conduit threaded into a box or enclosure and engaged wrenchtight is acceptable as a grounding connection (see Figure 121). See NEC® Section 502.30(B).

In general, equipment in Class II, Division 1 locations should be dust-ignition-proof, while equipment in Division 2 locations need only be dust tight. Additionally, equipment is marked to indicate its operating temperature above ambient (T1 = 450°C [842°F]; T6 = 85°C [185°F]), which should be below the ignition temperature of the dust. For organic dusts that dehydrate or carbonize, the temperature marking cannot exceed the ignition temperature or 165°C (329°F), whichever is lower. The ignition temperature of equipment approved prior to this requirement is assumed to be as shown in Table 307-7. NEC® Section 502.10 provides additional information on proper wiring methods in Class II locations.

Table 307-7 – Maximum Surface Temperatures

Group	Equipment That is Not Subject to Overloading		Equipment (Such as Motors or Power Transformers) That May be Overloaded			
			Normal Operation		Abnormal Operation	
Class II	°C	°F	°C	°F	°C	°F
E	200	392	200	392	200	392
F	200	392	150	302	200	392
G	165	329	120	248	165	329

Source: Table 500.8(D)(2) of the National Electrical Code®

Transformers and Capacitors

In Class II, Division 1 locations, all transformers and capacitors must be installed in vaults or must be approved as a complete assembly for Class II locations. In Division 2 areas, transformers and capacitors containing liquids that will burn must be installed in a vault. However, no transformer or capacitor may be installed where aluminum, magnesium, or other metals of similarly hazardous characteristics may be present.

This can be accomplished in one of the following ways:

(1) By installing permanent, effective seals with fittings that are easy to reach for repairs. See Figure 117.

(2) By arranging 10-foot or longer raceways horizontally between enclosures. See Figure 118.

(3) Arranging 5-foot or longer vertical raceways that extend downward from the dust-ignition proof enclosure to a general-purpose enclosure. See Figure 119.

Pipe-ventilated motors, generators, or other rotating electrical machinery must be enclosed in a dust tight enclosure that is connected to outside clean air.

In Class II Division 1 locations, ventilating pipe must be dust-tight—that is, constructed to minimize the entrance of dust. See NEC® Section 502.128(A).

Figure 120 illustrates a pipe-ventilated motor for Class II, Division 1 and 2 areas.

In Class II locations, all exposed non-current-carrying metal parts of the electrical system must be grounded. Bonding jumpers are used to prevent arcs across joints (Figure 121).

Figure 117 – Preventing dust from entering the dust-ignition-proof enclosure by sealing between enclosures.

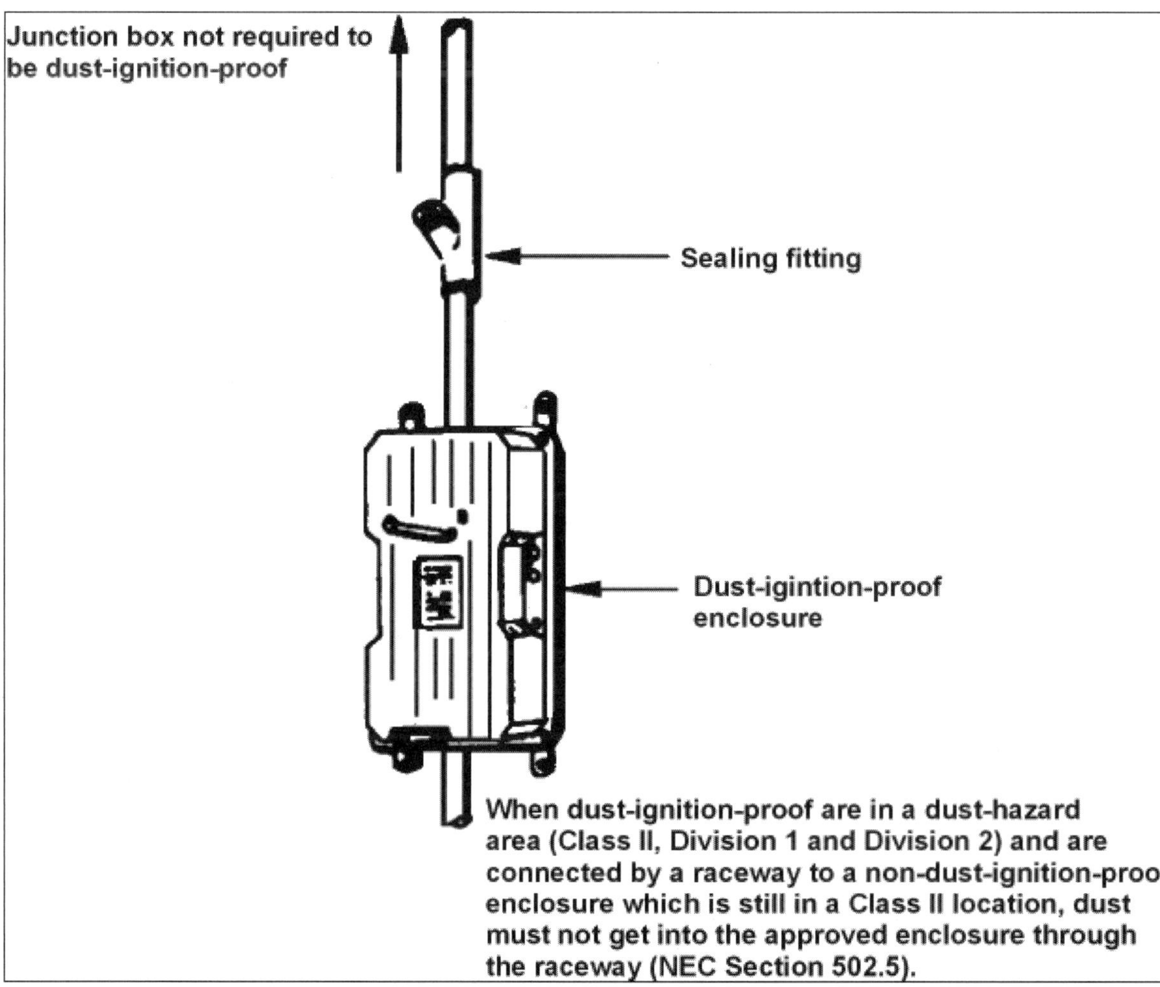

Figure 118 – Preventing dust from entering the dust-ignition-proof enclosures by horizontal distance (no seal).

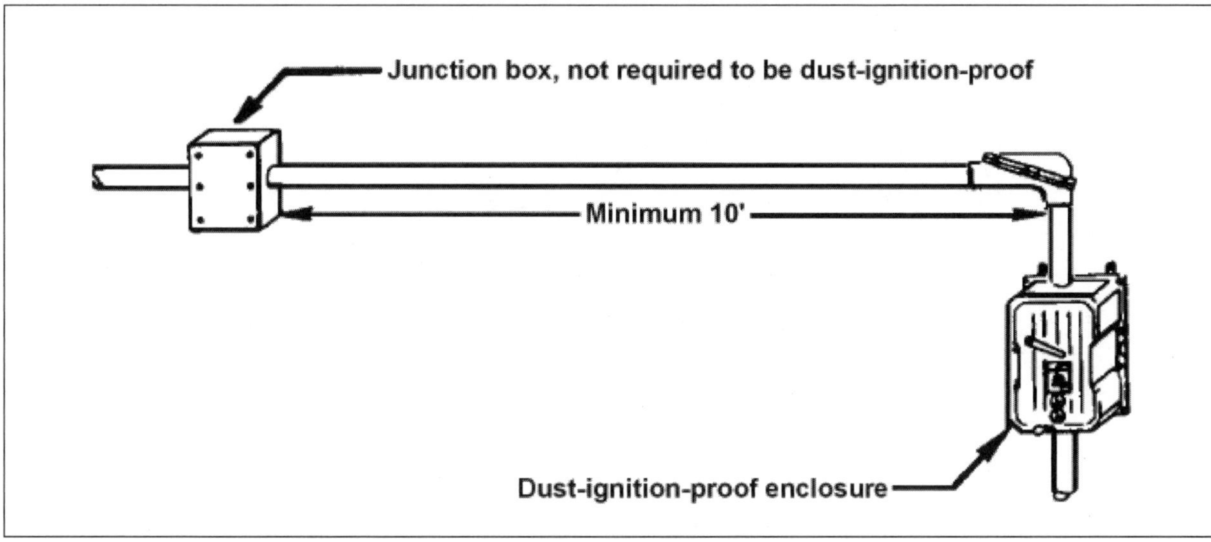

Figure 119 – Preventing dust from entering the dust-ignition-proof enclosure by vertical distance (no seal)

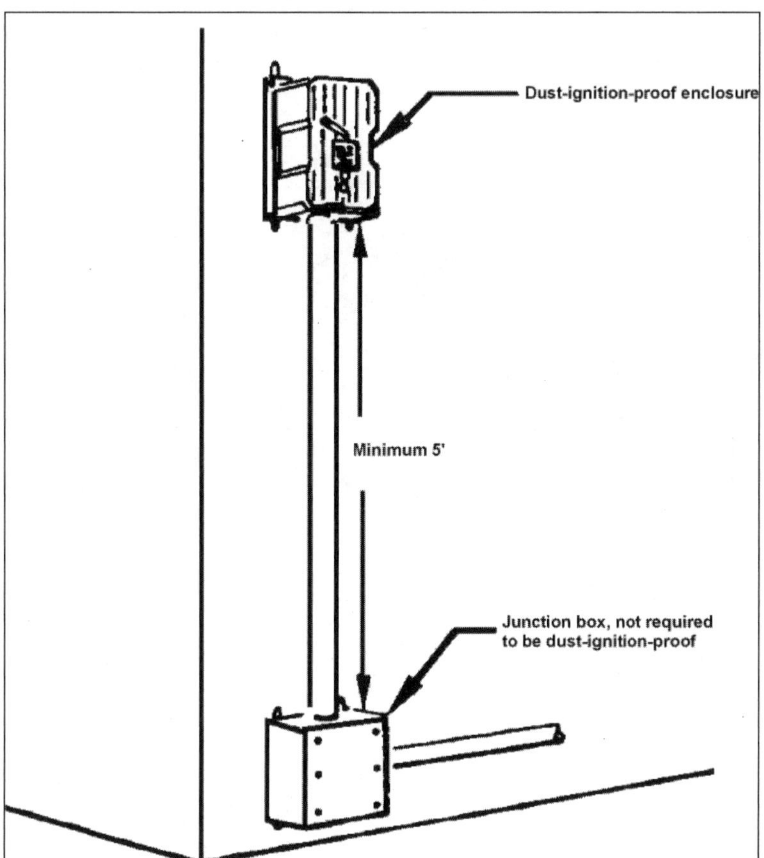

Figure 120 – Totally enclosed pipe-ventilated motor.

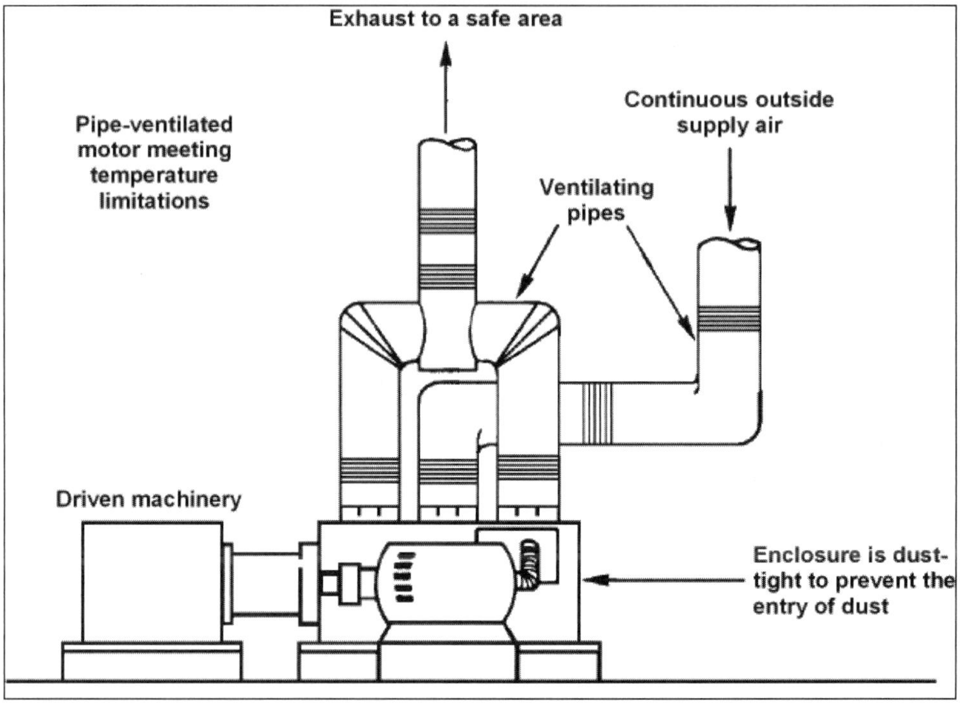

Figure 121 – Bonding in Class II hazardous (classified) locations.

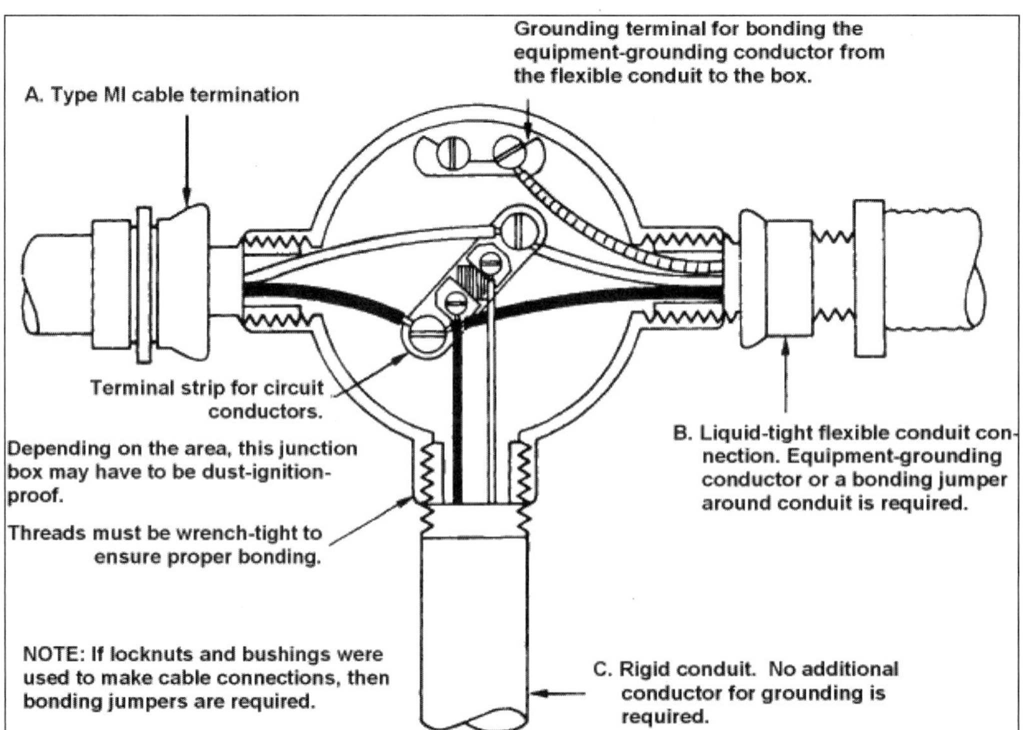

1910.307

Title: Hazardous (classified) locations.

Subpart: S

SubPart Title: Electrical - Design Safety Standards for Electrical Systems

Class III Hazardous Locations

Class III hazardous locations are areas where ignitable fibers and flyings are present. In general, equipment acceptable for use in Class II, Division 2 locations is also acceptable for installation in Class III locations. Equipment in Class III locations should be able to operate at full rating without causing excessive dehydration or carbonization of accumulated fibers or flyings. The maximum operating surface temperature is 165°C (329°F) for equipment that is not subject to overloading, and 120° (248°F) for equipment that may be overloaded. Figure 122 and Table 307-8 summarize some of the requirements for installations in Class III locations. Grounding requirements for Class III locations are the same as those for Class II locations. (See NEC® Sections 503.30 and 502.30.)

Notice that, unlike Class I and II locations, Class III locations do not necessarily require devices and equipment that are labeled for the location. They are required to be "dust tight," meeting certain performance requirements as outlined in Table 307-8 and set forth more completely in NEC® Article 503.

Table 307-8 – Summary of Class III Locations (See Figure 121)

A	In Division 1 and most Division 2 locations, wiring must be in conduit, electrical metallic tubing, dust tight wireways, or Type MC or MI cable, all with boxes, fittings and connections that will provide a dust-tight wiring system. See NEC® Article 503.10.
B	Switches, circuit breakers, motor controllers, and similar devices used in Class III hazardous locations must be within tight metal enclosures that are designed to minimize the entry of fibers and flyings and must not have any openings through which sparks or burning materials might escape. See NEC® 503.115.
C	Motors, generators, and other rotating electric machinery must be totally enclosed nonventilated, totally enclosed pipe-ventilated, or totally enclosed fan-cooled. The windings of totally enclosed non-ventilated motors are completely enclosed in a tight casing and are cooled by radiation and conduction through the frame. Enclosed pipe-ventilated motors have openings for a ventilating pipe that conveys air to the motor and then discharges it to a safe area. See Figure 120. In totally enclosed fan-cooled motors, the windings are cooled by an internal fan that circulates air inside the enclosure. Under certain conditions, self-cleaning textile motors and standard open-type machines may be used. See NEC® Section 503.125.
D	Lighting fixtures must have enclosures designed to minimize the entry of fibers, to prevent the escape of sparks or hot metal, and to have a maximum exposed surface temperature of less than 165°C. (329°F) See NEC® Section 503.130(A).

Cranes and Hoists

NEC® Article 503 includes requirements for cranes and hoists that are in Class III hazardous locations. Electric cranes, hoists, and similar equipment installed or located to operate over areas where combustible fibers are present must have an ungrounded power supply that is isolated from any other system. Also, such equipment must have a means of alarming and automatically de-energizing the contact conductors when a ground fault occurs. A ground-fault indicator that gives visual and audible alarm is also acceptable if the alarm is maintained until the circuit is opened. The contact conductors should be located so that they are guarded against tampering and contact by foreign objects. The current collectors must have protection to prevent the escape of sparks or hot particles, and they must be kept free of lint accumulations (see NEC® Section 503.155.) Live parts other than contacts and collectors for cranes and hoists may not be exposed in Class III locations.

Figure 122 – Class III hazardous locations. (See Table 307-8.)

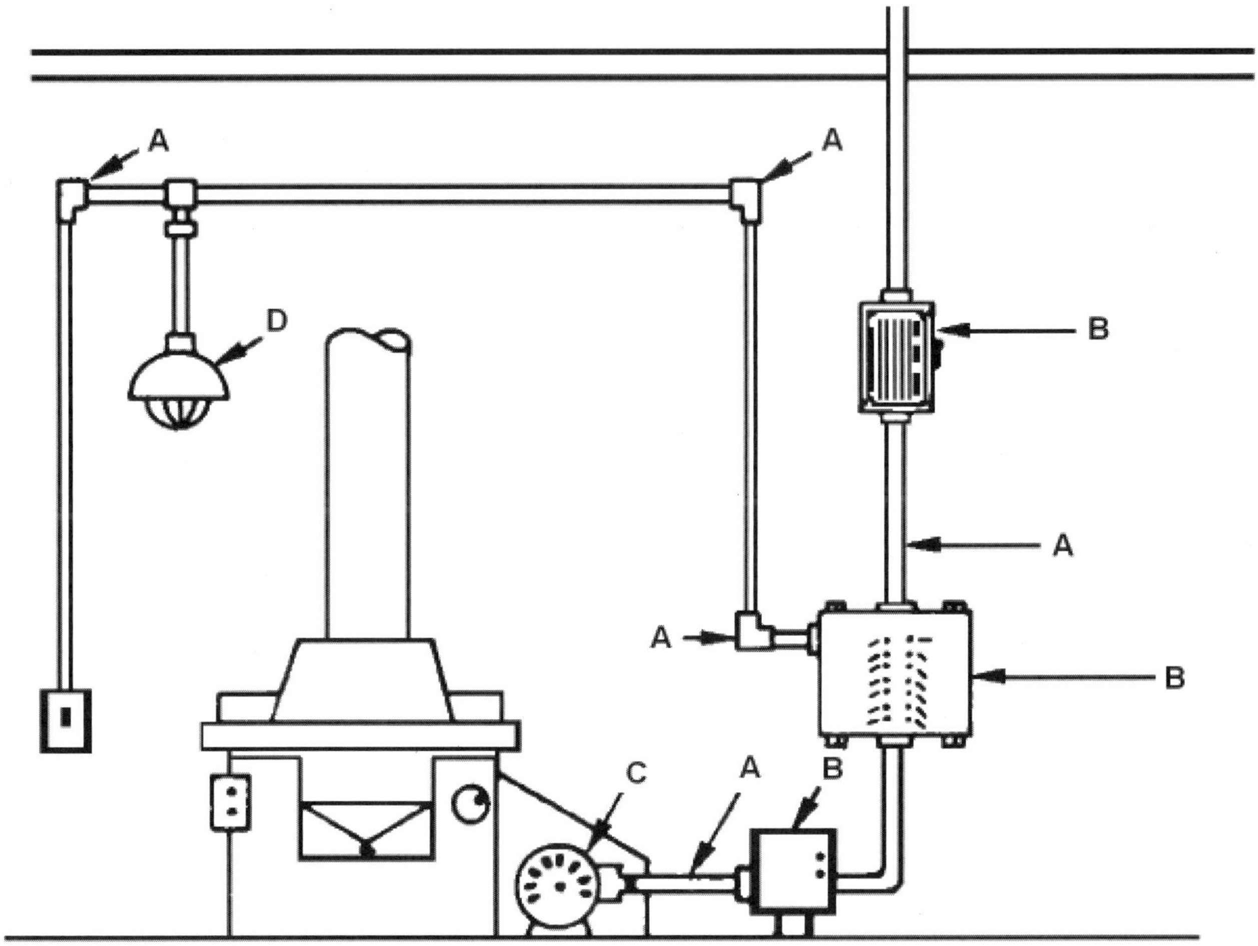

(d) **Conduits.** All conduits shall be threaded and shall be made wrench-tight. Where it is impractical to make a threaded joint tight, a bonding jumper shall be utilized.

Conduits in Hazardous Locations

The first sentence of (c) seems to restrict wiring installations in any Class I, II or III location to threaded rigid or intermediate metal conduit. However, the note immediately preceding this requirement in the OSHA standards permits wiring methods as set forth in NEC® (Section 503.10) as one means of compliance with OSHA's Electrical Installations paragraph.

(e) **Equipment in Division 2 locations**. Equipment that has been approved for a Division 1 location may be installed in a Division 2 location of the same class and group. General-purpose equipment or equipment in general-purpose enclosures may be installed in Division 2 locations if the employer can demonstrate that the equipment does not constitute a source of ignition under normal operating conditions.

(f) **Protection techniques.** The following are acceptable protection techniques for electric and electronic equipment in hazardous (classified) locations.

(f)(1)	**Explosionproof apparatus.** This protection technique is permitted for equipment in the Class I, Division 1 and 2 locations for which it is approved.
(f)(2)	**Dust ignitionproof.** This protection technique is permitted for equipment in the Class II, Division 1 and 2 locations for which it is approved.
(f)(3)	**Dust-tight**. This protection technique is permitted for equipment in the Class II, Division 2 and Class III locations for which it is approved.
(f)(4)	**Purged and pressurized.** This protection technique is permitted for equipment in any hazardous (classified) location for which it is approved.
(f)(5)	**Nonincendive circuit.** This protection technique is permitted for equipment in Class I, Division 2; Class II, Division 2; or Class III, Division 1 or 2 locations.
(f)(6)	**Nonincendive equipment.** This protection technique is permitted for equipment in Class I, Division 2; Class II, Division 2; or Class III, Division 1 or 2 locations.
(f)(7)	**Nonincendive component.** This protection technique is permitted for equipment in Class I, Division 2; Class II, Division 2; or Class III, Division 1 or 2 locations.
(f)(8)	**Oil immersion**. This protection technique is permitted for current-interrupting contacts in Class I, Division 2 locations as described in the Subpart.
(f)(9)	**Hermetically sealed**. This protection technique is permitted for equipment in Class I, Division 2; Class II, Division 2; and Class III, Division 1 or 2 locations.
(f)(10)	**Other protection techniques.** Any other protection technique that meets paragraph (c) of this section is acceptable in any hazardous (classified) location.
(g)	**Class I, Zone 0, 1, and 2 locations.**
(g)(1)	**Scope.** Employers may use the zone classification system as an alternative to the division classification system for electric and electronic equipment and wiring for all voltage in Class I, Zone 0, Zone 1, and Zone 2 hazardous (classified) locations where fire or explosion hazards may exist due to flammable gases, vapors, or liquids.
(g)(2)	**Location and general requirements.**
(g)(2)(i)	Locations shall be classified depending on the properties of the flammable vapors, liquids, or gases that may be present and the likelihood that a flammable or combustible concentration or quantity is present. Where pyrophoric materials are the only materials used or handled, these locations need not be classified.
(g)(2)(ii)	Each room, section, or area shall be considered individually in determining its classification.
(g)(2)(iii)	All threaded conduit shall be threaded with an NPT (National (American) Standard Pipe Taper) standard conduit cutting die that provides 3/4-in. taper per foot. The conduit shall be made wrench tight to prevent sparking when fault current flows through the conduit system and to ensure the explosionproof or flameproof integrity of the conduit system where applicable.
(g)(2)(iv)	Equipment provided with threaded entries for field wiring connection shall be installed in accordance with paragraph (g)(2)(iv)(A) or (g)(2)(iv)(B) of this section.
(g)(2)(iv)(A)	For equipment provided with threaded entries for NPT threaded conduit or fittings, listed conduit, conduit fittings, or cable fittings shall be used.

(g)(2)(iv)(B) For equipment with metric threaded entries, such entries shall be identified as being metric, or listed adaptors to permit connection to conduit of NPT-threaded fittings shall be provided with the equipment. Adapters shall be used for connection to conduit or NPT-threaded fittings.

(g)(3) **Protection techniques.** One or more of the following protection techniques shall be used for electric and electronic equipment in hazardous (classified) locations classified under the zone classification system.

(g)(3)(i) Flameproof "d" - This protection technique is permitted for equipment in the Class I, Zone 1 locations for which it is approved.

(g)(3)(ii) Purged and pressurized - This protection technique is permitted for equipment in the Class I, Zone 1 or Zone 2 locations for which it is approved.

(g)(3)(iii) Intrinsic safety - This protection technique is permitted for equipment in the Class I, Zone 0 or Zone 1 locations for which it is approved.

(g)(3)(iv) Type of protection "n" - This protection technique is permitted for equipment in the Class I, Zone 2 locations for which it is approved. Type of protection "n" is further subdivided into nA, nC, and nR.

(g)(3)(v) Oil Immersion "o" - This protection technique is permitted for equipment in the Class I, Zone 1 locations for which it is approved.

(g)(3)(vi) Increased safety "e" - This protection technique is permitted for equipment in the Class I, Zone 1 locations for which it is approved.

(g)(3)(vii) Encapsulation "m" - This protection technique is permitted for equipment in the Class I, Zone 1 locations for which it is approved.

(g)(3)(viii) Powder Filling "q" - This protection technique is permitted for equipment in the Class I, Zone 1 locations for which it is approved.

(g)(4) **Special precaution.** Paragraph (g) of this section requires equipment construction and installation that will ensure safe performance under conditions of proper use and maintenance.

(g)(4)(i) Classification of areas and selection of equipment and wiring methods shall be under the supervision of a qualified registered professional engineer.

(g)(4)(ii) In instances of areas within the same facility classified separately, Class I, Zone 2 locations may abut, but not overlap, Class I, Division 2 locations. Class I, Zone 0 or Zone 1 locations may not abut Class I, Division 1 or Division 2 locations.

(g)(4)(iii) A Class I, Division 1 or Division 2 location may be reclassified as a Class I, Zone 0, Zone 1, or Zone 2 location only if all of the space that is classified because of a single flammable gas or vapor source is reclassified.

Note to paragraph (g)(4) of this section: Low ambient conditions require special consideration. Electric equipment depending on the protection techniques described by paragraph (g)(3)(i) of this section may not be suitable for use at temperatures lower than -20 °C (-4 °F) unless they are approved for use at lower temperatures. However, at low ambient temperatures, flammable concentrations of vapors may not exist in a location classified Class I, Zone 0, 1, or 2 at normal ambient temperature.

(g)(5) **Listing and marking.**

(g)(5)(i) Equipment that is listed for a Zone 0 location may be installed in a Zone 1 or Zone 2 location of the same gas or vapor. Equipment that is listed for a Zone 1 location may be installed in a Zone 2 location of the same gas or vapor.

(g)(5)(ii)	Equipment shall be marked in accordance with paragraph (g)(5)(ii)(A) and (g)(5)(ii)(B) of this section, except as provided in (g)(5)(ii)(C).
(g)(5)(ii)(A)	Equipment approved for Class I, Division 1 or Class 1, Division 2 shall, in addition to being marked in accordance with (c)(2)(ii), be marked with the following:
(g)(5)(ii)(A)(1)	Class I, Zone 1 or Class I, Zone 2 (as applicable);
(g)(5)(ii)(A)(2)	Applicable gas classification groups; and
(g)(5)(ii)(A)(3)	Temperature classification; or
(g)(5)(ii)(B)	Equipment meeting one or more of the protection techniques described in paragraph (g)(3) of this section shall be marked with the following in the order shown:
(g)(5)(ii)(B)(1)	Class, except for intrinsically safe apparatus;
(g)(5)(ii)(B)(2)	Zone, except for intrinsically safe apparatus;
(g)(5)(ii)(B)(3)	Symbol "AEx;"
(g)(5)(ii)(B)(4)	Protection techniques;
(g)(5)(ii)(B)(5)	Applicable gas classification groups; and
(g)(5)(ii)(B)(6)	Temperature classification, except for intrinsically safe apparatus.
	Note to paragraph (g)(5)(ii)(B) of this section: An example of such a required marking is "Class I, Zone 0, AEx ia IIC T6." See Figure S-1 for an explanation of this marking.
(g)(5)(ii)(C)	Equipment that the employer demonstrates will provide protection from the hazards arising from the flammability of the gas or vapor and the zone of location involved and will be recognized as providing such protection by employees need not be marked.
	Note to paragraph (g)(5)(ii)(C) of this section: The National Electrical Code, NFPA 70, contains guidelines for determining the type and design of equipment and installations that will meet this provision.

For clarification purposes, OSHA added a list of all the specific protective techniques for electrical installations in hazardous locations classified under the division classification system found in Article 500 of the NEC®. The list is based on the 2002 National Electrical Code®.

§ 1910.308 Special Systems

[46 FR 4056, Jan. 16, 1981; 46 FR 40185, Aug. 7, 1981; 72 FR 7212, Feb. 14, 2007]

(a) **Systems over 600 volts, nominal.** This paragraph covers the general requirements for all circuits and equipment operated at over 600 volts.

(a)(1) **Aboveground wiring methods.**

(a)(1)(i) Aboveground conductors shall be installed in rigid metal conduit, in intermediate metal conduit, in electrical metallic tubing, in rigid nonmetallic conduit, in cable trays, as busways, as cablebus, in other identified raceways, or as open runs of metal-clad cable suitable for the use and purpose. In locations accessible to qualified persons only, open runs of Type MV cables, bare conductors, and bare busbars are also permitted. Busbars shall be either copper or aluminum. Open runs of insulated wires and cables having a bare lead sheath or a braided outer covering shall be supported in a manner designed to prevent physical damage to the braid or sheath.

Above-Ground Conductors

For installations over 600 volts, insulated conductors and cables which have a lead sheath or metal-braided outer covering must be supported so that the sheath or covering is not damaged. The metallic sheath or braid for open conductors is grounded to provide a path for fault current to operate circuit breakers in the event of a fault. Any damage to the shielding may cause the continuity of this ground path to be broken.

In addition, metal braids and lead sheaths provide protection for conductor insulation. If the lead sheath is damaged, moisture may penetrate the insulation, providing a path for a ground fault.

Figure 123 shows an example of a properly supported cable and an example of the damage which could result should the cable be improperly supported.

Figure 123 – Improperly supported cable and properly supported cable.

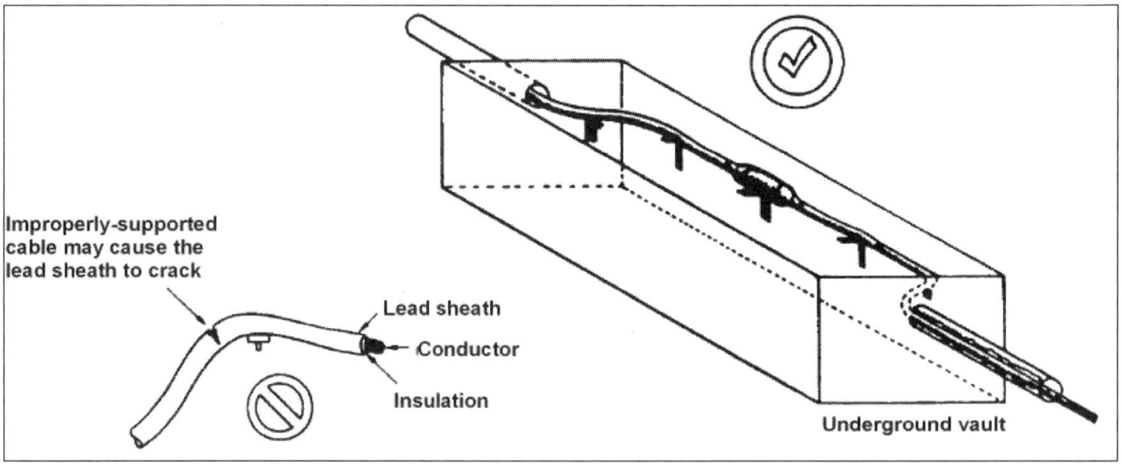

(a)(1)(ii) Conductors emerging from the ground shall be enclosed in approved raceways.

(a)(2) **Braid-covered insulated conductors - open installations.** The braid on open runs of braid-covered insulated conductors shall be flame retardant or shall have a flame-retardant saturant applied after installation. This treated braid covering shall be stripped back a safe distance at conductor terminals, according to the operating voltage.

(a)(3) **Insulation shielding.**

(a)(3)(i) Metallic and semiconductor insulation shielding components of shielded cables shall be removed for a distance dependent on the circuit voltage and insulation. Stress reduction means shall be provided at all terminations of factory-applied shielding.

(a)(3)(ii)	Metallic shielding components such as tapes, wires, or braids, or combinations thereof, and their associated conducting and semiconducting components shall be grounded.
(a)(4)	**Moisture or mechanical protection for metal-sheathed cables**. Where cable conductors emerge from a metal sheath and where protection against moisture or physical damage is necessary, the insulation of the conductors shall be protected by a cable sheath terminating device.
(a)(5)	**Interrupting and isolating devices.**
(a)(5)(i)	Circuit breaker installations located indoors shall consist of metal-enclosed units or fire-resistant cell-mounted units. In locations accessible only to qualified employees, open mounting of circuit breakers is permitted. A means of indicating the open and closed position of circuit breakers shall be provided.
(a)(5)(ii)	Where fuses are used to protect conductors and equipment, a fuse shall be placed in each ungrounded conductor. Two power fuses may be used in parallel to protect the same load, if both fuses have identical ratings, and if both fuses are installed in an identified common mounting with electrical connections that will divide the current equally. Power fuses of the vented type may not be used indoors, underground, or in metal enclosures unless identified for the use.
(a)(5)(iii)	Fused cutouts installed in buildings or transformer vaults shall be of a type identified for the purpose. Distribution cutouts may not be used indoors, underground, or in metal enclosures. They shall be readily accessible for fuse replacement.
(a)(5)(iv)	Where fused cutouts are not suitable to interrupt the circuit manually while carrying full load, an approved means shall be installed to interrupt the entire load. Unless the fused cutouts are interlocked with the switch to prevent opening of the cutouts under load, a conspicuous sign shall be placed at such cutouts reading: "WARNING - DO NOT OPERATE UNDER LOAD."
(a)(5)(v)	Suitable barriers or enclosures shall be provided to prevent contact with nonshielded cables or energized parts of oil-filled cutouts.
(a)(5)(vi)	Load interrupter switches may be used only if suitable fuses or circuits are used in conjunction with these devices to interrupt fault currents.
(a)(5)(vi)(A)	Where these devices are used in combination, they shall be coordinated electrically so that they will safely withstand the effects of closing, carrying, or interrupting all possible currents up to the assigned maximum short-circuit rating.
(a)(5)(vi)(B)	Where more than one switch is installed with interconnected load terminals to provide for alternate connection to different supply conductors, each switch shall be provided with a conspicuous sign reading: "WARNING - SWITCH MAY BE ENERGIZED BY BACKFEED."
(a)(5)(vii)	A means (for example, a fuseholder and fuse designed for the purpose) shall be provided to completely isolate equipment for inspection and repairs. Isolating means that are not designed to interrupt the load current of the circuit shall be either interlocked with an approved circuit interrupter or provided with a sign warning against opening them under load.

Interrupting and Isolating Devices

Interrupting and isolating devices are switches used to disconnect power in a circuit so service and maintenance work on equipment down stream of the switch can be accomplished safely. Figure 124 is a schematic illustration of two isolating switches used to de-energize a piece of equipment for maintenance and repair. In this case, two switches are used to sectionalize a high voltage grid to prevent back-feed current to the unit being serviced. Interrupting devices that are rated for high-voltage application, 600 volts or more, include air break, gas-filled and oil-immersed switches. These switches operate in an enclosure that is filled with air, sulphur-hexafluoride or oil to suppress arcing and flash-over during operation.

Isolating switches intended for use as load-interrupting devices must be rated for the short-circuit current load of the circuit they interrupt. Isolating switches can, however, be used in a circuit and not be rated as an interrupting device, but only if certain precautions are followed. If the switches are not designed to be opened under load, arcing and flash-over may occur and cause severe damage to the equipment and serious injury to the employee operating the switch. If an isolating means is used but is not designed to interrupt the circuit while the switch is carrying current, it must either be interlocked with a rated circuit interrupter or be provided with a sign warning against opening it under load. In either case, current through the switch must be stopped before the non-rated switch can be opened.

Figure 124 – Isolating switches.

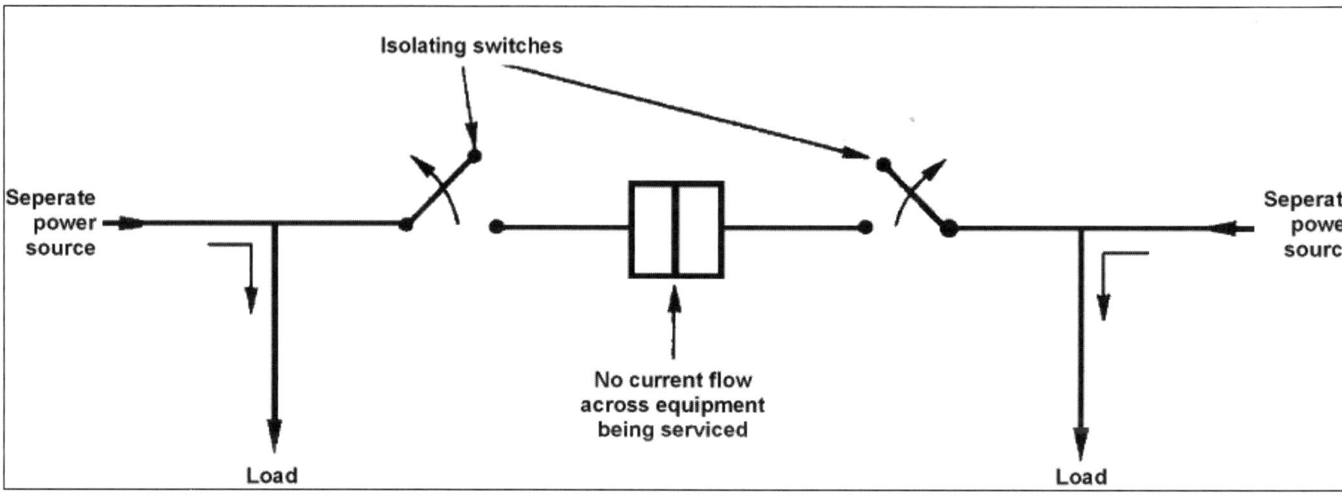

(a)(6)	**Mobile and portable equipment.**
(a)(6)(i)	A metallic enclosure shall be provided on the mobile machine for enclosing the terminals of the power cable. The enclosure shall include provisions for a solid connection for the grounding terminal to effectively ground the machine frame. The method of cable termination used shall prevent any strain or pull on the cable from stressing the electrical connections. The enclosure shall have provision for locking so only authorized qualified persons may open it and shall be marked with a sign warning of the presence of energized parts.
(a)(6)(ii)	All energized switching and control parts shall be enclosed in effectively grounded metal cabinets or enclosures. Circuit breakers and protective equipment shall have the operating means projecting through the metal cabinet or enclosure so these units can be reset without locked doors being opened. Enclosures and metal cabinets shall be locked so that only authorized qualified persons have access and shall be marked with a sign warning of the presence of energized parts. Collector ring assemblies on revolving-type machines (shovels, draglines, etc.) shall be guarded.
(a)(7)	**Tunnel installations.** This paragraph applies to installation and use of high-voltage power distribution and utilization equipment that is portable or mobile, such as substations, trailers, cars, mobile shovels, draglines, hoists, drills, dredges, compressors, pumps, conveyors, and underground excavators.
(a)(7)(i)	Conductors in tunnels shall be installed in one or more of the following:
(a)(7)(i)(A)	Metal conduit or other metal raceway;
(a)(7)(i)(B)	Type MC cable; or
(a)(7)(i)(C)	Other approved multiconductor cable.
(a)(7)(ii)	Multiconductor portable cable may supply mobile equipment.
(a)(7)(iii)	Conductors and cables shall also be so located or guarded as to protect them from physical damage. An equipment grounding conductor shall be run with circuit conductors inside the metal raceway or inside the multiconductor cable jacket. The equipment grounding conductor may be insulated or bare.

(a)(7)(iv)	Bare terminals of transformers, switches, motor controllers, and other equipment shall be enclosed to prevent accidental contact with energized parts.
(a)(7)(v)	Enclosures for use in tunnels shall be drip-proof, weatherproof, or submersible as required by the environmental conditions.
(a)(7)(vi)	Switch or contactor enclosures may not be used as junction boxes or raceways for conductors feeding through or tapping off to other switches, unless special designs are used to provide adequate space for this purpose.
(a)(7)(vii)	A disconnecting means that simultaneously opens all ungrounded conductors shall be installed at each transformer or motor location.
(a)(7)(viii)	All nonenergized metal parts of electric equipment and metal raceways and cable sheaths shall be effectively grounded and bonded to all metal pipes and rails at the portal and at intervals not exceeding 305 m (1000 ft) throughout the tunnel.
(b)	**Emergency power systems.** This paragraph applies to circuits, systems, and equipment intended to supply power for illumination and special loads in the event of failure of the normal supply.
(b)(1)	**Wiring methods.** Emergency circuit wiring shall be kept entirely independent of all other wiring and equipment and may not enter the same raceway, cable, box, or cabinet or other wiring except either where common circuit elements suitable for the purpose are required, or for transferring power from the normal to the emergency source.

Wiring Methods for Emergency Power Systems

Emergency circuit wiring must be completely independent of and physically separated from all other wiring and equipment. Such wiring must not enter raceways, boxes, cables, or cabinets that contain other wiring. This ensures that any fault on the normal wiring circuits will not affect the performance of

Figure 125 – Normal and emergency lighting circuits where permitted in a common junction box.

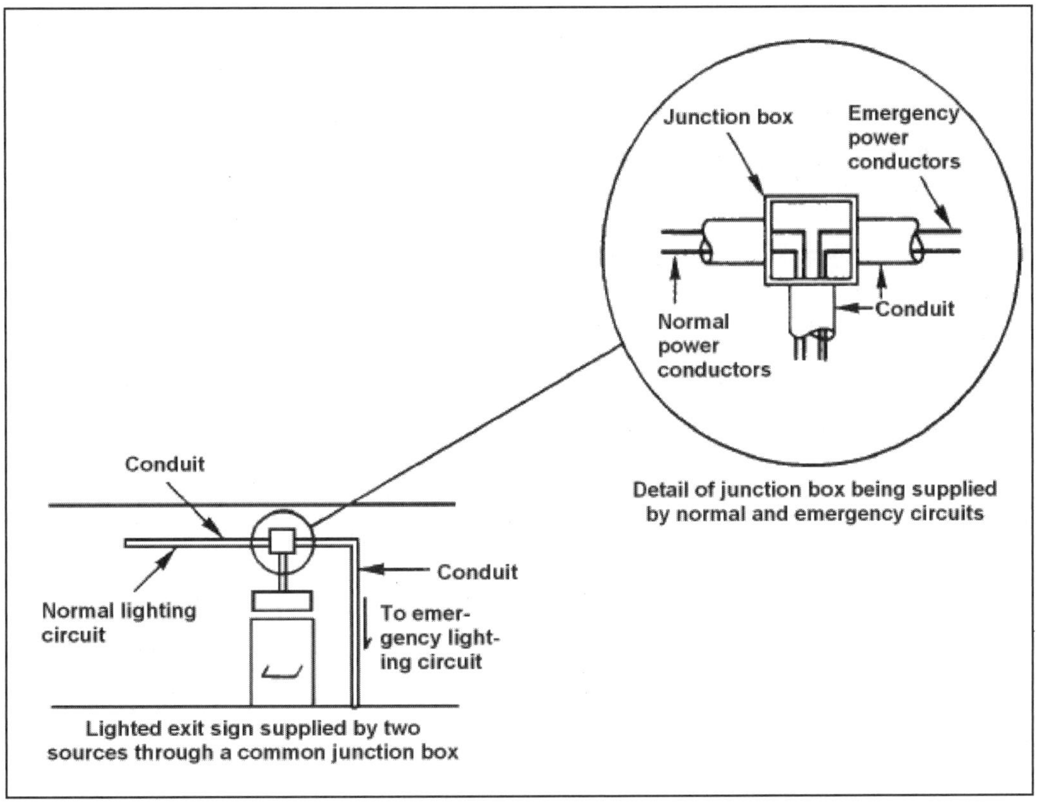

the emergency system.

There are two exceptions to this requirement. First, where power is transferred from the normal source to the emergency source, the transfer switch requires circuit wiring from both sources to enter the same box or cabinet. Second, if common circuit elements in equipment are suitable for emergency and normal lighting purposes, both systems may occupy the same enclosure. For example, the junction box for a lighted sign, which is supplied under normal conditions by a regular branch circuit, may also contain emergency system conductors. See Figure 125.

(b)(2)	**Emergency illumination.** Emergency illumination shall include all required means of egress lighting, illuminated exit signs, and all other lights necessary to provide illumination. Where emergency lighting is necessary, the system shall be so arranged that the failure of any individual lighting element, such as the burning out of a light bulb, cannot leave any space in total darkness.
(b)(3)	**Signs.**
(b)(3)(i)	A sign shall be placed at the service entrance equipment indicating the type and location of onsite emergency power sources. However, a sign is not required for individual unit equipment.
(b)(3)(ii)	Where the grounded circuit conductor connected to the emergency source is connected to a grounding electrode conductor at a location remote from the emergency source, there shall be a sign at the grounding location that shall identify all emergency and normal sources connected at that location.
(c)	**Class 1, Class 2, and Class 3 remote control, signaling, and power-limited circuits.**
(c)(1)	**Classification.** Class 1, Class 2, and Class 3 remote control, signaling, or power-limited circuits are characterized by their usage and electrical power limitation that differentiates them from light and power circuits. These circuits are classified in accordance with their respective voltage and power limitations as summarized in paragraphs (c)(1)(i) through (c)(1)(iii) of this section.
(c)(1)(i)	A Class 1 power-limited circuit shall be supplied from a source having a rated output of not more than 30 volts and 1000 volt-amperes.
(c)(1)(ii)	A Class 1 remote control circuit or a Class 1 signaling circuit shall have a voltage not exceeding 600 volts; however, the power output of the source need not be limited.
(c)(1)(iii)	The power source for a Class 2 or Class 3 circuit shall be listed equipment marked as a Class 2 or Class 3 power source, except as follows:
(c)(1)(iii)(A)	Thermocouples do not require listing as a Class 2 power source; and
(c)(1)(iii)(B)	A dry cell battery is considered an inherently limited Class 2 power source, provided the voltage is 30 volts or less and the capacity is less than or equal to that available from series-connected No. 6 carbon zinc cells.

Class 1, Class 2, and Class 3 Remote Control, Signaling, and Power-Limited Circuits

Class 1 Circuits

Class 1 power-limited circuits are supplied from a power source that has a rated output of not more than 30 volts and a power limitation of 1000 volt-amps. Class 1 power-limited circuits have a current limiter on the power source that supplies them. This limiter is an overcurrent protection device that restricts the amount of supply current to the circuit in the event of an overload, short circuit or ground fault. These Class 1 circuits may be supplied from a transformer or other type of power supply such as generators or batteries. See Figure 126.

Class 1 remote control or signaling circuits are permitted to operate at up to 600 volts and have no limitation on the power rating of the source. Class 1 systems generally must meet most wiring requirements for power and light circuits. Class 1 remote control circuits are commonly used in motor controllers that operate mechanical processes, elevators, conveyors, and equipment that is controlled from one or more remote locations. Class 1 signaling circuits are used in nurses' call systems in hospitals, electric clocks, bank alarm systems, and factory call systems. See Figure 127.

Figure 126 – Example of a Class 1 power-limited circuit

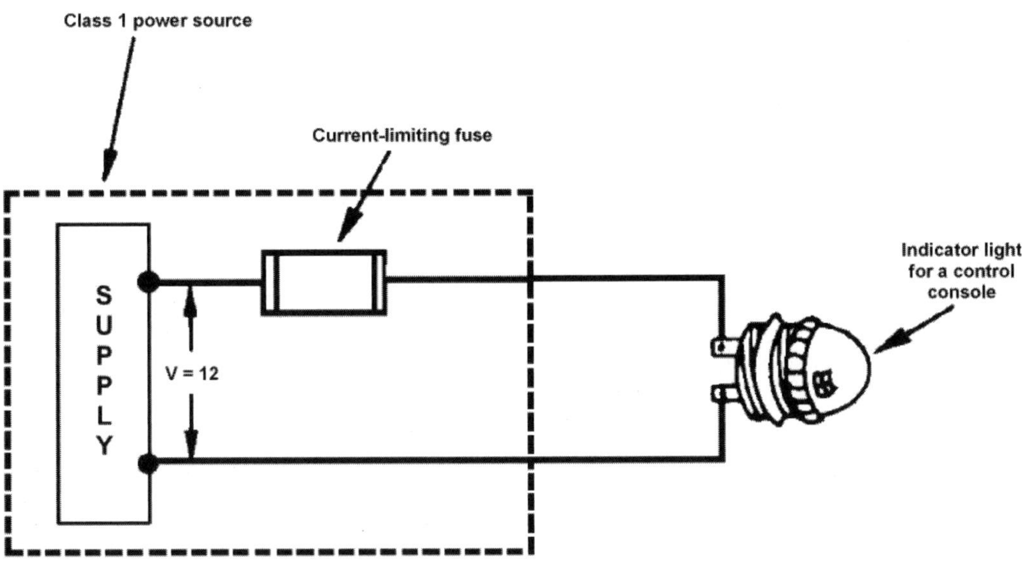

Figure 127 – Example of a Class 1 remote control circuit

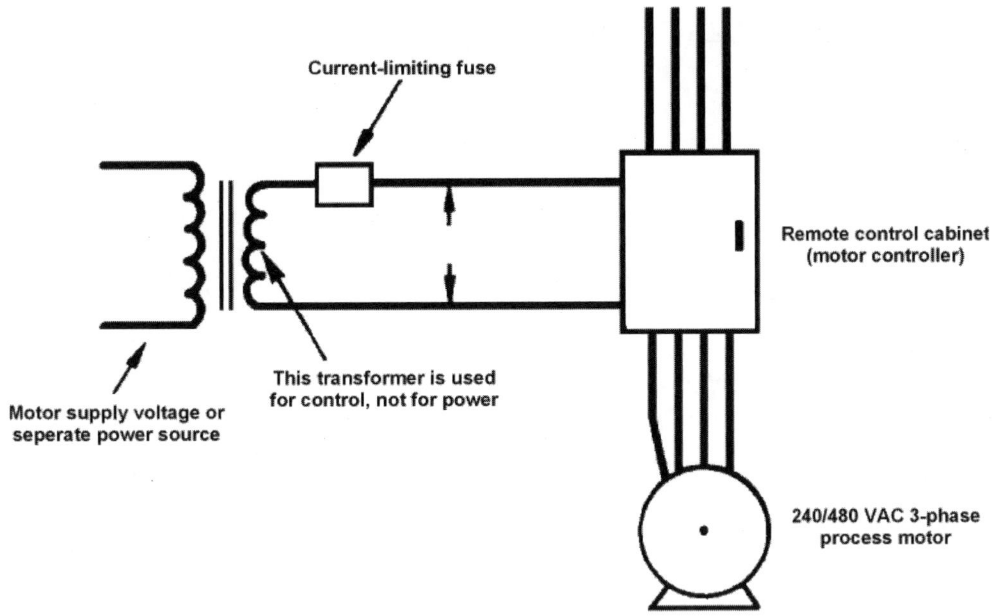

Class 2 and Class 3 Circuits

The power for Class 2 and Class 3 circuits must be inherently limited, which requires no overcurrent protection, or must be limited by combining a power source and overcurrent protection. These circuits can be inherently limited by power-limited transformers that have high-impedance windings to limit power output. Batteries also are common power sources that are inherently limiting.

Heating system thermostats are commonly Class 2 systems. Figure 128 shows a current-limiting 24-volt, Class 2 circuit for a boiler thermostat control.

Figure 128 – Thermostat control circuit, Class 2

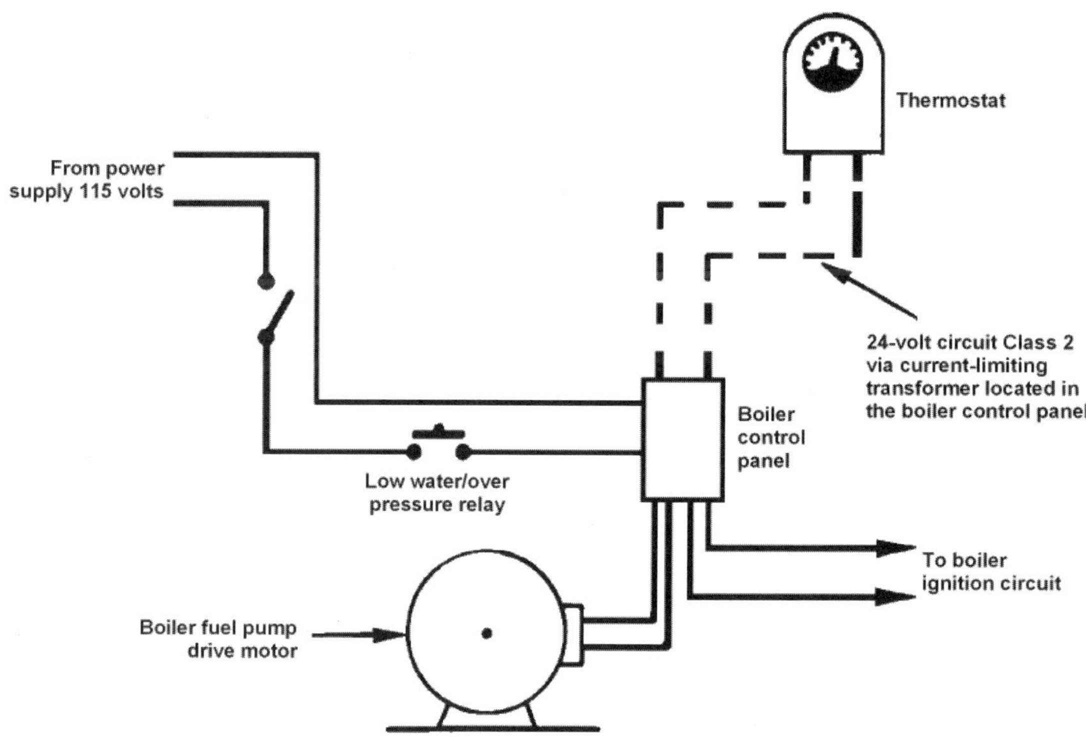

The majority of small bell, buzzer, and annunciator systems are Class 2 circuits. Class 2 also includes small intercommunicating telephone systems in which the voice circuit is supplied by a battery and the ringing circuit by a transformer.

Because power is limited in Class 2 and 3 wiring, less stringent requirements apply than to power, light and Class 1 systems, and typically it can be installed without conduit or raceway. In recent years, there has been a proliferation of special-purpose cables for Class 2 and 3 systems, including plenum and shaft cable, power-limited fire alarm cables, cables for 1- and 2-family dwellings, etc. Specific restrictions applying to each are set forth in NEC® Sections 725.51 through 725.71.

Table 308-1 presents a comparison of Class 1, Class 2, and Class 3 circuits, including common uses. This table does not include most telephone and similar communications wiring installations, which are covered by OSHA Section 1910.308(e) and NEC® Article 800. Optical fiber cables are included in NEC® under Article 770.

limited fire alarm circuits, and medium power network-powered broadband communications cables unless a barrier or other equivalent form of protection against contact is employed.

(d) **Fire alarm systems.**

(d)(1) **Classifications.** Fire alarm circuits shall be classified either as nonpower limited or power limited.

(d)(2) **Power sources.** The power sources for use with fire alarm circuits shall be either power limited or nonlimited as follows:

(d)(2)(i) The power source of nonpower-limited fire alarm (NPLFA) circuits shall have an output voltage of not more than 600 volts, nominal; and

(d)(2)(ii) The power source for a power-limited fire alarm (PLFA) circuit shall be listed equipment marked as a PLFA power source.

(d)(3) **Separation from conductors of other circuits.**

(d)(3)(i) Nonpower-limited fire alarm circuits and Class 1 circuits may occupy the same enclosure, cable, or raceway provided all conductors are insulated for maximum voltage of any conductor within the enclosure, cable, or raceway. Power supply and fire alarm circuit conductors are permitted in the same enclosure, cable, or raceway only if connected to the same equipment.

Non-Power-Limited Conductor Location

Non-power-limited fire protective signaling circuits may include circuits that are part of a central station signaling system, a sprinkler water flow alarm, or a local fire alarm in a building. Voltages for these circuits range up to 600 volts. These circuits are permitted to be located in the same enclosure, cable, or raceway as Class 1 circuits if the insulation on all of the wires within that enclosure is rated for the highest voltage of any conductor therein.

Power supply conductors are not usually permitted in the same enclosure, cable, or raceway as fire-protective conductors because a fault or overcurrent condition in the power supply conductor could damage the fire-protective circuits. This would cause the fire-protective signal circuit to malfunction and perhaps not transmit a needed alarm or fire signal. However, power supply conductors and fire-protective signaling circuits may occupy the same enclosure if they are connected to the same equipment.

(d)(3)(ii) Power-limited circuit cables and conductors may not be placed in any cable, cable tray, compartment, enclosure, outlet box, raceway, or similar fitting with conductors of electric light, power, Class 1, nonpower-limited fire alarm circuit conductors, or medium power network-powered broadband communications circuits.

(d)(3)(iii) Power-limited fire alarm circuit conductors shall be separated at least 50.8 mm (2 in.) from conductors of any electric light, power, Class 1, nonpower-limited fire alarm, or medium power network-powered broadband communications circuits unless a special and equally protective method of conductor separation is employed.

(d)(3)(iv) Conductors of one or more Class 2 circuits are permitted within the same cable, enclosure, or raceway with conductors of power-limited fire alarm circuits provided that the insulation of Class 2 circuit conductors in the cable, enclosure, or raceway is at least that needed for the power-limited fire alarm circuits.

1910.308

Title: Special Systems.

Subpart: S

SubPart Title: Electrical - Design Safety Standards for Electrical Systems

Table 308-1 – Comparison of Class 1, 2, and 3 Remote Control, Signaling, and Power-Limited Circuits

Group	Common AC/DC Source	Power-Limiting Device	Maximum† Power Output (Volt-Amperes)	Voltage† Limitation Volts	Common Uses
Class 1 Circuits					
Power limited	AC or DC	Overcurrent Device	1,000	30	
Remote Control And Signaling Circuits	AC or DC	Overcurrent Device	No Power Limitations	600	Remote Motor Control Circuits Bank Alarm Systems, Call Systems in Hospitals
Class 2 Circuits					
Inherently Limited	AC or DC with High-Resistance Windings	Transformer	100††	150	Oil Burner Control Circuit Small Bell, Buzzer, or Annunciator System
Overcurrent Device Limited	AC	Overcurrent Device	250	30	Small Interconnected Telephone systems
	DC	250	250	60	
Class 3 Circuits					
Inherently Limited	AC or DC	Transformer With High-Resistance Windings	100†	100	
Overcurrent Device Limited	AC or DC	Overcurrent Device	250	150	

† Maximum power output and voltage limitations contained in this table are at different circuit voltages. Both power output and voltage limitation must be satisfied. For more information refer to Article 725 of the National Electrical Code®.

†† Maximum nameplate rating of power source.

Power-Limited Conductor Location

Since power-limited conductors are usually light gauge wires with low voltage rating and operate at lower voltages and power ratings than Class 1 circuits, power circuits, and non-power-limited circuits, special measures must be taken to keep these conductors physically separate. Power-limited cables must be separated from non-power-limited wiring by methods such as barriers, conduit enclosures, tubing, distance (50.8mm (2 in.)) for open wiring, 6.35mm (0.25 inch) in enclosures) or insulation as required for non-power-limited applications. See NEC® Section 725.136 for details.

Because of the differing operating voltages and insulation levels of power-limited signaling circuits, their conductors are only permitted to be located where not subject to damage or interference from other types of circuits. Class 3 circuits and power-limited fire-protective signal circuits operate at similar voltages and power levels. Therefore, the conductors and cables of two or more power-limited fire protective signaling circuits or Class 3 circuits may occupy the same enclosure. However, power-limited fire-protective signaling circuits and Class 2 circuits differ in operating voltages and insulation, and they are permitted in the same enclosure only when the insulation of the Class 2 conductors is comparable to that of the fire-protective circuit. See Figure 129.

Figure 129 – Well-laid-out fire control panel has barriers to minimize wiring errors. *Source: The Protectowire Co., Inc., Hanover, Massachusetts.*

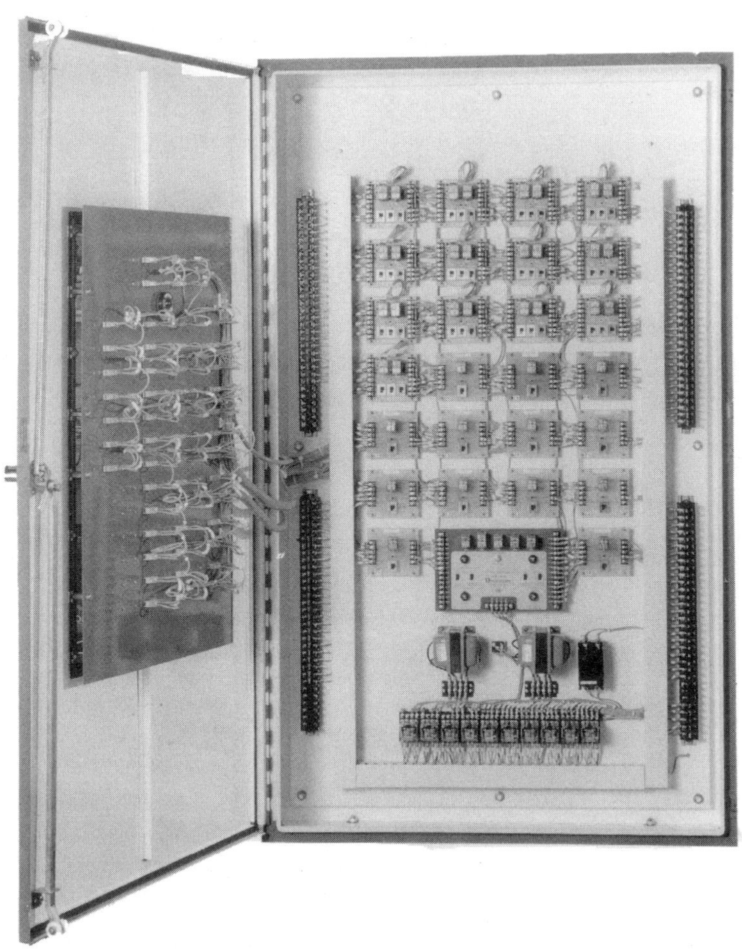

(d)(4) **Identification.** Fire alarm circuits shall be identified at terminal and junction locations in a manner which will prevent unintentional interference with the signaling circuit during testing and servicing. Power-limited fire alarm circuits shall be durably marked as such where plainly visible at terminations.

Identification

Identification for fire protective signaling circuits is required so that these systems are not interfered with during maintenance operations. Because these are essential systems, this requirement is intended to protect signaling circuits while work is being performed on other systems or while the signaling circuit itself is being serviced, avoiding damage to the circuit and false alarms. See Figure 130. This fire alarm system is identified.

For power-limited fire-protective signaling circuits, the marking must indicate that the circuit is a power-limited fire-protective signaling circuit. This rule is intended to ensure that the power-limited fire circuits, which operate at lower power and voltage levels, are not confused with other circuits operating at higher voltages.

Figure 130 – Identification of fire protective signaling circuits. *Source: Bert Cohn*

(e) **Communications systems.** This paragraph applies to central-station-connected and non-central-station-connected telephone circuits, radio and television receiving and transmitting equipment, including community antenna television and radio distribution systems, telegraph, district messenger, and outside wiring for fire and burglar alarm, and similar central station systems. These installations need not comply with the provisions of § 1910.303 through § 1910.308(d) except for § 1910.304(c)(1) and § 1910.307.

1910.308

Title: Special Systems.

Subpart: S

SubPart Title: Electrical - Design Safety Standards for Electrical Systems

(e)(1)	**Protective devices.**
(e)(1)(i)	A listed primary protector shall be provided on each circuit run partly or entirely in aerial wire or aerial cable not confined within a block.
(e)(1)(ii)	A listed primary protector shall be also provided on each aerial or underground circuit when the location of the circuit within the block containing the building served allows the circuit to be exposed to accidental contact with electric light or power conductors operating at over 300 volts to ground.
(e)(1)(iii)	In addition, where there exists a lightning exposure, each interbuilding circuit on premises shall be protected by a listed primary protector at each end of the interbuilding circuit.

Figure 131 – Protection of communication system from accidental contact with power conductors.

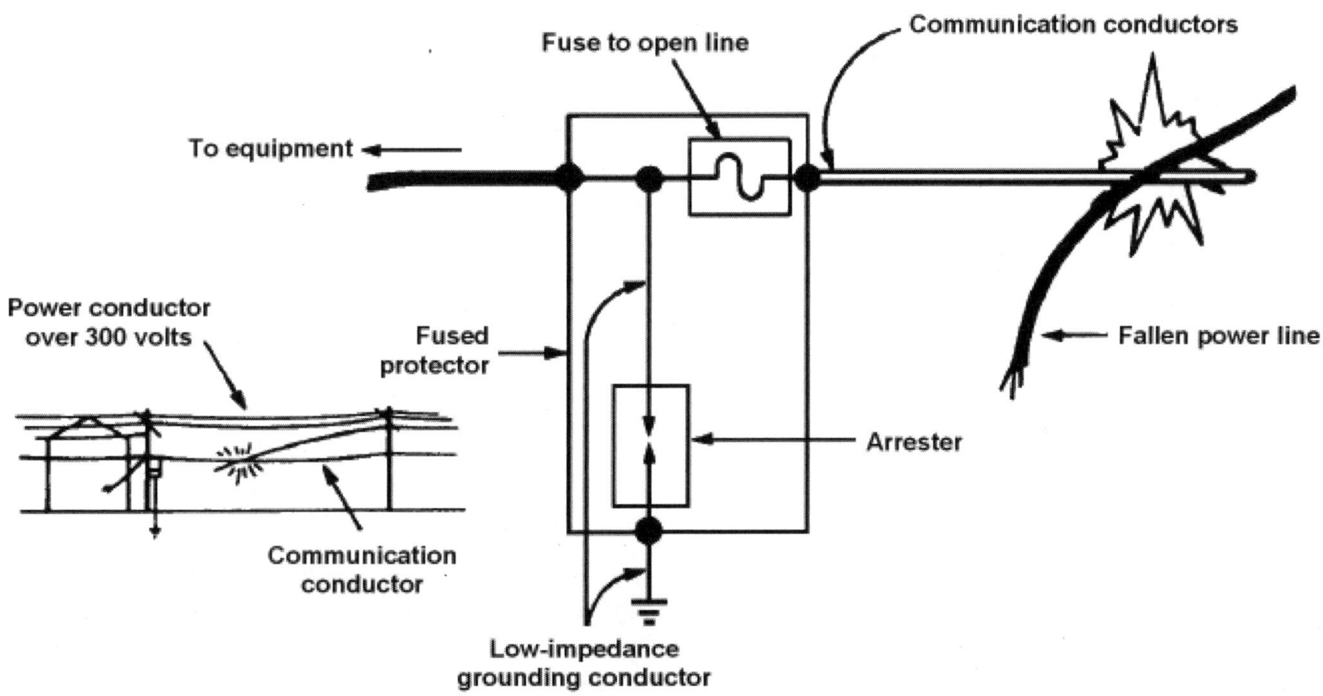

If a power line operating at over 300 volts accidentally contacts the communication conductor, the protective arrester allows the excess current to travel to ground instead of along the communication conductor. The fuse then opens the circuit to the equipment.

Communication Systems—Protective Devices

Communication circuits near light and power conductors that operate at over 300 volts must be provided with a protective device that will stop current flow from the power lines through the communication lines. Fused arresters on each communication line are commonly used protective devices. If high voltage power lines on poles come down in a storm and accidentally contact the communication lines that share the same pole, these protective devices will not allow the high voltage to be impressed on the communication circuit. Protectors are required by the National Electrical Code® to be located as near as practicable to the power conductors and inside, on, or adjacent to the buildings being serviced.

Figure 131 illustrates protection from accidental contact with power lines operating at over 300 volts with a low-impedance protective grounding conductor and arrester.

(e)(2) **Conductor location.**

(e)(2)(i) Lead-in or aerial-drop cables from a pole or other support, including the point of initial attachment to a building or structure, shall be kept away from electric light, power, Class 1, or nonpower-limited fire alarm circuit conductors so as to avoid the possibility of accidental contact.

(e)(2)(ii) A separation of at least 1.83 m (6 ft) shall be maintained between communications wires and cables on buildings and lightning conductors.

(e)(2)(iii) Where communications wires and cables and electric light or power conductors are supported by the same pole or run parallel to each other in-span, the following conditions shall be met:

(e)(2)(iii)(A) Where practicable, communication wires and cables on poles shall be located below the electric light or power conductors; and

(e)(2)(iii)(B) Communications wires and cables may not be attached to a crossarm that carries electric light or power conductors.

(e)(2)(iv) Indoor communications wires and cables shall be separated at least 50.8 mm (2 in.) from conductors of any electric light, power, Class 1, nonpower-limited fire alarm, or medium power network-powered broadband communications circuits, unless a special and equally protective method of conductor separation, identified for the purpose, is employed.

(e)(3) **Equipment location.** Outdoor metal structures supporting antennas, as well as self-supporting antennas such as vertical rods or dipole structures, shall be located as far away from overhead conductors of electric light and power circuits of over 150 volts to ground as necessary to prevent the antenna or structure from falling into or making accidental contact with such circuits.

(e)(4) **Grounding.**

(e)(4)(i) If exposed to contact with electric light and power conductors, the metal sheath of aerial cables entering buildings shall be grounded or shall be interrupted close to the entrance to the building by an insulating joint or equivalent device. Where protective devices are used, they shall be grounded in an approved manner.

Lead-in Conductor Grounding

Where aerial communication conductors enter buildings and are located near light and power conductors, the metal sheath covering these cables must be effectively grounded or its continuity must be interrupted before reaching the building. Either method will prevent the presence of power line voltage on equipment in case a power line accidentally contacts the communication conductors. In addition, protective devices or arresters must also be properly grounded because of unusual current surges and power levels that may occur. Proper grounding means that grounding electrode conductors must have approved insulation, be no smaller than no. 18 copper, be run in straight lines, be protected from physical damage, be connected to the building grounding electrode system or equivalent at the nearest accessible location, and be connected with bolted clamps or bonded.

(e)(4)(ii) Masts and metal structures supporting antennas shall be permanently and effectively grounded without splice or connection in the grounding conductor.

(e)(4)(iii) Transmitters shall be enclosed in a metal frame or grill or separated from the operating space by a barrier, all metallic parts of which are effectively connected to ground. All external metal handles and controls accessible to the operating personnel shall be effectively grounded. Unpowered equipment and enclosures are considered to be grounded where connected to an attached coaxial cable with an effectively grounded metallic shield.

(f) **Solar photovoltaic systems**. This paragraph covers solar photovoltaic systems that can be interactive with other electric power production sources or can stand alone with or without electrical energy storage such as batteries. These systems may have ac or dc output for utilization.

(f)(1) **Conductors of different systems**. Photovoltaic source circuits and photovoltaic output circuits may not be contained in the same raceway, cable tray, cable, outlet box, junction box, or similar fitting as feeders or branch circuits of other systems, unless the conductors of the different systems are separated by a partition or are connected together.

(f)(2) **Disconnecting means**. Means shall be provided to disconnect all current-carrying conductors of a photovoltaic power source from all other conductors in a building or other structure. Where a circuit grounding connection is not designed to be automatically interrupted as part of the ground-fault protection system, a switch or circuit breaker used as disconnecting means may not have a pole in the grounded conductor.

The use of solar photovoltaic systems as alternative power methods has increased dramatically over the past few years as technology and the availability of equipment has become better. Article 690 of the NEC® covers the installation and use of such systems in detail, and continues to grow as these systems become more and more popular.

(g) **Integrated electrical systems.**

(g)(1) **Scope**. Paragraph (g) of this section covers integrated electrical systems, other than unit equipment, in which orderly shutdown is necessary to ensure safe operation. An integrated electrical system as used in this section shall be a unitized segment of an industrial wiring system where all of the following conditions are met:

(g)(1)(i) An orderly shutdown process minimizes employee hazard and equipment damage;

(g)(1)(ii) The conditions of maintenance and supervision ensure that only qualified persons will service the system; and

(g)(1)(iii) Effective safeguards are established and maintained.

(g)(2) **Location of overcurrent devices in or on premises**. Overcurrent devices that are critical to integrated electrical systems need not be readily accessible to employees as required by § 1910.304(f)(1)(iv) if they are located with mounting heights to ensure security from operation by nonqualified persons.

In some industrial processes, the loss of electric power to critical equipment is unacceptable, and an orderly shutdown process is essential to preventing injury or extreme equipment damage. Integrated electrical systems are often used in nuclear facilities, smelters, mills and similar locations, but must be installed and maintained under strict engineering supervision.

The following Sections have been reserved for future use by OSHA:

1910.309	Reserved
1910.310	Reserved
1910.311	Reserved
1910.312	Reserved
1910.313	Reserved
1910.314	Reserved
1910.315	Reserved
1910.316	Reserved
1910.317	Reserved
1910.318	Reserved
1910.319	Reserved
1910.320	Reserved
1910.321	Reserved
1910.322	Reserved
1910.323	Reserved
1910.324	Reserved
1910.325	Reserved
1910.326	Reserved
1910.327	Reserved
1910.328	Reserved
1910.329	Reserved
1910.330	Reserved

Subpart S
Electrical - General

Authority: Authority: Secs. 4, 6, 8, Occupational Safety and Health Act of 1970 (29 U.S.C. 653, 655, 657); Secretary of Labor's Order No. 8-76 (41 FR 25059), 1-90 (55 FR 9033), 5-2002 (67 FR 65008), 5-2007 (72 FR 31160), as applicable; 29 CFR part 1911.

Source: 46 FR 4056, Jan. 16, 1981, unless otherwise noted.

Part 2
Electrical Safety-Related Work Practices

§ 1910.331 Electrical Safety-Related Work Practices

[55 FR 32016, Aug. 6, 1990; 59 FR 4476, Jan 31, 1994]

Scope.

Division (b) of Subpart S, Electrical Safety Related Work Practices was first published for public comment in 1987. The final rules were published on August 6, 1990 and became effective on December 4, 1990 (except for article 1910.332, which became effective on August 6, 1990). This performance-oriented standard sets forth new levels for the training of employees and requires the development of specific safe work procedures. The standard is based on the 1985 edition of NFPA 70E, part II, Electrical Safety Requirements for Employee Workplaces (now updated to Chapter 1 of NFPA 70E, Standard for Electrical Safety in the Workplace, 2009 Edition) and ANSI Z244.1, Control of Hazardous Energy - Lockout/Tagout and Alternative Methods. The standard covers safety-related work practices for both qualified and unqualified persons.

Qualified persons — those who have skills and knowledge related to construction and operation of the electrical equipment and installations ***and*** have been trained to recognize avoid the electrical hazards involved with working on or near electricity and electrical equipment.

Unqualified persons — those workers with little or no such skills, knowledge and training who work on or n195ear the following four types of insta195llations.

(a) **Covered work by both qualified and unqualified persons.** The provisions of §§ 1910.331 through 1910.335 cover electrical safety-related work practices for both qualified persons (those who have training in avoiding the electrical hazards of working on or near exposed energized parts) and unqualified persons (those with little or no such training) working on, near, or with the following installations:

(a)(1) **Premises wiring.** Installations of electric conductors and equipment within or on buildings or other structures, and on other premises such as yards, carnival, parking, and other lots, and industrial substations;

(a)(2) **Wiring for connection to supply.** Installations of conductors that connect to the supply of electricity; and

(a)(3) **Other wiring.** Installations of other outside conductors on the premises.

(a)(4) **Optical fiber cable.** Installations of optical fiber cable where such installations are made along with electric conductors.

Title: Scope

Subpart: S

SubPart Title: Electrical - Safety-Related Work Practices

Note: See § 1910.399 for the definition of "qualified person." See § 1910.332 for training requirements that apply to qualified and unqualified persons.

Installations covered by the standard

(1) **Premises wiring**—interior and exterior wiring, including power lighting, signaling and control wiring and all associated hardware.

(2) **Wiring for connection to the electrical source or supply.**

(3) **Other wiring**—other exterior wiring on the premises.

(4) **Optical fiber cable**—where such installations are made near or adjacent to electrical conductors.

At the time OSHA published the final rule, the fourth edition of NFPA 70E had been published. This was followed by the fifth edition, published in 1995. Although both the OSHA regulations and NFPA 70E stress the need for improved training and development of specific safe work practices, they generated many questions regarding compliance. This was especially true regarding the use of personal protective equipment (PPE) when performing various tasks that exposed an employee to an electrical hazard.

Many of these questions were answered by the sixth edition of NFPA 70E. Part II of the standard provided much needed guidance regarding the selection of PPE, and was further extensively expanded upon by Chapter 1 of the 2004 and now 2009 Editions. NFPA 70E also goes beyond OSHA requirements in many areas, providing for a more effective electrical safety program. Every safety manager and electrical maintenance supervisor should have a copy of NFPA 70E and become familiar with it, especially the requirements stated in Chapter 1 - Safety-Related Work Practices.

(b) **Other covered work by unqualified persons.** The provisions of §§ 1910.331 through 1910.335 also cover work performed by unqualified persons, on, near, or with the installations listed in paragraphs (c)(1) through (c)(4) of this section.

Although the following paragraphs exclude certain work or industries from coverage, all unqualified persons working in these industries *are covered* by 1910.331 through 1910.335. As explained below, other standards cover these industries and may include additional requirements for unqualified persons unique to the specific industry.

(c) **Excluded work by qualified persons.** The provisions of §§ 1910.331 through 1910.335 do not apply to work performed by qualified persons on or directly associated with the following installations:

(c)(1) **Generation, transmission, and distribution of electric energy** (including communication and metering) located in buildings used for such purposes or located outdoors.

Note 1: Work on or directly associated with installations of utilization equipment used for purposes other than generating, transmitting, or distributing electric energy (such as installations which are in office buildings, warehouses, garages, machine shops, or recreational buildings, or other utilization installations which are not an integral part of a generating installation, substation, or control center) is covered under paragraph (a)(1) of this section.

Note 2: For work on or directly associated with utilization installations, an employer who complies with the work practices of 1910.269 (electric power generation, transmission, and distribution) will be deemed to be in compliance with 1910.333(c) and 1910.335. However, the requirements of 1910.332, 1910.333(a), 1910.333(b), and 1910.334 apply to all work on or directly associated with utilization installations, regardless of whether the work is performed by qualified or unqualified persons.

Note 3: Work on or directly associated with generation, transmission, or distribution installations includes:

(1) Work performed directly on such installations, such as repairing overhead or underground distribution lines or repairing a feed-water pump for the boiler in a generating plant.

(2) Work directly associated with such installations, such as line-clearance tree trimming and replacing utility poles.

(3) Work on electric utilization circuits in a generating plant provided that:

(A) Such circuits are commingled with installations of power generation equipment or circuits, and

(B) The generation equipment or circuits present greater electrical hazards than those posed by the utilization equipment or circuits (such as exposure to higher voltages or lack of overcurrent protection).

This work is covered by 1910.269 of this Part.

See OSHA Subpart R, Special Industries, § 1910.269 and the National Electrical Safety Code, for work practices relating to operation and maintenance of electric power generation, control, transmission and distribution lines and equipment.

(c)(2) **Communications installations.** Installations of communication equipment to the extent that the work is covered under 1910.268.

See OSHA Subpart R, Special Industries, § 1910.268, for work conditions, practices, installations, etc. at telecommunications centers and telecommunications field installations.

(c)(3) **Installations in vehicles.** Installations in ships, watercraft, railway rolling stock, aircraft or automotive vehicles other than mobile homes and recreational vehicles.

(c)(4) **Railway installations.** Installations of railways for generation, transformation, transmission, or distribution of power used exclusively for operation of rolling stock or installations of railways used exclusively for signaling and communication purposes.

Installations not covered — these provisions do not apply to work practices and related activities for ships, watercraft, railway rolling stock, aircraft or automotive vehicles other than mobile homes and recreational vehicles.

§ 1910.332 Training

[55 FR 32016, Aug. 6, 1990]

(a) **Scope.** The training requirements contained in this section apply to employees who face a risk of electric shock that is not reduced to a safe level by the electrical installation requirements of §§ 1910.303 through 1910.308.

Note: Employees in occupations listed in Table S-4 face such a risk and are required to be trained. Other employees who also may reasonably be expected to face a comparable risk of injury due to electric shock or other electrical hazards must also be trained.

Table S-4

Typical Occupational Categories of Employees Facing a Higher Than Normal Risk of Electrical Accidents

Occupation
Blue collar supervisors. [1]
Electrical and electronic engineers. [1]
Electrical and electronic equipment assemblers. [1]
Electrical and electronic technicians. [1]
Electricians.
Industrial machine operators. [1]
Material handling equipment operators. [1]
Mechanics and repairers. [1]
Painters. [1]
Riggers and roustabouts. [1]
Stationary engineers. [1]
Welders.

[1] Workers in these groups do not need to be trained if their work or the work of those they supervise does not bring them or the employees they supervise close enough to exposed parts of electric circuits operating at 50 volts or more to ground for a hazard to exist.

(b) **Content of training.**

(b)(1) **Practices addressed in this standard.** Employees shall be trained in and familiar with the safety-related work practices required by §§ 1910.331 through 1910.335 that pertain to their respective job assignments.

(b)(2) **Additional requirements for unqualified persons.** Employees who are covered by paragraph (a) of this section but who are not qualified persons shall also be trained in and familiar with any electrically related safety practices not specifically addressed by §§ 1910.331 through 1910.335 but which are necessary for their safety.

(b)(3) **Additional requirements for qualified persons.** Qualified persons (i.e. those permitted to work on or near exposed energized parts) shall, at a minimum, be trained in and familiar with the following:

(b)(3)(i) The skills and techniques necessary to distinguish exposed live parts from other parts of electric equipment.

(b)(3)(ii) The skills and techniques necessary to determine the nominal voltage of exposed live parts, and

(b)(3)(iii) The clearance distances specified in § 1910.333(c) and the corresponding voltages to which the qualified person will be exposed.

Note 1: For the purpose of §§ 1910.331 through 1910.335, a person must have the training required by paragraph (b)(3) of this section in order to be considered a qualified person.

Note 2: Qualified persons whose work on energized equipment involves either direct contact or contact by means of tools or materials must also have the training needed to meet § 1910.333(c)(2).

(c) **Type of training.** The training required by this section shall be of the classroom or on-the-job type. The degree of training provided shall be determined by the risk to the employee.

Training

The training requirements in this standard apply to any and all employees who are faced with a risk of electrical shock, electrocution, burns, arc flash and blast, or other electrical hazard(s). Table S-4 in the standard lists job titles of employees who have a higher than normal risk of shock or other electrical hazard.

All employees shall be trained in the safety-related work practices required by articles §§1910.331-1910.335 as they pertain to their respective job assignments.

In addition, *unqualified personnel* shall be trained in *any* safety-related practice necessary for the safe performance of their job.

Qualified personnel shall, at a minimum, be trained in the following areas:

(1) The ability to distinguish between exposed live parts and other parts of electrical equipment.

(2) The skills and techniques necessary to determine the nominal voltage of exposed live parts.

(3) The required working clearances or safe working distances as defined in 1910.333(c) (Table S-5 Approach Distances for Qualified Employees - Alternating Current.)

These three items provide the minimum training requirements under OSHA for a worker to be considered as being a "qualified person."

The 2009 Edition of NFPA 70E adds a fourth requirement in Section 110.6(D)(1)(b):

"The decision-making process necessary to determine the degree and extent of the hazard and the personal protective equipment and job planning necessary to perform the task safely."

This requirement indicates that qualified persons must have the knowledge necessary to make decisions about how to perform their work safely.

If a qualified worker interacts with energized equipment, either directly or indirectly (as through the use of tools and/or other materials), then he or she must be trained in the use of special precautionary techniques, such as personal protective equipment, insulation and shielding materials, and insulated tools.

The 2009 Edition of NFPA 70E (Section 110.6(D)(1)(e)) now requires that qualified persons be able to *demonstrate* some of these skills, as well.

All training sessions should be documented (also now required by the 2009 Edition of NFPA 70E, Section 110.6(E)) in detail as to the date, subject content, and the name of all employees who participated. At the end of each session, the employer should have every employee sign a statement acknowledging their participation in the training. Updates or refresher sessions should be scheduled as determined to be necessary by the employer. It is also important to note that as of this publication, proposals have been accepted to add to the next addition of NFPA 70E a requirement for retraining of personnel in safe work practices a minimum of every three years. The training can be conducted either in the classroom or on the job with actual hands-on-type instruction, or a combination of both.

The depth of training necessary for each employee is determined by the risk faced by that employee.

It is important that management realize the difference between technical training and the safety-related work practice training. There is usually considerable pressure to make sure maintenance personnel are technically qualified, since this directly affects their efficiency. However, OSHA's primary concern is not with task efficiency but how safely a worker can perform a task.

Many electrical maintenance persons receive regular training in the National Electrical Code® and related standards. Such classes do not always cover the safety-related work practice issues involved. Therefore, the content of the training provided should be closely monitored so that workers receive the proper safety-related work practice training specific to their job tasks.

It is critical that employees and employers alike recognize that training alone does not make a person qualified. It is a combination of their skills, knowledge and training, *specific* to the task the worker is assigned. It is then the employers' responsibility to document who their qualified persons are, and what tasks those persons are qualified for. NFPA 70E makes it clear that a person can be considered qualified with respect to certain equipment and methods, yet remain unqualified for others (NFPA 70E, 2009 Edition, 110.6(D)(1)(a)).

Table S-4 should not be the primary guiding factor in determining who needs safety-related work practice training. This table was necessary to show that, as a minimum, there was a substantial need for training. However, accident data confirms that every employee needs some degree of electrical safety-related work practice training, including office workers. When addressing the needs of such people as office workers, it may be more appropriate to refer to the training as *electrical safety-awareness* training so as not to overwhelm the employee. Training for such employees could easily be accomplished in as little as two hours and repeated on a regular basis. It is the responsibility of the employer to determine who should receive training and to what extent.

§ 1910.333 Selection and Use of Work Practices

[55 FR 32016, Aug. 6, 1990; 55 FR 42053, Nov. 1, 1990; as amended at 59 FR 4476, Jan. 31, 1994]

(a) **General.** Safety-related work practices shall be employed to prevent electric shock or other injuries resulting from either direct or indirect electrical contacts, when work is performed near or on equipment or circuits which are or may be energized. The specific safety-related work practices shall be consistent with the nature and extent of the associated electrical hazards.

(a)(1) **Deenergized parts.** Live parts to which an employee may be exposed shall be deenergized before the employee works on or near them, unless the employer can demonstrate that deenergizing introduces additional or increased hazards or is infeasible due to equipment design or operational limitations. Live parts that operate at less than 50 volts to ground need not be deenergized if there will be no increased exposure to electrical burns or to explosion due to electric arcs.

Note 1: Examples of increased or additional hazards include interruption of life support equipment, deactivation of emergency alarm systems, shutdown of hazardous location ventilation equipment, or removal of illumination for an area.

Note 2: Examples of work that may be performed on or near energized circuit parts because of infeasibility due to equipment design or operational limitations include testing of electric circuits that can only be performed with the circuit energized and work on circuits that form an integral part of a continuous industrial process in a chemical plant that would otherwise need to be completely shut down in order to permit work on one circuit or piece of equipment.

Note 3: Work on or near deenergized parts is covered by paragraph (b) of this section.

1910.333(a)(1) states the required method of preventing electrical injury or fatalities, and that is to deenergize live parts to which an employee may be exposed before the employee works on or near them. To deviate from this requirement, the employer must demonstrate that deenergizing introduces additional or increased hazards or is infeasible due to equipment design or operational limitations. Notes 1 and 2 provide some examples of increased hazard or infeasibility. Normally, lost production is not an acceptable reason.

(a)(2) **Energized parts.** If the exposed live parts are not deenergized (i.e., for reasons of increased or additional hazards or infeasibility), other safety-related work practices shall be used to protect employees who may be exposed to the electrical hazards involved. Such work practices shall protect employees against contact with energized circuit parts directly with any part of their body or indirectly through some other conductive object. The work practices that are used shall be suitable for the conditions under which the work is to be performed and for the voltage level of the exposed electric conductors or circuit parts. Specific work practice requirements are detailed in paragraph (c) of this section.

Selection and Use of Work Practices

The Fundamental goal of § 1910.333 is to develop safety-related work practices to prevent shock, electrocution or other injuries resulting from direct or indirect contact while working with electrical circuits and equipment.

In addition to the contact hazards (shock / electrocution) typically addressed by safety programs, careful attention must be given to the prevention of injury or death due to arc burns and arc blast. Numerous studies indicate that a very high percentage of serious injury or deaths are occurring as a result of an arc flash/blast. Some data indicates it is as high as 80 percent among qualified persons performing electrical work.

Many of the workers killed and injured could have been saved had they been wearing the proper personal protective equipment. However, the proper PPE cannot be selected unless the hazard is recognized. An effective electrical safety program will require thorough hazard analyzes and selection of PPE.

While OSHA only mandates that the employer address the hazard in a manner which protects employee from injury and/or death, NFPA 70E specifies specific methods of determining the shock, electrocution and burn hazards, as discussed in the next few pages.

Additionally, it is important for engineering and design personnel to recognize that these hazards exist and strive to remove serious arc flash and blast hazards through engineering designs, such as arc-resistant equipment, fiber-optic relaying and low voltage controls.

The last sentence in 1910.333(a)(1) states that live parts that operate at less than 50 volts to ground need not be deenergized if there will be no increased exposure to electrical burns or to explosion due to electric arcs. This is an area frequently overlooked. Batteries, such those used in material handling equipment, and DC power supplies for plating systems are typically under 50 volts, for example. However, they have the capability of providing significant amounts of current.

(b) **Working on or near exposed deenergized parts.**

(b)(1) **Application.** This paragraph applies to work on exposed deenergized parts or near enough to them to expose the employee to any electrical hazard they present. Conductors and parts of electric equipment that have been deenergized but have not been locked out or tagged in accordance with paragraph (b) of this section shall be treated as energized parts, and paragraph (c) of this section applies to work on or near them.

(b)(2) **Lockout and tagging.** While any employee is exposed to contact with parts of fixed electric equipment or circuits which have been deenergized, the circuits energizing the parts shall be locked out or tagged or both in accordance with the requirements of this paragraph. The requirements shall be followed in the order in which they are presented (i.e., paragraph (b)(2)(i) first, then paragraph (b)(2)(ii), etc.)

Note 1: As used in this section, fixed equipment refers to equipment fastened in place and connected by permanent wiring methods.

Note 2: Lockout and tagging procedures that comply with paragraphs (c) through (f) of § 1910.147 will also be deemed to comply with paragraph (b)(2) of this section provided that:

(1) The procedures address the electrical safety hazards covered by this Subpart; and

(2) The procedures also incorporate the requirements of paragraphs (b)(2)(iii)(D) and (b)(2)(iv)(B) of this section.

(b)(2)(i) **Procedures.** The employer shall maintain a written copy of the procedures outlined in paragraph (b)(2) and shall make it available for inspection by employees and by the Assistant Secretary of Labor and his or her authorized representatives.

Note: The written procedures may be in the form of a copy of paragraph (b) of this section.

(b)(2)(ii) **Deenergizing equipment.**

(b)(2)(ii)(A) Safe procedures for deenergizing circuits and equipment shall be determined before circuits or equipment are deenergized.

(b)(2)(ii)(B) The circuits and equipment to be worked on shall be disconnected from all electric energy sources. Control circuit devices, such as push buttons, selector switches, and interlocks, may not be used as the sole means for deenergizing circuits or equipment. Interlocks for electric equipment may not be used as a substitute for lockout and tagging procedures.

(b)(2)(ii)(C) Stored electric energy which might endanger personnel shall be released. Capacitors shall be discharged and high capacitance elements shall be short-circuited and grounded, if the stored electric energy might endanger personnel.

Note: If the capacitors or associated equipment are handled in meeting this requirement, they shall be treated as energized.

(b)(2)(ii)(D) Stored non-electrical energy in devices that could reenergize electric circuit parts shall be blocked or relieved to the extent that the circuit parts could not be accidentally energized by the device.

(b)(2)(iii) **Application of locks and tags.**

(b)(2)(iii)(A) A lock and a tag shall be placed on each disconnecting means used to deenergize circuits and equipment on which work is to be performed, except as provided in paragraphs (b)(2)(iii)(C) and (b)(2)(iii)(E) of this section. The lock shall be attached so as to prevent persons from operating the disconnecting means unless they resort to undue force or the use of tools.

(b)(2)(iii)(B)	Each tag shall contain a statement prohibiting unauthorized operation of the disconnecting means and removal of the tag.
(b)(2)(iii)(C)	If a lock cannot be applied, or if the employer can demonstrate that tagging procedures will provide safety equivalent to that of a lock, a tag may be used without a lock.
(b)(2)(iii)(D)	A tag used without a lock, as permitted by paragraph (b)(2)(iii)(C) of this section, shall be supplemented by at least one additional safety measure that provides a level of safety equivalent to that obtained by the use of a lock. Examples of additional safety measures include the removal of an isolating circuit element, blocking of a controlling switch, or opening of an extra disconnecting device.
(b)(2)(iii)(E)	A lock may be placed without a tag only under the following conditions:
(b)(2)(iii)(E)(1)	Only one circuit or piece of equipment is deenergized, and
(b)(2)(iii)(E)(2)	The lock period does not extend beyond the work shift, and
(b)(2)(iii)(E)(3)	Employees exposed to the hazard associated with reenergizing the circuit or equipment are familiar with this procedure.
(b)(2)(iv)	**Verification of deenergized condition.** The requirements of this paragraph shall be met before any circuits or equipment can be considered and worked as deenergized.
(b)(2)(iv)(A)	A qualified person shall operate the equipment operating controls or otherwise verify that the equipment cannot be restarted.
(b)(2)(iv)(B)	A qualified person shall use test equipment to test the circuit elements and electrical parts of equipment to which employees will be exposed and shall verify that the circuit elements and equipment parts are deenergized. The test shall also determine if any energized condition exists as a result of inadvertently induced voltage or unrelated voltage backfeed even though specific parts of the circuit have been deenergized and presumed to be safe. If the circuit to be tested is over 600 volts, nominal, the test equipment shall be checked for proper operation immediately after this test.
(b)(2)(v)	**Reenergizing equipment.** These requirements shall be met, in the order given, before circuits or equipment are reenergized, even temporarily.
(b)(2)(v)(A)	A qualified person shall conduct tests and visual inspections, as necessary, to verify that all tools, electrical jumpers, shorts, grounds, and other such devices have been removed, so that the circuits and equipment can be safely energized.
(b)(2)(v)(B)	Employees exposed to the hazards associated with reenergizing the circuit or equipment shall be warned to stay clear of circuits and equipment.
(b)(2)(v)(C)	Each lock and tag shall be removed by the employee who applied it or under his or her direct supervision. However, if this employee is absent from the workplace, then the lock or tag may be removed by a qualified person designated to perform this task provided that:
(b)(2)(v)(C)(1)	The employer ensures that the employee who applied the lock or tag is not available at the workplace, and
(b)(2)(v)(C)(2)	The employer ensures that the employee is aware that the lock or tag has been removed before he or she resumes work at that workplace.
(b)(2)(v)(D)	There shall be a visual determination that all employees are clear of the circuits and equipment.

1910.333

Title: Selection and use of work practices.

Subpart: S

SubPart Title: Electrical - Safety-Related Work Practices

Working on Deenergized and Energized Systems

As stated in 1910.333(a)(2), the required method is to deenergize all of the electrical power supplying the circuits or equipment and to apply lockout procedures to the electrical disconnect(s). It is the function of the lock to secure the disconnecting means or other energy-isolating device in the open (deenergized) or safe position. It is not sufficient to lockout push buttons, equipment interlocks and other control circuits. It is necessary to physically interrupt the power or current carrying conductors leading to circuits or equipment. Potential sources of electrical energy storage, such as capacitors and inductors, need to be dissipated and grounded. Sources of non-electrical energy also need to be blocked or released as discussed below.

The lockout/tagout procedures are developed to prevent an accidental startup or release of energy into the circuit or equipment while it is being serviced. OSHA's definitive standard on lockout/tagout procedures is presented in CFR1910.147, The Control of Hazardous Energy (Lockout/Tagout). This generic lockout/tagout standard covers all forms of potential energy release including:

- electrical
- compressed air
- chemical sources
- pressurized hydraulic systems
- steam and other thermal storage systems
- gravitational potential energy storage devices such as in a punch press
- inertial energy devices such as in a flywheel

The idea is that each employee working on electrical circuits and equipment will have his or her own lock with a single key (maintained in his/her possession) to insure that power is not accidentally restored to the system before he or she is ready and is free and clear of the equipment. When multiple people are working on a circuit or equipment, either together or independently, each applies their own lock so that each individual has control and knowledge of when the system is to be restarted. The tag is used to identify the owner of the lock and to inform or warn that the circuit (or equipment) is being worked on. Figure 132A shows a typical group lockout/tagout device. Locks and tags are to be removed only by their owner or the worker's supervisor under certain conditions if the employer certifies that the employee is not available at the work site. Other means of complying with the lockout rules are shown in Figures 132B and 132C.

Figure 132A – Group lockout/tagout of electrical disconnecting means. *Source: Master Lock Co., Milwaukee, Wis.*

Figure 132B – Cable lockout device for several electrical disconnects. Source: Brady USA Inc., Signmark Division, Milwaukee, Wis.

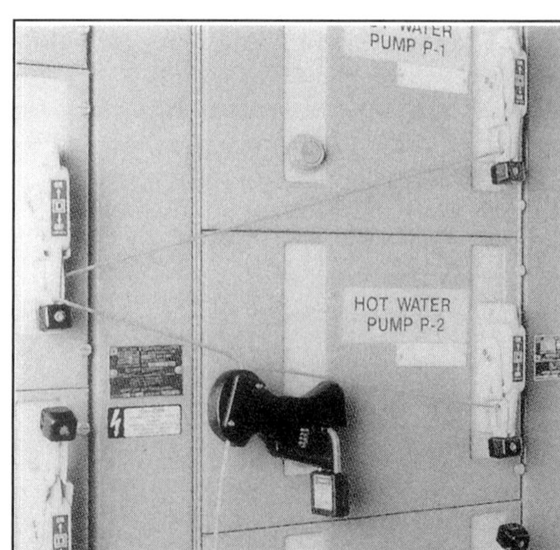

Figure 132C – Lockout procedures affixed to a pump motor. *Source: Brady USA, Inc., Signmark Division, [Milwaukee, Wis.*

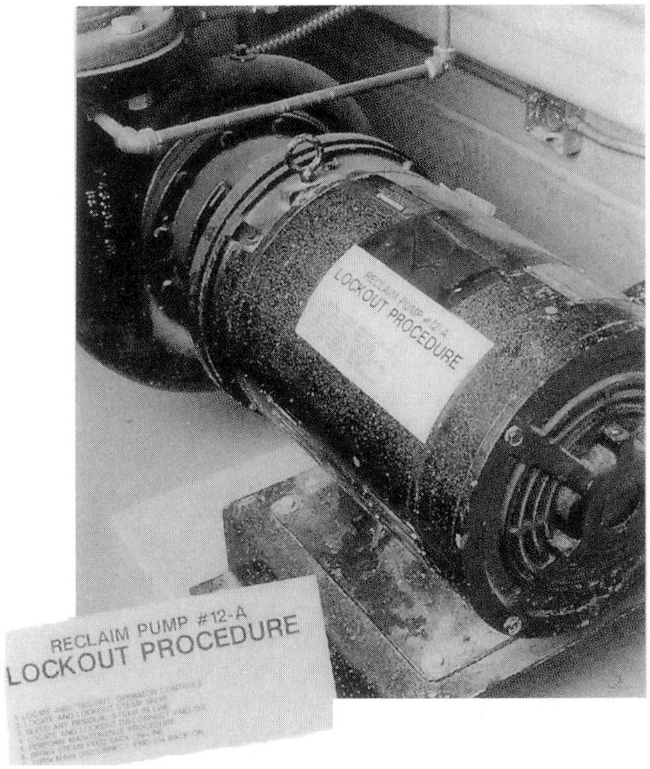

A lock without a tag can be used if only one circuit or one piece of equipment is to be deenergized, the lockout period is to be wholly contained within one work shift, and the employees exposed to such hazards are familiar with the procedures.

It is the responsibility of the employer to develop a detailed set of lockout/tagout procedures for each type of machine or electrical circuit in service. These procedures need to be available for inspection by employees and OSHA. All affected employees need to be trained in their use. It is the employer's responsibility to assure that these designated procedures are being followed by their employees. Although not required in Subpart S, NFPA 70E, 2009 Edition (Section 120.2(C)(3)) also requires and annual audit of these procedures in harmony with 1910.147.

The detail necessary to be specified for these procedures is suggested by the following procedural outline:

1. Warn affected personnel of impending shutdown.
2. Prepare for equipment shutdown.
3. Equipment shutdown.
4. Isolation of hazardous energy sources.
5. Lockout/tagout(s) applied.
6. Verification of deenergized circuit(s) (and removal of other energy sources).
7. Conduct service on the equipment.
8. A qualified person determines that all tools, electrical jumpers, shorts, temporary grounds, and other devices have been removed.
9. Affected personnel are warned of impending startup.
10. Lockout/tagouts are removed by person(s) who applied visual inspection to ensure that all personnel are clear.
11. Restore energy sources.
12. Equipment startup.

Major problem areas are the details of how to coordinate group lockout/tagout projects; lockout/tagout procedures which extend over multiple shifts; employees who are present on shutdown and install a lock but are absent at startup time; and communication between or among multiple work groups, including outside contractors. The employer is responsible for developing and maintaining all of these procedures and is responsible for ensuring that everyone, including all outside personnel such as contractors, is knowledgeable about, and participates in, the lockout/tagout program. The employer is also responsible for providing all necessary training. In addition, the employer is responsible for auditing the lockout/tagout procedures and correcting any deficiencies found. Periodic review and provisions for retraining are good items to include in the overall plan.

NFPA 70E (2009), Article 120, provides more detailed guidance on "establishing an electrically safe work condition," which follows along the methodology of the OSHA regulations, but provide additional detail for instituting specific lockout/tagout procedures as required here. Article 120 includes six basic steps to establishing a deenergized condition, including verifying that the circuits are indeed deenergized (at zero voltage). Such determination shall be made by a qualified person who will 1) operate the system controls and verify that the system cannot be restarted or reenergized, and 2) verify by using appropriate test instruments. NFPA 70E (2009) specifies that the qualified person will perform both a phase-to-phase and a phase-to-ground measurement with the appropriate test instrument (voltage meter). NFPA 70E (2009), Section 110.9(A) also provides direction on the proper rating, design and inspection of such test equipment consistent with 1910.334(C). The measuring or test equipment must be tested both before and after the deenergized circuit measurement is made to ensure that the test equipment is operating properly.

An alternate method

As stated above, lockout is the preferred method of controlling employee exposure to electrical energy hazards. Tagout is an alternate method that can be used by an employer. Under OSHA's requirements, a tag can be used in place of a lock only under two conditions: (1) A lock cannot be applied due to the design of the equipment, or (2) The employer can demonstrate that a tag provides the same level of safety as a lock. The tag must warn that the switch is not to be opened, closed, energized, etc.

The tagout procedure does not physically prevent the flow of power and does not provide the physical protection of a lockout. For that reason, if no lock can be applied, at least one additional safety measure must be used. For example, the removal of an isolating circuit element, i.e., the removal of a fuse from its fuse holder, will provide an additional level of safety.

All the training and auditing procedures discussed above in the lockout procedure are also applicable for tagout procedures.

All circuits which have been deenergized but not locked out or tagged out will be considered to be energized.

(c) **Working on or near exposed energized parts.**

(c)(1) **Application.** This paragraph applies to work performed on exposed live parts (involving either direct contact or by means of tools or materials) or near enough to them for employees to be exposed to any hazard they present.

(c)(2) **Work on energized equipment.** Only qualified persons may work on electric circuit parts or equipment that have not been deenergized under the procedures of paragraph (b) of this section. Such persons shall be capable of working safely on energized circuits and shall be familiar with the proper use of special precautionary techniques, personal protective equipment, insulating and shielding materials, and insulated tools.

If systems or equipment cannot be deenergized

OSHA allows only qualified personnel to work on or near electrical circuits or equipment that has not been deenergized. Working on or near energized systems presents the personal hazard of direct or indirect contact (indirect contact means contact with the energized part through a tool or other conductive material). In addition to the contact hazard, there is also the possibility of an arc-flash/blast hazard. The qualified person shall be trained to work safely on energized systems and shall be familiar with the proper use of special precautionary techniques, personal protective equipment, insulating and shielding material, and insulated tools.

Section 110.7 of the 2009 Edition of NFPA 70E requires that a documented electrical safety program identify a hazard/risk evaluation procedure to be used before work is started on or near live parts. This means if the system or equipment cannot be placed in an electrically safe condition, a hazard/risk evaluation must be performed to identify all electrical hazards to which an employee will be exposed. In most cases the electrical hazards can be placed into three categories:

- Contact hazard (current flow through the body due to direct or indirect contact with live parts)
- Arc Flash/Blast
- High-current hazard (common to systems or equipment operating at less than 50 volts but with high current available)

In some cases only a contact hazard may exist, whereas in other cases all three hazards could be present. Once the hazards are identified, the proper work procedures and PPE must be selected.

Section 130.3 of NFPA 70E requires that a Flash Hazard Analysis be done before a person approaches any exposed electrical conductor or circuit part that has not been placed in an electrically safe condition. This Flash Hazard Analysis will establish the Flash Protection Boundary and the personal protective equipment that persons within the Flash Protection Boundary will use. A Flash Hazard Analysis should be performed under the supervision of a qualified engineer, and determines the incident energy (generally considered as the heat energy expelled during an arc flash incident, measured in cal/cm^2) that will be present during an arc flash incident. The onset of second degree burning to the human skin occurs at 1.2 cal/cm^2. The definitive standard for the performance of a Flash Hazard Analysis is IEEE Std. 1584, *IEEE Guide for Performing Arc Flash Hazard Calculations*.

(c)(3) **Overhead lines.** If work is to be performed near overhead lines, the lines shall be deenergized and grounded, or other protective measures shall be provided before work is started. If the lines are to be deenergized, arrangements shall be made with the person or organization that operates or controls the electric circuits involved to deenergize and ground them. If protective measures, such as guarding, isolating, or insulating, are provided, these precautions shall prevent employees from contacting such lines directly with any part of their body or indirectly through conductive materials, tools, or equipment.

* Note: The work practices used by qualified persons installing insulating devices on overhead power transmission or distribution lines are covered by 1910.269 of this Part, not by 1910.332 through 1910.335 of this Part. Under paragraph (c)(2) of this section, unqualified persons are prohibited from performing this type of work.

(c)(3)(i) **Unqualified persons.**

(c)(3)(i)(A) When an unqualified person is working in an elevated position near overhead lines, the location shall be such that the person and the longest conductive object he or she may contact cannot come closer to any unguarded, energized overhead line than the following distances:

(c)(3)(i)(A)(1) For voltages to ground 50kV or below 10 ft. (305 cm);

(c)(3)(i)(A)(2) For voltages to ground over 50kV - 10 ft. (305 cm) plus 4 in. (10 cm) for every 10kV over 50kV.

(c)(3)(i)(B) When an unqualified person is working on the ground in the vicinity of overhead lines, the person may not bring any conductive object closer to unguarded, energized overhead lines than the distances given in paragraph (c)(3)(i)(A) of this section.

Note: For voltages normally encountered with overhead power lines, objects which do not have an insulating rating for the voltage involved are considered to be conductive.

A competent person must perform a survey of the work site to determine where overhead lines are present and their nominal operating voltage. In some cases these overhead lines may not have anything to do with the actual workplace. An example is overhead utility lines that run along property lines. However, employees could become involved with these lines as they perform some work task.

If overhead lines are present in the work area, 1910.333(c)(3)(i)(A)(1) states that unqualified persons are required to come no closer than 10 feet for voltages to ground of 50 kV or below. For voltages to ground over 50 kV, a person must come no closer than 10 feet plus 4 inches for every 10 kV over 50 kV.

A qualified person (as determined by the employer, based on minimums outlined in 1910.332 and NFPA 70E, Section 110.6(C)) familiar with the system supplying the overhead lines should be consulted to confirm the voltage to ground. Frequently the system voltage is expressed in phase-to-phase voltage as compared to voltage to ground. For example, a system supplying overhead lines may be referred to as a 13.8 kV system. This is normally the phase-to-phase voltage. The phase-to-ground voltage would be 13.8 kV divided by 1.73 (for a 3-phase system) or approximately 7.9 kV.

When applying these approach distances, it is necessary to consider the work task and the materials and equipment involved. The distances specified would be measured from the end of any conductive object the worker may be handling. In some cases the overhead lines may be guarded using insulating blankets or sheathing, allowing the unqualified person to approach closer.

(c)(3)(ii) **Qualified persons.** When a qualified person is working in the vicinity of overhead lines, whether in an elevated position or on the ground, the person may not approach or take any conductive object without an approved insulating handle closer to exposed energized parts than shown in Table S-5 unless:

(c)(3)(ii)(A) The person is insulated from the energized part (gloves or gloves with sleeves rated for the voltage involved are considered to be insulation of the person from the energized part on which work is performed), or

(c)(3)(ii)(B) The energized part is insulated from all other conductive objects at a different potential and from the person, or

(c)(3)(ii)(C) The person is insulated from all conductive objects at a potential different from that of the energized part.

Table S-5 Approach Distances for Qualified Employees—Alternating Current

Voltage Range (phase to phase)	Minimum approach distance
300V and less	Avoid contact
Over 300V, not over 750V	1 ft. 0 in. (30.5 cm)
Over 750V, not over 2kV	1 ft. 6 in. (46 cm)
Over 2kV, not over 15kV	2 ft. 0 in. (61 cm)
Over 15kV, not over 37kV	3 ft. 0 in. (91 cm)
Over 37kV, not over 87.5kV	3 ft. 6 in. (107 cm)
Over 87.5kV, not over 121kV	4 ft. 0 in. (122 cm)
Over 121kV, not over 140kV	4 ft. 6 in. (137 cm)

1910.333

Title: Selection and use of work practices.

Subpart: S

SubPart Title: Electrical - Safety-Related Work Practices

It is important to realize there is a major difference between OSHA and NFPA 70E regarding the application of Approach Distances and Table S-5. OSHA applies this requirement when a qualified person is working in the vicinity of overhead lines. NFPA 70E applies it to all work on live parts. Table 130.2(C) in NFPA 70E also contains more voltage ranges and conditions than Table S-5.

As can be seen from a comparison of the two tables (see Table 333–1), there are other major differences between OSHA and NFPA 70E. Columns 2 and 3 with the major heading "Limited Approach Boundary" applies to *unqualified persons*. Column 2 applies in the majority of instances to overhead lines. Column 3 applies to the typical fixed circuit part that would be exposed in the event a *qualified* person is working in an energized control enclosure.

Section 130.2 of NFPA 70E places some important responsibilities on both the qualified and unqualified persons to assure that these approach boundaries are not violated. This section reads as follows:

130.2(D) Where one or more unqualified persons are working at or close to the Limited Approach Boundary, the designated person in charge of the work space where the electrical hazard exists shall cooperate with the designated person in charge of the unqualified person(s) to ensure that all work can be done safely. *This shall include advising the unqualified person(s) of the electrical hazard and warning him or her to stay outside of the Limited Approach Boundary.* (Italics added for emphasis.)

Figure 133 – Example of danger tape used to establish work zone to restrict approach by unqualified persons.

1910.333

Title: Selection and use of work practices.

Subpart: S

SubPart Title: Electrical - Safety-Related Work Practices

Proper training must be provided to ALL employees so that these approach boundaries are respected. So as not to place undue responsibility on the qualified person, it is required that a work zone be created about equipment or systems whenever energized work is to be performed. The work zone should be marked off with some appropriate method, such as warning tape (Figure 133), establishing an area that is equal to or greater than the required limited approach boundary. Other employees can then be easily instructed not to cross over the work zone barrier unless they are qualified and have the permission of the person in charge of the area where the electrical hazard exists.

Where the need exists, an unqualified person can cross over the limited approach boundary under the direct control and continual supervision of the qualified person in charge of the area where the electrical hazard exists. The qualified person is responsible for the safety of the unqualified person while inside the limited approach boundary, as well as instructing the unqualified person on the hazards they are exposed to before entering the area. The qualified person must ensure the protection of the unqualified person from the hazards thorough the use of personal protective equipment, distance, or other methods, and must ensure that the unqualified person does not approach any closer than the Restricted Approach Boundary (see Table 333-1).

Table 333-1 Approach Boundaries to Live Parts for Shock Protection

(All dimensions are distance from live part to employee.)

(1)	(2)	(3)	(4)	(5)
Nominal System Voltage Range, Phase to Phase	Limited Approach Boundary		Restricted Approach Boundary ; Includes Inadvertent Movement Adder	Prohibited Approach Boundary
	Exposed Movable Conductor	Exposed Fixed Circuit Part		
< 50V	Not specified	Not specified	Not specified	Not specified
50 to 300 V	3.0m (10 ft.)	1.0m (3' 6")	Avoid contact	Avoid contact
301 to 750 V	3.0m (10 ft.)	1.0m (3' 6")	.3m (1 ft.)	25mm (0 ft. 1 in.)
751 to 15 kV	3.0m (10 ft.)	1.5m (5 ft.)	.7m (2' 2")	.2m (0 ft. 7 in.)
15.1 kV to 36 kV	3.0m (10 ft.)	1.8m (6 ft.)	.8m (2' 7")	.3m (0 ft. 10 in.)
36.1 kV to 46 kV	3.0m (10 ft.)	2.5m (8 ft.)	.8m (2' 9")	.4m (1 ft. 5 in.)
46.1 kV to 72.5 kV	3.0m (10 ft.)	2.5m (8 ft.)	1.0m (3' 3")	.1mm (2' 2")
72.6 kV to 121 kV	3.3m (10' 8")	2.5m (8 ft.)	1.0m (3' 4")	.8mm (2' 9")
138 kV to 145 kV	3.4m (11 ft.)	3.0m (10 ft.)	1.2m (3' 10")	1.0m (3' 4")
161 kV to 169 kV	3.6m (11' 8")	3.6m (11' 8")	1.3m (4' 3")	1.1m (3' 9")
230 kV to 242 kV	4.0m (13 ft.)	4.0m (13 ft.)	1.7m (5' 8")	1.6m (5' 2")
345 kV to 362 kV	4.7m (15' 4")	4.7m (15' 4")	2.8m (9' 2")	2.6m (8' 8")
500 kV to 550 kV	5.8m (19 ft.)	5.8m (19 ft.)	3.6m (11' 10")	3.5m (11' 4")
765 kV to 800 kV	7.2m (23' 9")	7.2m (23' 9")	4.9m (15' 11")	4.7m (15' 5")

Source: 2012 Edition of NFPA 70E, Table 130.4(C)(a)

(c)(3)(iii)	**Vehicular and mechanical equipment.**
(c)(3)(iii)(A)	Any vehicle or mechanical equipment capable of having parts of its structure elevated near energized overhead lines shall be operated so that a clearance of 10 ft. (305 cm) is maintained. If the voltage is higher than 50kV, the clearance shall be increased 4 in. (10 cm) for every 10kV over that voltage. However, under any of the following conditions, the clearance may be reduced:
(c)(3)(iii)(A)(1)	If the vehicle is in transit with its structure lowered, the clearance may be reduced to 4 ft. (122 cm). If the voltage is higher than 50 kV, the clearance shall be increased 4 in. (10 cm) for every 10 kV over that voltage.
(c)(3)(iii)(A)(2)	If insulating barriers are installed to prevent contact with the lines, and if the barriers are rated for the voltage of the line being guarded and are not a part of or an attachment to the vehicle or its raised structure, the clearance may be reduced to a distance within the designed working dimensions of the insulating barrier.
(c)(3)(iii)(A)(3)	If the equipment is an aerial lift insulated for the voltage involved, and if the work is performed by a qualified person, the clearance (between the uninsulated portion of the aerial lift and the power line) may be reduced to the distance given in Table S-5.
(c)(3)(iii)(B)	Employees standing on the ground may not contact the vehicle or mechanical equipment or any of its attachments, unless:
(c)(3)(iii)(B)(1)	The employee is using protective equipment rated for the voltage; or
(c)(3)(iii)(B)(2)	The equipment is located so that no part of its structure can come closer to the line than permitted in paragraph (c)(3)(iii) of this section.
(c)(3)(iii)(C)	**If any vehicle or mechanical equipment capable of having parts of its** structure elevated near energized overhead lines is intentionally grounded, employees working on the ground near the point of grounding may not stand at the grounding location whenever there is a possibility of overhead line contact. Additional precautions, such as the use of barricades or insulation shall be taken to protect employees from hazardous ground potentials, depending on earth resistivity and fault currents, which can develop with the first few feet outward from the grounding point.

Working On or Near Overhead Lines

When work is to be performed near overhead power lines, it is desirable to first deenergize and ground the overhead lines. However, if this procedure is not feasible, protective measures such as guarding, isolating, or insulating shall be used to prevent employees from having direct or indirect contact (through conductive materials, tools, or equipment) with the lines.

When applying insulation to overhead lines as a method of protection and contacting the overhead line with such equipment as cranes, backhoes and similar equipment is possible, consideration should be given to the likelihood that such contact could break the line. When the overhead line contacts the earth, or other conductive equipment, there will be significant voltage gradients in the vicinity as a result of the high current flow. If other workers are present in the area, this could result in serious injury or death.

The methods of installing insulation materials on power lines by qualified persons are not covered by these articles. Unqualified personnel are prohibited from engaging in these types of activities by §1910.333(c)(2).

Unqualified employees working on the ground or in an elevated position near overhead lines, the location shall be such that the employee and the longest conductive object the employee might contact cannot come closer to any unguarded, energized overhead power line than the following:

(1) For voltages of 50kV to ground or less—10 feet

(2) For voltages over 50kV to ground—10 feet plus 4 inches for every 10kV over 50kV.

At these high transmission voltages, any object which does not have an insulation rating for the voltages involved is considered to be a conductor.

Figure 134 – Metallic crane booms may come in contact with overhead lines.

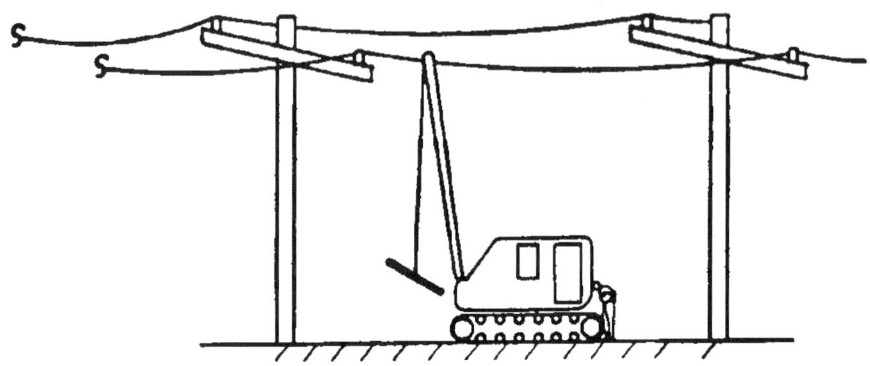

Approach distances for a qualified person working near overhead lines are presented in Table S-5, Approach Distances for Qualified Employees - Alternating Current. The distances in the table apply unless the qualified person is appropriately insulated from the line or is using tools with approved insulating handles.

Vehicular and mechanical equipment capable of having parts of its structure elevated (such as an aerial lift truck, truck crane, dumpster truck, well-drilling rig, etc.) near overhead lines (Figure 134) shall maintain a minimum clearance of 10 feet for voltages of 50kV or less to ground and 10 feet plus 4 inches for every 10kV over 50kV. There are several circumstances where these clearances may be somewhat reduced. For example, if the vehicle is in transit with its structure lowered, the minimum clearance may be lowered to 4 feet plus 4 inches for every 10kV over 50kV.

Also allowed are insulating barriers that prevent the elevated structure from contacting the lines. If the equipment is an aerial lift insulated for the voltage involved, the minimum distance for clearance can be reduced to those given in Table S-5.

It is important for all employers to recognize the potential for very serious accidents to occur from this type of hazard. Each employer should develop a comprehensive set of work rules and provide adequate training for drivers or operators of all vehicles with structures capable of being elevated. It is so easy for driver-operators to become complacent in their operation of the elevation mechanism. Most of the time there are no power lines around. However, it only takes one mistake for a tragic accident to occur.

(c)(4) **Illumination.**

(c)(4)(i) Employees may not enter spaces containing exposed energized parts, unless illumination is provided that enables the employees to perform the work safely.

(c)(4)(ii) Where lack of illumination or an obstruction precludes observation of the work to be performed, employees may not perform tasks near exposed energized parts. Employees may not reach blindly into areas which may contain energized parts.

Illumination

Illumination must be adequate before an employee may enter any space which contains exposed energized parts. Employees may not reach blindly into areas which may contain energized parts.

(c)(5) **Confined or enclosed work spaces**. When working in confined or enclosed spaces (such as manholes or vaults) that contain exposed energized parts, the employer shall provide, and the employee shall use, protective shields, protective barriers, or insulating materials as necessary to avoid inadvertent contact with these parts. Doors, hinged panels and the like shall be secured to prevent their swinging into an employee and causing the employee to contact exposed energized parts.

Confined Work Spaces

When an employee finds it necessary to work in a confined or enclosed space with exposed energized parts, the employer shall provide and the employee shall use, appropriate protective shields or insulating materials and barriers so that the employee will not contact directly, or indirectly, the energized systems.

(c)(6) **Conductive materials and equipment.** Conductive materials and equipment that are in contact with any part of an employee's body shall be handled in a manner that will prevent them from contacting exposed energized conductors or circuit parts. If an employee must handle long dimensional conductive objects (such as ducts and pipes) in areas with exposed live parts, the employer shall institute work practices (such as the use of insulation, guarding, and material handling techniques) which will minimize the hazard.

Conductive Materials and Equipment

When conductive materials or equipment must be handled by employees in areas where there are exposed energized parts, appropriate work practices will be developed to minimize the hazard. For example, if long lengths of metallic conduit or piping need to be handled or carried through an area with exposed parts, appropriate handling techniques shall be developed and enforced by the employer.

(c)(7) **Portable ladders.** Portable ladders shall have non-conductive siderails if they are used where the employee or the ladder could contact exposed energized parts.

(c)(8) **Conductive apparel.** Conductive articles of jewelry and clothing (such as watch bands, bracelets, rings, key chains, necklaces, metalized aprons, cloth with conductive thread, or metal headgear) may not be worn if they might contact exposed energized parts. However, such articles may be worn if they are rendered nonconductive by covering, wrapping, or other insulating means.

Employees may wear any of the items listed above as long as they are not exposed to energized parts. If the task requires them to be exposed to live parts, they must be removed to protect against electrocution and arc-flash hazards. It is not recommended that such items be placed in shirt or pants pockets while performing the work. Metal frame eye glasses may be worn if they are restrained.

(c)(9) **Housekeeping duties.** Where live parts present an electrical contact hazard, employees may not perform housekeeping duties at such close distances to the parts that there is a possibility of contact, unless adequate safeguards (such as insulating equipment or barriers) are provided. Electrically conductive cleaning materials (including conductive solids such as steel wool, metalized cloth, and silicon carbide as well as conductive liquid solutions) may not be used in proximity of energized parts unless procedures are followed which will prevent electrical contact.

(c)(10) **Interlocks.** Only a qualified person following the requirements of paragraph (c) of this section may defeat an electrical safety interlock, and then only temporarily while he or she is working on the equipment. The interlock system shall be returned to its operable condition when this work is completed.

§ 1910.334 Use of Equipment

[55 FR 32016, Aug. 6, 1990; 55 FR 46054, Nov. 1, 1990]

(a)	**Portable electric equipment.** This paragraph applies to the use of cord- and plug-connected equipment, including flexible cord sets (extension cords).
(a)(1)	**Handling.** Portable equipment shall be handled in a manner which will not cause damage. Flexible electric cords connected to equipment may not be used for raising or lowering the equipment. Flexible cords may not be fastened with staples or otherwise hung in such a fashion as to damage the outer jacket or insulation.
(a)(2)	**Visual inspection.**
(a)(2)(i)	Portable cord- and plug-connected equipment and flexible cord sets (extension cords) shall be visually inspected before use on any shift for external defects (such as loose parts, deformed and missing pins, or damage to outer jacket or insulation) and for evidence of possible internal damage (such as pinched or crushed outer jacket). Cord- and plug-connected equipment and flexible cord sets (extension cords) which remain connected once they are put in place and are not exposed to damage need not be visually inspected until they are relocated.
(a)(2)(ii)	If there is a defect or evidence of damage that might expose an employee to injury, the defective or damaged item shall be removed from service, and no employee may use it until repairs and tests necessary to render the equipment safe have been made.
(a)(2)(iii)	When an attachment plug is to be connected to a receptacle (including any on a cord set), the relationship of the plug and receptacle contacts shall first be checked to ensure that they are of proper mating configurations.
(a)(3)	**Grounding-type equipment.**
(a)(3)(i)	A flexible cord used with grounding-type equipment shall contain an equipment grounding conductor.
(a)(3)(ii)	Attachment plugs and receptacles may not be connected or altered in a manner which would prevent proper continuity of the equipment grounding conductor at the point where plugs are attached to receptacles. Additionally, these devices may not be altered to allow the grounding pole of a plug to be inserted into slots intended for connection to the current-carrying conductors.
(a)(3)(iii)	Adapters which interrupt the continuity of the equipment grounding connection may not be used.
(a)(4)	**Conductive work locations.** Portable electric equipment and flexible cords used in highly conductive work locations (such as those inundated with water or other conductive liquids), or in job locations where employees are likely to contact water or conductive liquids, shall be approved for those locations.
(a)(5)	**Connecting attachment plugs.**
(a)(5)(i)	Employees' hands may not be wet when plugging and unplugging flexible cords and cord- and plug-connected equipment, if energized equipment is involved.
(a)(5)(ii)	Energized plug and receptacle connections may be handled only with insulating protective equipment if the condition of the connection could provide a conducting path to the employee's hand (if, for example, a cord connector is wet from being immersed in water).
(a)(5)(iii)	Locking-type connectors shall be properly secured after connection.

Cord- and Plug-Connected Equipment

This section is consistent with NFPA 70E, 2009 Edition, Section 110.9.

Portable electrical equipment shall be handled carefully to prevent damage to the circuit parts. For example, the cord will not be used to raise or lower the equipment from one level to another. Flexible cords cannot be stapled or otherwise treated such that the outer jacket or insulation might be damaged.

214All portable cord- and plug-connected equipment and flexible cord sets (extension cords) shall be visually inspected before use on any work shift for external defects or any suggestion of an insulation defect. Once connected and put into operation, they need not be reinspected until they are relocated. Defective cords shall be repaired and tested before being put back into service.

1910.334

Title: Use of equipment.

Subpart: S

SubPart Title: Electrical - Safety-Related Work Practices

Mating plugs and receptacles should be carefully inspected before they are plugged together to ensure that they mate each other properly. Plug-receptacle sets have used a key-and-keyway design to achieve proper mating. The key on the plug can become so worn that it is possible to rotate the plug and mate the receptacle four different ways. In one case a worker plugged the phase A pin on the plug into the grounded slot on the receptacle and he literally had a ground fault in his hands. He was killed instantly. A careful inspection should have caught the wear of the key and kept the cord out of service.

Employees may not plug and/or unplug flexible cords and cord- and plug-connected equipment when they have wet hands or are standing in water or in other conductive liquids (milk for instance). Also an employee handling an energized plug and/or receptacle whose condition (wet with water or other liquid) might provide a conducting path to the hands, must utilize insulating protective equipment.

(b) **Electric power and lighting circuits.**

(b)(1) **Routine opening and closing of circuits.** Load rated switches, circuit breakers, or other devices specifically designed as disconnecting means shall be used for the routine opening, reversing, or closing of circuits under load conditions. Cable connectors not of the load-break type, fuses, terminal lugs, and cable splice connections may not be used for such purposes, except in an emergency.

(b)(2) **Reclosing circuits after protective device operation.** After a circuit is deenergized by a circuit protective device, the circuit may not be manually reenergized until it has been determined that the equipment and circuit can be safely energized. The repetitive manual reclosing of circuit breakers or reenergizing circuits through replaced fuses is prohibited.

Note: When it can be determined from the design of the circuit and the overcurrent devices involved that the automatic operation of a device was caused by an overload rather than a fault condition, no examination of the circuit or connected equipment is needed before the circuit is reenergized.

(b)(3) **Overcurrent protection modification.** Overcurrent protection of circuits and conductors may not be modified, even on a temporary basis, beyond that allowed by § 1910.304(e), the installation safety requirements for overcurrent protection.

Electric Power and Lighting Circuits

Routine energizing or deenergizing of circuits shall be done by switches, circuit breakers or other devices specifically designed to be the disconnecting means. Cable connectors not designed for such duty, fuses, terminal lugs, and cable splice connections may not be used for making and breaking power circuits except in an emergency situation.

A circuit may not be reenergized immediately after a fuse has blown or a circuit breaker tripped until it has been determined that the circuit and its connected utilization equipment can be safely energized. Repetitive resetting of a circuit breaker is not permitted until the reason for the tripping of the overcurrent device is determined. (See also NFPA 70E, 2009 Edition, Sections 130.6(J) and (K)).

(c) **Test instruments and equipment.**

(c)(1) **Use.** Only qualified persons may perform testing work on electric circuits or equipment.

(c)(2) **Visual inspection.** Test instruments and equipment and all associated test leads, cables, power cords, probes, and connectors shall be visually inspected for external defects and damage before the equipment is used. If there is a defect or evidence of damage that might expose an employee to injury, the defective or damaged item shall be removed from service and no employee may use it until repairs and tests necessary to render the equipment safe have been made.

(c)(3) **Rating of equipment.** Test instruments and equipment and their accessories shall be rated for the circuits and equipment to which they will be connected and shall be designed for the environment in which they will be used.

1910.334

Title: Use of equipment.

Subpart: S

SubPart Title: Electrical - Safety-Related Work Practices

Test instruments, like any other electrical equipment, must be approved (1910.303(a)), in other words listed by an OSHA Nationally Recognized Testing Laboratory (NRTL). Additionally, the use of equipment which has a Category rating (see UL 61010-1) is highly recommended. Meters and other test equipment, when tested under the UL or IEC standards are given Category ratings of I-IV (abbreviated CAT I-IV). Standard 61010-1 defines Categories I through IV, dividing the power distribution system into categories based on the fact that dangerous high-energy transients, such as lightning strikes, will be attenuated as they travel through the impedance of the system. A higher CAT number refers to an electrical environment with higher available power and higher-energy transients. Therefore, a meter designed to meet a CAT III standard is resistant to much higher-energy transients than one designed to CAT II standards. Within a Category, a higher voltage rating denotes a higher transient withstand rating, so for example, a CAT III-1000 V meter has superior protection compared to a CAT III-600 V rated meter. Misunderstanding often occurs when personnel select a CAT II-1000 V rated meter, thinking that it is superior to a CAT III-600 V meter. Category I test equipment is generally designed for electronic applications, while Category II is for use with single-phase receptacle loads. Category III equipment is for three-phase distribution. Utility connections (within the rating of the equipment) and outdoor applications are covered by Category IV. CAT III and CAT IV equipment is highly recommended for most commercial and industrial applications.

(d) **Occasional use of flammable or ignitable materials.** Where flammable materials are present only occasionally, electric equipment capable of igniting them shall not be used, unless measures are taken to prevent hazardous conditions from developing. Such materials include, but are not limited to: flammable gases, vapors, or liquids; combustible dust; and ignitable fibers or flyings.

Note: Electrical installation requirements for locations where flammable materials are present on a regular basis are contained in 1910.307.

§ 1910.335 Safeguards for Personnel Protection

[55 FR 32016, Aug. 6, 1990]

(a) **Use of protective equipment.**

(a)(1) **Personal protective equipment.**

(a)(1)(i) Employees working in areas where there are potential electrical hazards shall be provided with, and shall use, electrical protective equipment that is appropriate for the specific parts of the body to be protected and for the work to be performed.

Note: Personal protective equipment requirements are contained in Subpart I of this part.

(a)(1)(ii) Protective equipment shall be maintained in a safe, reliable condition and shall be periodically inspected or tested, as required by 1910.137.

Table 335-1 shows the required test intervals for common types of rubber insulating PPE.

Table 335-1. Rubber Insulating Equipment Test Intervals

Type of equipment	When to test
Rubber insulating line hose.	Upon indication that insulating value is suspect.
Rubber insulating covers.	Upon indication that insulating value is suspect.
Rubber insulating blankets	Before first issue and every 12 months thereafter.[1]
Rubber insulating gloves	Before first issue and every 6 months thereafter.[1]
Rubber insulating sleeves	Before first issue and every 12 months thereafter.[1]

1. If the insulating equipment has been electrically tested but not issued for service, it may not be placed into service unless it has been electrically tested within the previous 12 months.

Table 335-2 shows the six categories of dielectric gloves and protective sleeves. Maximum use voltage and retest voltages for both ac and dc application are shown. Note that Class 00 gloves, which may be selected for use on nominal 480 volt ac systems, typically found in industrial facilities, has a very small safety margin. Many of these systems can go over 500 volts at certain times during the day depending on loading and regulation characteristics. Thus many electricians do not feel comfortable using this class glove and prefer to use a Class 0 glove.

Table 335-2. Maximum-Use and Test Voltages—Gloves and Sleeves

Label Color	Class	Maximum Use Voltage		Retest Voltage	
		AC	DC	AC	DC
Beige	00	500	750	2500	10000
Red	0	1000	1500	5000	20000
White	1	7500	11250	10000	40000
Yellow	2	17000	25500	20000	50000
Green	3	26500	39750	30000	60000
Orange	4	36000	54000	40000	70000

Title: Safeguards for personnel protection.

Subpart: S

SubPart Title: Electrical - Safety-Related Work Practices

1910.132 and 1910.137 cover the details of classification, testing and more, as related to PPE, but other details, such as types of rubber (Type I (natural rubber) and Type II (synthetic rubber)) are covered in the appropriate ASTM standards that apply to the specific product in question (see Table 335-3).

Table 335-3. Rubber Insulating Glove Standards

ASTM Standard	Subject
D-120, 2002a (R 2006)	Standard Specification for Rubber Insulating Gloves
F-496, 2006	Standard Specification for In-Service Care of Insulating Gloves and Sleeves
D-1051, 2007	Standard Specification for Rubber Insulating Sleeves
F-696, 2006	Standard Specification for Leather Protectors for Rubber Insulating Gloves and Mittens

Figure 135 – Example of rubber insulating glove label.

(a)(1)(iii) If the insulating capability of protective equipment may be subject to damage during use, the insulating material shall be protected. (For example, an outer covering of leather is sometimes used for the protection of rubber insulating material).

Most care, storage and use instructions are also provided in standards outside the OSHA regulations found here. For example, rubber insulating gloves are required to be used with leather protectors. While 1910.335(a)(1)(iii) above requires that the insulating material be protected from damage, it is actually ASTM 496-02a which gives detailed instruction on how. The ASTM standard requires that protector gloves "shall be worn over insulating gloves to prevent mechanical damage", and proceeds to provide in depth specifications and instructions on their use and care (ASTM 496-02a, Section 8.7). Section 8.7.4 of the ASTM standard also contains an exception to the use of protectors; where unusual dexterity is required, the protectors may be omitted, *only* where the possibility of physical damage to the gloves is unlikely *and* provided the voltage class of the glove used is one class above the voltage exposure. The voltage rating of Class 00 gloves is reduced to 250V for this exception. Additionally, rubber insulating gloves that have been used without protectors shall not be used with protectors until given an inspection and electrical retest. While this exception exists, it is extremely difficult to guarantee conditions where the rubber insulating glove will not be subject to damage of any kind, and it is not recommended that they be used in energized conditions without the protectors.

(a)(1)(iv) Employees shall wear nonconductive head protection wherever there is a danger of head injury from electric shock or burns due to contact with exposed energized parts.

1910.335

Title: Safeguards for personnel protection.

Subpart: S

SubPart Title: Electrical - Safety-Related Work Practices

(a)(1)(v)	Employees shall wear protective equipment for the eyes or face wherever there is danger of injury to the eyes or face from electric arcs or flashers or from flying objects resulting from electrical explosion.
(a)(2)	**General protective equipment and tools.**
(a)(2)(i)	When working near exposed energized conductors or circuit parts, each employee shall use insulated tools or handling equipment if the tools or handling equipment might make contact with such conductors or parts. If the insulating capability of insulated tools or handling equipment is subject to damage, the insulating material shall be protected.
(a)(2)(i)(A)	Fuse handling equipment, insulated for the circuit voltage, shall be used to remove or install fuses when the fuse terminals are energized.
(a)(2)(i)(B)	Ropes and handlines used near exposed energized parts shall be nonconductive.
(a)(2)(ii)	Protective shields, protective barriers, or insulating materials shall be used to protect each employee from shock, burns, or other electrically related injuries while that employee is working near exposed energized parts which might be accidentally contacted or where dangerous electric heating or arcing might occur. When normally enclosed live parts are exposed for maintenance or repair, they shall be guarded to protect unqualified persons from contact with the live parts.

Personal Protective Equipment

The employer shall provide, and the employee shall use, appropriate protective equipment when a potential electrical hazard exists. This equipment will be properly maintained and tested periodically to ensure its reliable condition. Procedures for testing are outlined in OSHA §1910.137.

Employees shall use protective equipment appropriate for the areas of the body to be protected, i.e., rubber gloves and sleeves for the hands and arms, nonconductive headgear for the head, eye and face protection to protect the face from arcs and sparks, etc.

Before the proper PPE can be selected, a hazard analysis must be done as mandated by 1910.132(d). This hazard analysis includes a complete review of the electrical distribution system, including not only contact hazards but also the probability of a flash hazard, which may require information such as available system fault-current. It is also important to note that while the standards and regulations may allow certain practices, they remain minimum standards, and a hazard analysis may determine that minimum activities are not sufficient to protect employees from harm.

The use of Personal Protective Equipment, such as voltage-rated gloves and fire-rated clothing has been a controversial issue ever since OSHA published their safety-related work practice requirements in August of 1990. Although much training has been done over the years since this standard was published, few workplaces are in compliance with the intended objectives. One of the problems is the difference of opinions as to where and when such PPE is required.

NFPA 70E provides much needed guidance. Chapter 1 of NFPA 70E covers in detail the proper selection and use of Personal Protective Equipment, including the use of insulated tools.

In most cases there will be two hazards to deal with. One is the contact or shock/electrocution hazard and the second is a possible arc-flash/blast hazard. The flash hazard requires the greatest amount of effort. NFPA 70E permits two methods of providing flash protection. One method is to do a detailed flash hazard analysis as required by 130.3 (see also Appendix D). The other method is to use Table 130.7(C)(9) of NFPA 70E, provided that the applicable system is within the fault current and clearing times specified in the notes of the Table. Tables 335-3A through 335-3I reproduce some of the information found in Table 130.7(C)(9) of NFPA 70E. It is important to keep in mind that the tables contained here are for informational purposes only, and the actual tables found in NFPA 70E must be consulted before selecting any protection approach. To gain a more complete understanding of flash protection concepts, it is important that the reader obtain a copy of NFPA 70E and thoroughly study Chapter 1.

Section 110.7(D) of NFPA 70E states that "An electrical safety program shall identify the procedures for working on or near live parts before work is started." An example of an electrical safety program procedure can be found in Appendix E of NFPA 70E.

As can be seen from looking at Tables 335-3A through 335-3I, a hazard/risk evaluation process was completed to determine the hazard/risk category for different tasks. If tasks are to be performed that are not listed in the tables or if the employer chooses not to use the tables, then a hazard/risk evaluation process must be completed. Appendix F of NFPA 70E, 2009 Edition, contains an example of a hazard/risk evaluation process.

Normally 100% burn protection is not practical. In most cases the protection is selected based upon the concept of a curable burn. Research has shown that an incident energy level of 1.2 cal/cm^2 would cause the onset of a second degree burn. When protective clothing is tested for burn protection it is given an ATPV (Arc Thermal Performance Value) rating. This rating indicates the highest incident energy which did not cause fabric break open and did not exceed the second-degree burn criteria.

Synthetic materials, such as polyester, nylon, acetate, rayon, and synthetic-cotton blends shall not be used. These materials will melt into the skin when exposed to high temperatures and aggravate the burn injury. Not all flame-resistant clothing is suitable for electrical protection.

Table 335-3A. Hazard Risk Category Classifications
Panelboards or other equipment rated 240 V and below
Maximum of 25 kA short circuit current available; maximum of 0.03 sec (2 cycle) fault clearing time.
Working Distance: 18 inches; Approximate Flash Protection Boundary for set parameters: 19 inches

Task (assumes equipment is energized; work is done within the flash protection boundary)	Hazard/ Risk Category	V-rated Gloves	V-rated Tools
Perform infrared thermography and other non-contact inspections outside the restricted approach boundary	0	N	N
Circuit breaker (CB) or fused switch operation with covers on	0	N	N
CB or fused switch operation with covers off	0	N	N
Work on energized parts, including voltage testing	1	Y	Y
Removal/install CBs or fused switches	1	Y	Y
Removal of bolted covers (to expose bare, energized parts)	1	N	N
Opening hinged covers (to expose bare, energized parts)	0	N	N
Work on energized parts of equipment fed directly by a branch circuit of the panelboard	1	Y	Y

Data Source: 2012 Edition NFPA 70E - Table 130.7(C)(15)(a)

Table 335-3B. Hazard Risk Category Classifications
Panelboards or Other Equipment Rated >240 V and up to 600 V
Maximum of 25 kA short circuit current available; maximum of 0.03 sec (2 cycle) fault clearing time.
Working Distance: 18 inches; Approximate Flash Protection Boundary for set parameters: 30 inches

Task (assumes equipment is energized; work is done within the flash protection boundary)	Hazard/ Risk Category	V-rated Gloves	V-rated Tools
Perform infrared thermography and other non-contact inspections outside the restricted approach boundary	1	N	N
Circuit breaker (CB) or fused switch operation with covers on	0	N	N
CB or fused switch operation with covers off	1	Y	N
Work on energized parts, including voltage testing	2	Y	Y
Removal/install CBs or fused switches	2	Y	Y
Removal of bolted covers (to expose bare, energized parts) – Note 3	1	N	N
Opening hinged covers (to expose bare, energized parts) – Note 3	0	N	N
Work on energized parts of equipment fed directly by a branch circuit of the panelboard or switchboard	2	Y	Y

Data Source: 2012 Edition NFPA 70E - Table 130.7(C)(15)(a)

Table 335-3C1. Hazard Risk Category Classifications
600 V Class Motor Control Centers (MCCs)
Maximum of 65 kA short circuit current available; maximum of 0.03 sec (2 cycle) fault clearing time.
Working Distance: 18 inches; Approximate Flash Protection Boundary for set parameters: 53 inches

Task (assumes equipment is energized; work is done within the flash protection boundary)	Hazard/ Risk Category	V-rated Gloves	V-rated Tools
Perform infrared thermography and other non-contact inspections outside the restricted approach boundary	1	N	N
CB or fused switch starter operation with enclosure doors closed	0	N	N
Reading a panel meter while operating a meter switch	0	N	N
CB or fused switch or starter operation with enclosure doors open	1	N	N
Work on energized parts, including voltage testing	2	Y	Y
Work on control circuits with energized parts 120 V or below exposed	0	Y	Y
Work on control circuits with energized parts >120 V exposed	2	Y	Y
Application of safety grounds, after voltage test	2	Y	N
Work on energized parts fed directly by a branch circuit of the MCC	2	Y	Y

Data Source: 2012 Edition NFPA 70E - Table 130.7(C)(15)(a)

Title: Safeguards for personnel protection.

Subpart: S

SubPart Title: Electrical - Safety-Related Work Practices

Table 335-3C2. Hazard Risk Category Classifications
600 V Class Motor Control Centers (MCCs)
Maximum of 42 kA short circuit current available; maximum of 0.33 sec (20 cycle) fault clearing time.
Working Distance: 18 inches; Approximate Flash Protection Boundary for set parameters: 165 inches

Task (assumes equipment is energized; work is done within the flash protection boundary)	Hazard/ Risk Category	V-rated Gloves	V-rated Tools
Insertion or removal of individual starter "buckets" from MCC - Note 3	4	Y	N
Removal of bolted covers (to expose bare, energized parts) – Note 3	4	N	N
Opening hinged covers (to expose bare, energized parts) – Note 3	1	N	N

Data Source: 2012 Edition NFPA 70E - Table 130.7(C)(15)(a)

Table 335-3D. Hazard Risk Category Classifications
600 V Class Switchgear (with power circuit breakers or fused switches)
Maximum of 35 kA short circuit current available; maximum of up to 0.5 sec (30 cycle) fault clearing time.
Working Distance: 18 inches; Approximate Flash Protection Boundary for set parameters: 233 inches

Task (assumes equipment is energized; work is done within the flash protection boundary)	Hazard/Risk Category	V-rated Gloves	V-rated Tools
Perform infrared thermography and other non-contact inspections outside the restricted approach boundary	2	N	N
CB or fused switch operation with enclosure doors closed	0	N	N
Reading a panel meter while operating a meter switch	0	N	N
CB or fused switch or starter operation with enclosure doors open	1	N	N
Work on energized parts, including voltage testing	2	Y	Y
Work on control circuits with energized parts 120 V or below exposed	0	Y	Y
Work on control circuits with energized parts >120 V exposed	2	Y	Y
Insertion or removal (racking) of CBs from cubicles, doors open or closed	4	N	N
Application of safety grounds, after voltage test	2	Y	N
Removal of bolted covers (to expose bare, energized parts)	4	N	N
Opening hinged covers (to expose bare, energized parts)	2	N	N

Data Source: 2012 Edition NFPA 70E - Table 130.7(C)(15)(a)

Table 335-3E. Hazard Risk Category Classifications
Other 600 V Class (277 V through 600 V, nominal) Equipment - Note 2 (except as indicated)
Maximum of 65 kA short circuit current available; maximum of 0.03 sec (2 cycle) fault clearing time.
Working Distance: 18 inches; Approximate Flash Protection Boundary for set parameters: 53 inches

Task (assumes equipment is energized; work is done within the flash protection boundary)	Hazard/Risk Category	V-rated Gloves	V-rated Tools
Lighting or small power transformers (600V, maximum)	-	-	-
Removal of bolted covers (to expose bare, energized parts)	2	N	N
Opening hinged covers (to expose bare, energized parts)	1	N	N
Work on energized parts, including voltage testing	2	Y	Y
Application of safety grounds, after voltage test	2	Y	N
Revenue meters (kW-hour, at primary voltage and current)	-	-	-
Insertion or removal	2	Y	N
Cable trough or tray cover removal or installation	1	N	N
Miscellaneous equipment cover removal or installation	1	N	N
Work on energized parts, including voltage testing	2	Y	Y
Application of safety grounds, after voltage test	2	Y	N
Insertion or removal of plug-in devices into or from busways	2	Y	N

Data Source: 2012 Edition NFPA 70E - Table 130.7(C)(15)(a)

Table 335-3F. Hazard Risk Category Classifications
NEMA E2 (fused contactor) Motor Starters, 2.3kV through 7.2kV

Maximum of 35 kA short circuit current available; maximum of up to 0.2 sec (12 cycle) fault clearing time.
Working Distance: 36 inches
Approximate Flash Protection Boundary for set parameters: 422 inches

Task (assumes equipment is energized; work is done within the flash protection boundary)	Hazard/ Risk Category	V-rated Gloves	V-rated Tools
Perform infrared thermography and other non-contact inspections outside the restricted approach boundary	3	N	N
Contactor operation with enclosure doors closed	0	N	N
Reading a panel meter while operating a meter switch	0	N	N
Contactor operation with enclosure doors open	2	N	N
Work on energized parts, including voltage testing	4	Y	Y
Work on control circuits with energized parts 120 V or below exposed	0	Y	Y
Work on control circuits with energized parts >120 V exposed	3	Y	Y
Insertion or removal (racking) of starters from cubicle, doors open or closed	4	N	N
Application of safety grounds, after voltage test	3	Y	N
Removal of bolted covers (to expose bare, energized parts)	4	N	N
Opening hinged covers (to expose bare, energized parts)	3	N	N
Insertion or removal (racking) of starters from cubicles of arc-resistant construction, tested in accordance with IEEE C37.20.7, doors closed only	0	N	N

Data Source: 2012 Edition NFPA 70E - Table 130.7(C)(15)(a)

Table 335-3G. Hazard Risk Category Classifications
Metal Clad Switchgear, 1 kV through 38kV

Maximum of 35 kA short circuit current available; maximum of up to 0.2 sec (12 cycle) fault clearing time.
Working Distance: 36 inches; Approximate Flash Protection Boundary for set parameters: 422 inches

Task (assumes equipment is energized; work is done within the flash protection boundary)	Hazard/ Risk Category	V-rated Gloves	V-rated Tools
Perform infrared thermography and other non-contact inspections outside the restricted approach boundary	3	N	N
CB operation with enclosure doors closed	2	N	N
Reading a panel meter while operating a meter switch	0	N	N
CB operation with enclosure doors open	4	N	N
Work on energized parts, including voltage testing	4	Y	Y
Work on control circuits with energized parts 120 V or below exposed	2	Y	Y
Work on control circuits with energized parts >120 V exposed	4	Y	Y
Insertion or removal (racking) of CBs from cubicle, doors open or closed	4	N	N
Application of safety grounds, after voltage test	4	Y	N
Removal of bolted covers (to expose bare, energized parts)	4	N	N
Opening hinged covers (to expose bare, energized parts)	3	N	N
Opening voltage transformer or control power transformer compartments	4	N	N

Data Source: 2012 Edition NFPA 70E - Table 130.7(C)(15)(a)

1910.335

Title: Safeguards for personnel protection.

Subpart: S

SubPart Title: Electrical - Safety-Related Work Practices

Table 335-3H. Hazard Risk Category Classifications
Arc-Resistant Switchgear
Type 1 or 2 (for clearing times of <0.5 sec with a perspective fault current not to exceed the arc-resistant rating of the equipment)

Maximum of 35 kA short circuit current available; maximum of up to 0.2 sec (12 cycle) fault clearing time. Working Distance: 36 inches; Approximate Flash Protection Boundary for set parameters: 422 inches

Task (assumes equipment is energized; work is done within the flash protection boundary)	Hazard/ Risk Category	V-rated Gloves	V-rated Tools
CB operation with enclosure doors closed	0	N	N
Insertion or removal (racking) of CBs from cubicles, doors closed	0	N	N
Insertion or removal of CBs from cubicles with door open	4	N	N
Work on control circuits with energized parts 120 V or below exposed	2	Y	Y
Insertion or removal (racking) of ground and test device with door closed	0	N	N
Insertion or removal (racking) of voltage transformers on or off the bus door closed	0	N	N

Data Source: 2012 Edition NFPA 70E - Table 130.7(C)(15)(a)

Table 335-3I. Hazard Risk Category Classifications
Other Equipment 1 kV Through 38 kV

Maximum of 35 kA short circuit current available; maximum of up to 0.2 sec (12 cycle) fault clearing time. Working Distance: 36 inches; Approximate Flash Protection Boundary for set parameters: 422 inches

Task (assumes equipment is energized; work is done within the flash protection boundary)	Hazard/ Risk Category	V-rated Gloves	V-rated Tools
Metal-enclosed interrupter switchgear, fused or unfused	-	-	-
Switch operation of arc-resistant-type construction, tested in accordance with IEEE C37.20.7, doors closed only	0	N	N
Switch operation, doors closed	2	N	N
Work on energized parts, including voltage testing	4	Y	Y
Removal of bolted covers (to expose bare, energized parts)	4	N	N
Opening hinged covers (to expose bare, energized parts)	3	N	N
Outdoor disconnect switch operation (hookstick operated)	3	Y	Y
Outdoor disconnect switch operation (gang-operated, from grade)	2	Y	N
Insulated cable examination, in manhole or other confined space	4	Y	N
Insulated cable examination, in open area	2	Y	N
Data Source: 2012 Edition NFPA 70E - Table 130.7(C)(15)(a)			

It is important to note that the PPE selection tables found in NFPA 70E, and reproduced in this guide, are based on these specific system parameters. Fault currents and clearing times be documented when using these PPE selection tables. The use of the tables should be considered a temporary solution until a detailed flash hazard analysis is performed for any facility, and are only valid if the actual system data documented matches the parameters set out.

Once the hazard/risk category is determined, the proper personal protective clothing and equipment can be selected from Table 130.7(C)(16) in NFPA 70E. Table 335-4 provides a summary of the information found in this table.

Table 335-4. Protective Clothing and Personal Protective Equipment (PPE) Matrix	
Hazard/Risk Category	**Protective Clothing and PPE** *Safety Glasses or Goggles & Hearing Protection required for all Levels*
Hazard/Risk Category 0 Protective Clothing, Nonmelting (according to ASTM F 1506-00) or Untreated Natural Fiber	Natural Fiber Shirt (long sleeve) Natural Fiber Pants (long) Leather gloves (as needed) [Rubber insulating gloves with leather protectors meet this requirement]
Hazard/Risk Category 1 FR Clothing, Minimum Arc Rating of 4 cal/cm^2	Arc-rated long-sleeve shirt (Minimum 4 cal/cm^2 rating) Arc-rated pants (Minimum 4 cal/cm^2 rating) Arc-rated coverall [May substitute for the arc-rated shirt & pants with minimum 4 cal/cm^2 rating] Arc-rated face shield or arc flash suit hood (Minimum 4 cal/cm^2 rating) [Must provide wrap-around guarding to protect the face, forehead, ears, and neck], with Class E hard hat Arc-rated jacket, parka, or rainwear [As needed] Leather gloves (as needed) [Rubber insulating gloves with leather protectors meet this requirement] Leather work shoes
Hazard/Risk Category 2 FR Clothing, Minimum Arc Rating of 8 cal/cm^2	Arc-rated long-sleeve shirt (Minimum 8 cal/cm^2 rating) Arc-rated pants (Minimum 8 cal/cm^2 rating) Arc-rated coverall [May substitute for the arc-rated shirt & pants with minimum 4 cal/cm^2 rating] Arc-rated flash suit hood (Minimum 8 cal/cm^2 rating) [or face shield with a minimum arc rating of 8 cal/cm^2 combined with a balaclava (sock hood) with a minimum arc rating of 8 cal/cm^2] with Class E hard hat Arc rated jacket, parka, or rainwear [As needed] Leather gloves [Rubber insulating gloves with leather protectors meet this requirement] Leather work shoes

Hazard/Risk Category 3 FR Clothing, Minimum Arc Rating of 25 cal/cm^2	Arc-rated long-sleeve shirt and arc-rated pants Arc-rated coverall Arc-rated arc flash suit jacket Arc-rated arc flash suit pants [Any combination of these items which provides a 25 cal/cm^2 total rating] Arc-rated flash suit hood with Class E hard hat Arc rated jacket, parka, or rainwear [As needed] Arc-rated gloves Leather work shoes
Hazard/Risk Category 4 FR Clothing, Minimum Arc Rating of 40 cal/cm^2	Arc-rated long-sleeve shirt and arc-rated pants Arc-rated coverall Arc-rated arc flash suit jacket Arc-rated arc flash suit pants [Any combination of these items which provides a 40 cal/cm^2 total rating] Arc-rated flash suit hood with Class E hard hat Arc rated jacket, parka, or rainwear [As needed] Arc-rated gloves Leather work shoes
Data Source: 2012 Edition NFPA 70E – Table 130.7(C)(16)	

(a)(2)	**General Protective equipment and tools.**
(a)(2)(i)	When working near exposed energized conductors or circuit parts, each employee shall use insulated tools or handling equipment if the tools or handling equipment might make contact with such conductors or parts. If the insulating capability of insulated tools or handling equipment is subject to damage, the insulating material shall be protected.
(a)(2)(i)(A)	Fuse handling equipment, insulated for the circuit voltage, shall be used to remove or install fuses when the fuse terminals are energized.
(a)(2)(i)(B)	Ropes and handlines used near exposed energized parts shall be nonconductive.
(a)(2)(ii)	Protective shields, protective barriers, or insulating materials shall be used to protect each employee from shock, burns, or other electrically related injuries while that employee is working near exposed energized parts which might be accidentally contacted or where dangerous electric heating or arcing might occur. When normally enclosed live parts are exposed for maintenance or repair, they shall be guarded to protect unqualified persons from contact with the live parts.

(b)	**Alerting techniques.** The following alerting techniques shall be used to warn and protect employees from hazards which could cause injury due to electric shock, burns, or failure of electric equipment parts:
(b)(1)	**Safety signs and tags.** Safety signs, safety symbols, or accident prevention tags shall be used where necessary to warn employees about electrical hazards which may endanger them, as required by § 1910.145.
(b)(2)	**Barricades.** Barricades shall be used in conjunction with safety signs where it is necessary to prevent or limit employee access to work areas exposing employees to uninsulated energized conductors or circuit parts. Conductive barricades may not be used where they might cause an electrical contact hazard.
(b)(3)	**Attendants.** If signs and barricades do not provide sufficient warning and protection from electrical hazards, an attendant shall be stationed to warn and protect employees.

Alerting Techniques

Safety signs, tags and symbols shall be used anywhere an accident might be prevented. OSHA §1910.145 requires potential electrical hazards that may endanger employees to be identified by appropriate safety signs.

Barricades will be used in conjunction with signs to prevent employees from entering areas where electrical hazards exist. Employees should be trained not to cross barricades unless they are trained to recognize and handle the potential electrical hazard.

The primary purpose of these "alerting techniques" is to keep unqualified persons away from the electrical hazards. OSHA requirements are not precise on how far unqualified persons must be kept away. The 2012 edition of NFPA 70E provides much needed guidance in this area.

For shock protection, NFPA 70E requires unqualified persons to approach exposed live parts no closer than the Limited Approach Boundary specified in Table 130.4(C)(a). (This table has been reproduced earlier in this document (Table 333-3).) In some cases the distance may even be greater than that specified in Table 130.4(C)(a) due to an arc-flash/blast hazard. Section 130.5 requires a Flash Hazard Analysis be performed before a person can approach exposed electrical conductors or circuit parts that have not been placed in an electrically safe condition. In some cases the Flash Protection Boundary may be greater than the Limited Approach Boundary.

The following Sections have been reserved for future use by OSHA

1910.336	Reserved
1910.337	Reserved
1910.338	Reserved
1910.339	Reserved
1910.340	Reserved
1910.341	Reserved
1910.342	Reserved
1910.343	Reserved
1910.344	Reserved
1910.345	Reserved
1910.346	Reserved
1910.347	Reserved
1910.348	Reserved
1910.349	Reserved
1910.350	Reserved
1910.351	Reserved
1910.352	Reserved
1910.353	Reserved
1910.354	Reserved
1910.355	Reserved
1910.356	Reserved
1910.357	Reserved
1910.358	Reserved
1910.359	Reserved
1910.360	Reserved
1910.361	Reserved
1910.362	Reserved
1910.363	Reserved
1910.364	Reserved
1910.365	Reserved
1910.366	Reserved
1910.367	Reserved
1910.368	Reserved
1910.369	Reserved
1910.370	Reserved
1910.371	Reserved
1910.372	Reserved
1910.373	Reserved
1910.374	Reserved
1910.375	Reserved
1910.376	Reserved
1910.377	Reserved
1910.378	Reserved
1910.379	Reserved
1910.380	Reserved
1910.381	Reserved
1910.382	Reserved
1910.383	Reserved
1910.384	Reserved
1910.385	Reserved
1910.386	Reserved
1910.387	Reserved
1910.388	Reserved
1910.389	Reserved
1910.390	Reserved
1910.391	Reserved
1910.392	Reserved
1910.393	Reserved
1910.394	Reserved
1910.395	Reserved
1910.396	Reserved
1910.397	Reserved
1910.398	Reserved

§ 1910.399 Definitions Applicable to This Subpart

[46 FR 4056, Jan. 16, 1981; 46 FR 40185, Aug. 7, 1981; as amended at 53 FR 12123, Apr. 12, 1988; 55 FR 32020, Aug. 6, 1990; 55 FR 46054, Nov. 1, 1990; 72 7215, Feb. 14, 2007]

(a) Definitions applicable to this subpart.

Acceptable. An installation or equipment is acceptable to the Assistant Secretary of Labor, and approved within the meaning of this Subpart S:

(1) If it is accepted, or certified, or listed, or labeled, or otherwise determined to be safe by a nationally recognized testing laboratory recognized pursuant to § 1910.7; or

(2) With respect to an installation or equipment of a kind that no nationally recognized testing laboratory accepts, certifies, lists, labels, or determines to be safe, if it is inspected or tested by another Federal agency, or by a State, municipal, or other local authority responsible for enforcing occupational safety provisions of the National Electrical Code®, and found in compliance with the provisions of the National Electrical Code® as applied in this Subpart; or

(3) With respect to custom-made equipment or related installations that are designed, fabricated for, and intended for use by a particular customer, if it is determined to be safe for its intended use by its manufacturer on the basis of test data which the employer keeps and makes available for inspection to the assistant Secretary and his authorized representatives. Refer to 1910.7 for definition of nationally recognized testing laboratory.

Accepted. An installation is "accepted" if it has been inspected and found by a nationally recognized testing laboratory to conform to specified plans or to procedures of applicable codes.

Accessible. (As applied to wiring methods.) Capable of being removed or exposed without damaging the building structure or finish, or not permanently closed in by the structure or finish of the building. (See "concealed" and "exposed".)

Accessible. (As applied to equipment.) Admitting close approach; not guarded by locked doors, elevation, or other effective means. (See "Readily accessible".)

Ampacity. The current, in amperes, that a conductor can carry continuously under the conditions of use without exceeding its temperature rating.

Appliances. Utilization equipment, generally other than industrial, normally built in standardized sizes or types, that is installed or connected as a unit to perform one or more functions such as clothes washing, air conditioning, food mixing, deep frying, etc.

Approved. Acceptable to the authority enforcing this Subpart. The authority enforcing this Subpart is the Assistant Secretary of Labor for Occupational Safety and Health. The definition of "acceptable" indicates what is acceptable to the Assistant Secretary of Labor, and therefore approved within the meaning of this Subpart.

Approved for the purpose. Approved for a specific purpose, environment, or application described in a particular standard requirement.

Suitability of equipment or materials for a specific purpose, environment or application may be determined by a nationally recognized testing laboratory, inspection agency or other organization concerned with product evaluation as part of its listing and labeling program. (See "Labeled" or "Listed".)

Armored cable *(Type AC)*. A fabricated assembly of insulated conductors in a flexible metallic enclosure.

Askarel. A generic term for a group of nonflammable synthetic chlorinated hydrocarbons used as electrical insulating media. Askarels of various compositional types are used. Under arcing conditions the gases produced, while consisting predominantly of noncombustible hydrogen chloride, can include varying amounts of combustible gases depending upon the askarel type.

Attachment plug (Plug cap) (Cap). A device which, by insertion in a receptacle, establishes connection between the conductors of the attached flexible cord and the conductors connected permanently to the receptacle.

Automatic. Self-acting, operating by its own mechanism when actuated by some impersonal influence, as, for example, a change in current strength, pressure, temperature, or mechanical configuration.

Bare conductor. See Conductor.

Barrier. A physical obstruction that is intended to prevent contact with equipment or live parts or to prevent unauthorized access to a work area.

Bathroom. An area including a basin with one or more of the following: a toilet, a tub, or a shower.

1910.399

Title: Definitions applicable to this subpart.

Subpart: S

SubPart Title: Electrical - Definitions

Bonding (Bonded). The permanent joining of metallic parts to form an electrically conductive path that ensures electrical continuity and the capacity to conduct safely any current likely to be imposed.

Bonding jumper. A conductor that assures the necessary electrical conductivity between metal parts required to be electrically connected.

Branch circuit. The circuit conductors between the final overcurrent device protecting the circuit and the outlet(s).

Building. A structure that stands alone or which is cut off from adjoining structures by fire walls with all openings therein protected by approved fire doors.

Cabinet. An enclosure designed either for surface or flush mounting, and provided with a frame, mat, or trim in which a swinging door or doors are or can be hung.

Cable tray system. A unit or assembly of units or sections and associated fittings forming a rigid structural system used to securely fasten or support cables and raceways. Cable tray systems include ladders, troughs, channels, solid bottom trays, and other similar structures.

Cablebus. An assembly of insulated conductors with fittings and conductor terminations in a completely enclosed, ventilated, protective metal housing.

Cell line: An assembly of electrically interconnected electrolytic cells supplied by a source of direct-current power.

Cell line attachments and auxiliary equipment: Cell line attachments and auxiliary equipment include, but are not limited to, auxiliary tanks, process piping, ductwork, structural supports, exposed cell line conductors, conduits and other raceways, pumps, positioning equipment, and cell cutout or bypass electrical devices. Auxiliary equipment also includes tools, welding machines, crucibles, and other portable equipment used for operation and maintenance within the electrolytic cell line working zone. In the cell line working zone, auxiliary equipment includes the exposed conductive surfaces of ungrounded cranes and crane-mounted cell-servicing equipment.

Center pivot irrigation machine. A multi-motored irrigation machine that revolves around a central pivot and employs alignment switches or similar devices to control individual motors.

Certified. Equipment is "certified" if it bears a label, tag or other record of certification that the equipment:

(1) Has been tested and found by a nationally recognized testing laboratory to meet nationally recognized standards or to be safe for use in a specified manner, or

(2) Is of a kind whose production is periodically inspected by a nationally recognized testing laboratory, and is accepted by the laboratory as safe for its intended use.

Circuit breaker. A device designed to open and close a circuit by nonautomatic means and to open the circuit automatically on a predetermined overcurrent without damage to itself when properly applied within its rating.

Class I locations. Class I locations are those in which flammable gases or vapors are or may be present in the air in quantities sufficient to produce explosive or ignitable mixtures. Class I locations include the following:

(1) Class I, Division 1. A Class I, Division 1 location is a location:

(i) In which ignitable concentrations of flammable gases or vapors may exist under normal operating conditions; or

(ii) In which ignitable concentrations of such gases or vapors may exist frequently because of repair or maintenance operations or because of leakage; or

(iii) In which breakdown or faulty operation of equipment or processes might release ignitable concentrations of flammable gases or vapors, and might also cause simultaneous failure of electric equipment.

Note to the definition of "Class 1, Division 1:" This classification usually includes locations where volatile flammable liquids or liquefied flammable gases are transferred from one container to another; interiors of spray booths and areas in the vicinity of spraying and painting operations where volatile flammable solvents are used; locations containing open tanks or vats of volatile flammable liquids; drying rooms or compartments for the evaporation of flammable solvents; locations containing fat and oil extraction equipment using volatile flammable solvents; portions of cleaning and dyeing plants where flammable liquids are used; gas generator rooms and other portions of gas manufacturing plants where flammable gas may escape; inadequately ventilated pump rooms for flammable gas or for volatile flammable liquids; the interiors of refrigerators and freezers in which volatile flammable materials are stored in open, lightly stoppered, or easily ruptured containers; and all other locations where ignitable concentrations of flammable vapors or gases are likely to occur in the course of normal operations.

(2) Class I, Division 2. A Class I, Division 2 location is a location:

(i) In which volatile flammable liquids or flammable gases are handled, processed, or used, but in which the hazardous liquids, vapors, or gases will normally be confined within closed containers or closed systems from which they can escape only in the event of accidental rupture or breakdown of such containers or systems, or as a result of abnormal operation of equipment; or

(ii) In which ignitable concentrations of gases or vapors are normally prevented by positive mechanical ventilation, and which might become hazardous through failure or abnormal operations of the ventilating equipment; or

(iii) That is adjacent to a Class I, Division 1 location, and to which ignitable concentrations of gases or vapors might occasionally be communicated unless such communication is prevented by adequate positive-pressure ventilation from a source of clean air, and effective safeguards against ventilation failure are provided.

Note to the definition of "Class 1, Division 2:" This classification usually includes locations where volatile flammable liquids or flammable gases or vapors are used, but which would become hazardous only in case of an accident or of some unusual operating condition. The quantity of flammable material that might escape in case of accident, the adequacy of ventilating equipment, the total area involved, and the record of the industry or business with respect to explosions or fires are all factors that merit consideration in determining the classification and extent of each location.

Piping without valves, checks, meters, and similar devices would not ordinarily introduce a hazardous condition even though used for flammable liquids or gases. Locations used for the storage of flammable liquids or a liquefied or compressed gases in sealed containers would not normally be considered hazardous unless also subject to other hazardous conditions.

Electrical conduits and their associated enclosures separated from process fluids by a single seal or barrier are classed as a Division 2 location if the outside of the conduit and enclosures is a nonhazardous location.

(3) *Class I, Zone 0.* A Class I, Zone 0 location is a location in which one of the following conditions exists:

(i) Ignitable concentrations of flammable gases or vapors are present continuously; or

(ii) Ignitable concentrations of flammable gases or vapors are present for long periods of time.

Note to the definition of "Class I, Zone 0:" As a guide in determining when flammable gases or vapors are present continuously or for long periods of time, refer to *Recommended Practice for Classification of Locations for Electrical Installations of Petroleum Facilities Classified as Class I, Zone 0, Zone 1 or Zone 2,* API RP 505-1997; *Electrical Apparatus for Explosive Gas Atmospheres, Classifications of Hazardous Areas,* IEC 79-10-1995; *Area Classification Code for Petroleum Installations, Model Code - Part 15,* Institute for Petroleum; and *Electrical Apparatus for Explosive Gas Atmospheres, Classifications of Hazardous (Classified) Locations,* ISA S12.24.01-1997.

(4) *Class I, Zone 1.* A Class I, Zone 1 location is a location in which one of the following conditions exists:

(i) Ignitable concentrations of flammable gases or vapors are likely to exist under normal operating conditions; or

(ii) Ignitable concentrations of flammable gases or vapors may exist frequently because of repair or maintenance operations or because of leakage; or

(iii) Equipment is operated or processes are carried on of such a nature that equipment breakdown or faulty operations could result in the release of ignitable concentrations of flammable gases or vapors and also cause simultaneous failure of electric equipment in a manner that would cause the electric equipment to become a source of ignition; or

(iv) A location that is adjacent to a Class I, Zone 0 location from which ignitable concentrations of vapors could be communicated, unless communication is prevented by adequate positive pressure ventilation from a source of clean air and effective safeguards against ventilation failure are provided.

(5) ***Class I, Zone 2.*** A Class I, Zone 2 location is a location in which one of the following conditions exists:

(i) Ignitable concentrations of flammable gases or vapors are not likely to occur in normal operation and if they do occur will exist only for a short period; or

(ii) Volatile flammable liquids, flammable gases, or flammable vapors are handled, processed, or used, but in which the liquids, gases, or vapors are normally confined within closed containers or closed systems from which they can escape only as a result of accidental rupture or breakdown of the containers or system or as the result of the abnormal operation of the equipment with which the liquids or gases are handled, processed, or used; or

(iii) Ignitable concentrations of flammable gases or vapors normally are prevented by positive mechanical ventilation, but which may become hazardous as the result of failure or abnormal operation of the ventilation equipment; or

(iv) A location that is adjacent to a Class I, Zone 1 location, from which ignitable concentrations of flammable gases or vapors could be communicated, unless such communication is prevented by adequate positive-pressure ventilation from a source of clean air, and effective safeguards against ventilation failure are provided.

Class II locations. Class II locations are those that are hazardous because of the presence of combustible dust. Class II locations include the following:

(1) Class II, Division 1. A Class II, Division 1 location is a location:

(i) In which combustible dust is or may be in suspension in the air under normal operating conditions, in quantities sufficient to produce explosive or ignitable mixtures; or

(ii) Where mechanical failure or abnormal operation of machinery or equipment might cause such explosive or ignitable mixtures to be produced, and might also provide a source of ignition through simultaneous failure of electric equipment, operation of protection devices, or from other causes, or

(iii) In which combustible dusts of an electrically conductive nature may be present.

Note to the definition of "Class II, Division 1:" This classification may include areas of grain handling and processing plants, starch plants, sugar-pulverizing plants, malting plants, hay-grinding plants, coal pulverizing plants, areas where metal dusts and powders are produced or processed, and other similar locations that contain dust producing machinery and equipment (except where the equipment is dust-tight or vented to the outside). These areas would have combustible dust in the air, under normal operating conditions, in quantities sufficient to produce explosive or ignitable mixtures. Combustible dusts that are electrically nonconductive include dusts produced in the handling and processing of grain and grain products, pulverized sugar and cocoa, dried egg and milk powders, pulverized spices, starch and pastes, potato and woodflour, oil meal from beans and seed, dried hay, and other organic materials which may produce combustible dusts when processed or handled. Dusts containing magnesium or aluminum are particularly hazardous and the use of extreme caution is necessary to avoid ignition and explosion.

(2) Class II, Division 2. A Class II, Division 2 location is a location where:

(i) Combustible dust will not normally be in suspension in the air in quantities sufficient to produce explosive or ignitable mixtures, and dust accumulations will normally be insufficient to interfere with the normal operation of electric equipment or other apparatus, but combustible dust may be in suspension in the air as a result of infrequent malfunctioning of handling or processing equipment; and

(ii) Resulting combustible dust accumulations on, in, or in the vicinity of the electric equipment may be sufficient to interfere with the safe dissipation of heat from electric equipment or may be ignitable by abnormal operation or failure of electric equipment.

Note to the definition of "Class II, Division 2:" This classification includes locations where dangerous concentrations of suspended dust would not be likely but where dust accumulations might form on or in the vicinity of electric equipment. These areas may contain equipment from which appreciable quantities of dust would escape under abnormal operating conditions or be adjacent to a Class II Division 1 location, as described above, into which an explosive or ignitable concentration of dust may be put into suspension under abnormal operating conditions.

Class III locations. Class III locations are those that are hazardous because of the presence of easily ignitable fibers or flyings but in which such fibers or flyings are not likely to be in suspension in the air in quantities sufficient to produce ignitable mixtures. Class III locations include the following:

(1) Class III, Division 1. A Class III, Division 1 location is a location in which easily ignitable fibers or materials producing combustible flyings are handled, manufactured, or used.

Note to the definition of "Class III, Division 1:" Such locations usually include some parts of rayon, cotton, and other textile mills; combustible fiber manufacturing and processing plants; cotton gins and cotton-seed mills; flax-processing plants; clothing manufacturing plants; woodworking plants, and establishments; and industries involving similar hazardous processes or conditions.

Easily ignitable fibers and flyings include rayon, cotton (including cotton linters and cotton waste), sisal or henequen, istle, jute, hemp, tow, cocoa fiber, oakum, baled waste kapok, Spanish moss, excelsior, and other materials of similar nature.

(2) Class III, Division 2. A Class III, Division 2 location is a location in which easily ignitable fibers are stored or handled, other than in the process of manufacture.

Collector ring. A collector ring is an assembly of slip rings for transferring electrical energy from a stationary to a rotating member.

Competent Person. One who is capable of identifying existing and predictable hazards in the surroundings or working conditions that are unsanitary, hazardous, or dangerous to employees and who has authorization to take prompt corrective measures to eliminate them.

Concealed. Rendered inaccessible by the structure or finish of the building. Wires in concealed raceways are considered concealed, even though they may become accessible by withdrawing them. [See "Accessible. (As applied to wiring methods.)"]

Conductor.

(1) Bare. A conductor having no covering or electrical insulation whatsoever.

(2) Covered. A conductor encased within material of composition or thickness that is not recognized by this subpart as electrical insulation.

(3) Insulated. A conductor encased within material of composition and thickness that is recognized by this subpart as electrical insulation.

Conduit body. A separate portion of a conduit or tubing system that provides access through one or more removable cover(s) to the interior of the system at a junction of two or more sections of the system or at a terminal point of the system. Boxes such as FS and FD or larger cast or sheet metal boxes are not classified as conduit bodies.

Controller. A device or group of devices that serves to govern, in some predetermined manner, the electric power delivered to the apparatus to which it is connected.

Covered conductor. See "Conductor".

Cutout. (Over 600 volts, nominal.) An assembly of a fuse support with either a fuseholder, fuse carrier, or disconnecting blade. The fuseholder or fuse carrier may include a conducting element (fuse link), or may act as the disconnecting blade by the inclusion of a nonfusible member.

Cutout box. An enclosure designed for surface mounting and having swinging doors or covers secured directly to and telescoping with the walls of the box proper. (See "Cabinet".)

Damp location. See "Location".

Dead front. Without live parts exposed to a person on the operating side of the equipment.

Deenergized. Free from any electrical connection to a source of potential difference and from electrical charge; not having a potential different from that of the earth.

1910.399

Title: Definitions applicable to this subpart.

Subpart: S

SubPart Title: Electrical - Definitions

Device. A unit of an electrical system which is intended to carry but not utilize electric energy.

Dielectric heating. Dielectric heating is the heating of a nominally insulating material due to its own dielectric losses when the material is placed in a varying electric field.

Disconnecting means. A device, or group of devices, or other means by which the conductors of a circuit can be disconnected from their source of supply.

Disconnecting (or Isolating) switch. (Over 600 volts, nominal.) A mechanical switching device used for isolating a circuit or equipment from a source of power.

Electrolytic cell line working zone. The cell line working zone is the space envelope wherein operation or maintenance is normally performed on or in the vicinity of exposed energized surfaces of electrolytic cell lines or their attachments.

Electrolytic cells. A tank or vat in which electrochemical reactions are caused by applying energy for the purpose of refining or producing usable materials.

Enclosed. Surrounded by a case, housing, fence or walls which will prevent persons from accidentally contacting energized parts.

Enclosure. The case or housing of apparatus, or the fence or walls surrounding an installation to prevent personnel from accidentally contacting energized parts, or to protect the equipment from physical damage.

Energized. Electrically connected to a source of potential difference.

Equipment. A general term including material, fittings, devices, appliances, fixtures, apparatus, and the like, used as a part of, or in connection with, an electrical installation.

Equipment grounding conductor. See "Grounding conductor, equipment".

Explosion-proof apparatus. Apparatus enclosed in a case that is capable of withstanding an explosion of a specified gas or vapor that may occur within it and of preventing the ignition of a specified gas or vapor surrounding the enclosure by sparks, flashes, or explosion of the gas or vapor within, and that operates at such an external temperature that it will not ignite a surrounding flammable atmosphere.

Exposed. (As applied to live parts.) Capable of being inadvertently touched or approached nearer than a safe distance by a person. It is applied to parts not suitably guarded, isolated, or insulated. (See "Accessible" and "Concealed".)

Exposed. (As applied to wiring methods.) On or attached to the surface or behind panels designed to allow access. [See "Accessible. (As applied to wiring methods.)"]

Exposed. (For the purposes of § 1910.308(e).) Where the circuit is in such a position that in case of failure of supports or insulation, contact with another circuit may result.

Externally operable. Capable of being operated without exposing the operator to contact with live parts.

Feeder. All circuit conductors between the service equipment, the source of a separate derived system, or other power supply source and the final branch-circuit overcurrent device.

Fitting. An accessory such as a locknut, bushing, or other part of a wiring system that is intended primarily to perform a mechanical rather than an electrical function.

Fountain. Fountains, ornamental pools, display pools, and reflection pools.

Note to the definition of "fountain:" This definition does not include drinking fountains.

Fuse. (Over 600 volts, nominal.) An overcurrent protective device with a circuit opening fusible part that is heated and severed by the passage of overcurrent through it. A fuse comprises all the parts that form a unit capable of performing the prescribed functions. It may or may not be the complete device necessary to connect it into an electrical circuit.

Ground. A conducting connection, whether intentional or accidental, between an electrical circuit or equipment and the earth, or to some conducting body that serves in place of the earth.

Grounded. Connected to earth or to some conducting body that serves in place of the earth.

1910.399

Title: Definitions applicable to this subpart.

Subpart: S

SubPart Title: Electrical - Definitions

Grounded, effectively. Intentionally connected to earth through a ground connection or connections of sufficiently low impedance and having sufficient current-carrying capacity to prevent the buildup of voltages that may result in undue hazards to connected equipment or to persons.

Grounded conductor. A system or circuit conductor that is intentionally grounded.

Grounding conductor. A conductor used to connect equipment or the grounded circuit of a wiring system to a grounding electrode or electrodes.

Grounding conductor, equipment. The conductor used to connect the non-current-carrying metal parts of equipment, raceways, and other enclosures to the system grounded conductor and/or the grounding electrode conductor, or both, at the service equipment or at the source of a separately derived system.

Grounding electrode conductor. The conductor used to connect the grounding electrode to the equipment grounding conductor and/or to the grounded conductor, or to both, of the circuits at the service equipment or at the source of a separately derived system.

Ground-fault circuit-interrupter. A device intended for the protection of personnel that functions to deenergize a circuit or a portion of a circuit within an established period of time when a current to ground exceeds some predetermined value that is less than that required to operate the overcurrent protective device of the supply circuit.

Guarded. Covered, shielded, fenced, enclosed, or otherwise protected by means of suitable covers, casings, barriers, rails, screens, mats, or platforms to remove the likelihood of approach to a point of danger or contact by persons or objects.

Health care facilities. Buildings or portions of buildings in which medical, dental, psychiatric, nursing, obstetrical, or surgical care are provided.

> **Note to the definition of "health care facilities:"** Health care facilities include, but are not limited to, hospitals, nursing homes, limited care facilities, clinics, medical and dental offices, and ambulatory care centers, whether permanent or movable.

Heating equipment. For the purposes of § 1910.306(g), the term "heating equipment" includes any equipment used for heating purposes if heat is generated by induction or dielectric methods.

Hoistway. Any shaftway, hatchway, well hole, or other vertical opening or space that is designed for the operation of an elevator or dumbwaiter.

Identified (as applied to equipment). Approved as suitable for the specific purpose, function, use, environment, or application, where described in a particular requirement.

> **Note to the definition of "identified:"** Some examples of ways to determine suitability of equipment for a specific purpose, environment, or application include investigations by a nationally recognized testing laboratory (through listing and labeling), inspection agency, or other organization recognized under the definition of "acceptable."

Induction heating. Induction heating is the heating of a nominally conductive material due to its own I^2R losses when the material is placed in a varying electromagnetic field.

Insulated. Separated from other conducting surfaces by a dielectric (including air space) offering a high resistance to the passage of current.

Insulated conductor. See Conductor, Insulated.

Interrupter switch. (Over 600 volts, nominal.) A switch capable of making, carrying and interrupting specified currents.

Irrigation Machine. An irrigation machine is an electrically driven or controlled machine, with one or more motors, not hand portable, and used primarily to transport and distribute water for agricultural purposes.

Isolated. (As applied to location.) Not readily accessible to persons unless special means for access are used.

Isolated power system. A system comprising an isolating transformer or its equivalent, a line isolation monitor, and its ungrounded circuit conductors.

Labeled. Equipment is "labeled" if there is attached to it a label, symbol, or other identifying mark of a nationally recognized testing laboratory:

(1) That makes periodic inspections of the production of such equipment, and

(2) Whose labeling indicates compliance with nationally recognized standards or tests to determine safe use in a specified manner.

Lighting outlet. An outlet intended for the direct connection of a lampholder, a lighting fixture, or a pendant cord terminating in a lampholder.

Line-clearance tree trimming. The pruning, trimming, repairing, maintaining, removing, or clearing of trees or cutting of brush that is within 10 feet (305 cm) of electric supply lines and equipment.

Listed. Equipment is "listed" if it is of a kind mentioned in a list that:

(1) Is published by a nationally recognized laboratory that makes periodic inspection of the production of such equipment, and

(2) States that such equipment meets nationally recognized standards or has been tested and found safe for use in a specified manner.

Live parts. Energized conductive components.

Location.

(1) **Damp location.** Partially protected locations under canopies, marquees, roofed open porches, and like locations, and interior locations subject to moderate degrees of moisture, such as some basements, some barns, and some cold-storage warehouses.

(2) **Dry location.** A location not normally subject to dampness or wetness. A location classified as dry may be temporarily subject to dampness or wetness, as in the case of a building under construction.

(3) **Wet location.** Installations underground or in concrete slabs or masonry in direct contact with the earth, and locations subject to saturation with water or other liquids, such as vehicle-washing areas, and locations exposed to weather and unprotected.

Medium voltage cable (Type MV). A single or multiconductor solid dielectric insulated cable rated 2001 volts or higher.

Metal-clad cable (Type MC). A factory assembly of one or more insulated circuit conductors with or without optical fiber members enclosed in an armor of interlocking metal tape, or a smooth or corrugated metallic sheath.

Mineral-insulated metal-sheathed cable (Type MI). Type MI, mineral-insulated metal-sheathed cable is a factory assembly of one or more conductors insulated with a highly compressed refractory mineral insulation and enclosed in a liquidtight and gastight continuous copper or alloy steel sheath.

Mobile X-ray. X-ray equipment mounted on a permanent base with wheels and/or casters for moving while completely assembled.

Motor control center. An assembly of one or more enclosed sections having a common power bus and principally containing motor control units.

Nonmetallic-sheathed cable (Types NM, NMC, and NMS). A factory assembly of two or more insulated conductors having an outer sheath of moisture resistant, flame-retardant, nonmetallic material.

Oil (filled) cutout. (Over 600 volts, nominal.) A cutout in which all or part of the fuse support and its fuse link or disconnecting blade are mounted in oil with complete immersion of the contacts and the fusible portion of the conducting element (fuse link), so that arc interruption by severing of the fuse link or by opening of the contacts will occur under oil.

Open wiring on insulators. Open wiring on insulators is an exposed wiring method using cleats, knobs, tubes, and flexible tubing for the protection and support of single insulated conductors run in or on buildings, and not concealed by the building structure.

Outlet. A point on the wiring system at which current is taken to supply utilization equipment.

Outline lighting. An arrangement of incandescent lamps or electric discharge lighting to outline or call attention to certain features such as the shape of a building or the decoration of a window.

Overcurrent. Any current in excess of the rated current of equipment or the ampacity of a conductor. It may result from overload (see definition), short circuit, or ground fault. A current in excess of rating may be accommodated by certain equipment and conductors for a given set of conditions. Hence the rules for overcurrent protection are specific for particular situations.

Overhaul means to perform a major replacement, modification, repair, or rehabilitation similar to that involved when a new building or facility is built, a new wing is added, or an entire floor is renovated.

Overload. Operation of equipment in excess of normal, full load rating, or of a conductor in excess of rated ampacity that, when it persists for a sufficient length of time, would cause damage or dangerous overheating. A fault, such as a short circuit or ground fault, is not an overload. (See Overcurrent.)

Panelboard. A single panel or group of panel units designed for assembly in the form of a single panel; including buses, automatic overcurrent devices, and with or without switches for the control of light, heat, or power circuits; designed to be placed in a cabinet or cutout box placed in or against a wall or partition and accessible only from the front. (See "Switchboard".)

Permanently installed decorative fountains and reflection pools. Pools that are constructed in the ground, on the ground, or in a building in such a manner that the fountain or pool cannot be readily disassembled for storage, whether or not served by electrical circuits of any nature. These units are primarily constructed for their aesthetic value and are not intended for swimming or wading.

Permanently installed swimming, wading, and therapeutic pools. Pools that are constructed in the ground or partially in the ground, and all other capable of holding water in a depth greater than 1.07 m (42 in.). The definition also applies to all pools installed inside of a building, regardless of water depth, whether or not served by electric circuits of any nature.

Portable X-ray. X-ray equipment designed to be hand-carried.

Power and control tray cable (Type TC). A factory assembly of two or more insulated conductors, with or without associated bare or covered grounding conductors under a nonmetallic sheath, approved for installation in cable trays, in raceways, or where supported by a messenger wire.

Power fuse. (Over 600 volts, nominal.) See Fuse.

Power-limited tray cable (Type PLTC). A factory assembly of two or more insulated conductors under a nonmetallic jacket.

Power outlet. An enclosed assembly which may include receptacles, circuit breakers, fuseholders, fused switches, buses and watt-hour meter mounting means; that is intended to supply and control power to mobile homes, recreational vehicles or boats, or to serve as a means for distributing power needed required to operate mobile or temporarily installed equipment.

Premises wiring. (Premises wiring system.) The interior and exterior wiring, including power, lighting, control, and signal circuit wiring together with all of their associated hardware, fittings, and wiring devices, both permanently and temporarily installed, that extends from the service point of utility conductors or source of power (such as a battery, a solar photovoltaic system, or a generator, transformer, or converter) to the outlets. Such wiring does not include wiring internal to appliances, fixtures, motors, controllers, motor control centers, and similar equipment.

Qualified person. One who has received training in and has demonstrated skills and knowledge in the construction and operation of electric equipment and installations and the hazards involved.

> **Note 1 to the definition of "qualified person:"** Whether an employee is considered to be a "qualified person" will depend upon various circumstances in the workplace. For example, it is possible and, in fact, likely for an individual to be considered "qualified" with regard to certain equipment in the workplace, but "unqualified" as to other equipment. (See § 1910.332 (b)(3) for training requirements that specifically apply to qualified persons.)

> **Note 2 to the definition of "qualified person:"** An employee who is undergoing on-the-job training and who, in the course of such training, has demonstrated an ability to perform duties safely at his or her level of training and who is under the direct supervision of a qualified person is considered to be a qualified person for the performance of those duties.

Raceway. An enclosed channel of metal or nonmetallic materials designed expressly for holding wires, cables, or busbars, with additional functions as permitted in this standard. Raceways include, but are not limited to, rigid metal conduit, rigid nonmetallic conduit, intermediate metal conduit, liquidtight flexible conduit, flexible metallic tubing, flexible metal conduit, electrical metallic tubing, electrical nonmetallic tubing, underfloor raceways, cellular concrete floor raceways, cellular metal floor raceways, surface raceways, wireways, and busways.

Readily accessible. Capable of being reached quickly for operation, renewal, or inspections, so that those needing ready access do not have to climb over or remove obstacles or to resort to portable ladders, chairs, etc. (See Accessible.)

Receptacle. A receptacle is a contact device installed at the outlet for the connection of an attachment plug. A single receptacle is a single contact device with no other contact device on the same yoke. A multiple receptacle is two or more contact devices on the same yoke.

Receptacle outlet. An outlet where one or more receptacles are installed.

Remote-control circuit. Any electric circuit that controls any other circuit through a relay or an equivalent device.

Sealable equipment. Equipment enclosed in a case or cabinet that is provided with a means of sealing or locking so that live parts cannot be made accessible without opening the enclosure. The equipment may or may not be operable without opening the enclosure.

Separately derived system. A premises wiring system whose power is derived from a battery, a solar photovoltaic system, or from a generator, transformer, or converter windings, and that has no direct electrical connection, including a solidly connected grounded circuit conductor, to supply conductors originating in another system.

Service. The conductors and equipment for delivering electric energy from the serving utility to the wiring system of the premises served.

Service cable. Service conductors made up in the form of a cable.

Service conductors. The conductors from the service point to the service disconnecting means.

Service drop. The overhead service conductors from the last pole or other aerial support to and including the splices, if any, connecting to the service-entrance conductors at the building or other structure.

Service-entrance cable. A single conductor or multiconductor assembly provided with or without an overall covering, primarily used for services, and is of the following types:

- **(1) Type SE**. Type SE, having a flame-retardant, moisture resistant covering; and
- **(2) Type USE.** Type USE, identified for underground use, having a moisture-resistant covering, but not required to have a flame-retardant covering. Cabled, single-conductor, Type USE constructions recognized for underground use may have a bare copper conductor cabled with the assembly. Type USE single, parallel, or cable conductor assemblies recognized for underground use may have a bare copper concentric conductor applied. These constructions do not require an outer overall covering.

Service-entrance conductors, overhead system. The service conductors between the terminals of the service equipment and a point usually outside the building, clear of building walls, where joined by tap or splice to the service drop.

Service entrance conductors, underground system. The service conductors between the terminals of the service equipment and the point of connection to the service lateral.

Service equipment. The necessary equipment, usually consisting of one or more circuit breakers or switches and fuses, and their accessories, connected to the load end of service conductors to a building or other structure, or an otherwise designated area, and intended to constitute the main control and cutoff of the supply.

Service point. The point of connection between the facilities of the serving utility and the premises wiring.

Shielded nonmetallic-sheathed cable (Type SNM). A factory assembly of two or more insulated conductors in an extruded core of moisture-resistant, flame-resistant nonmetallic material, covered with an overlapping spiral metal tape and wire shield and jacketed with an extruded moisture-, flame-, oil-, corrosion-, fungus-, and sunlight-resistant nonmetallic material.

Show window. Any window used or designed to be used for the display of goods or advertising material, whether it is fully or partly enclosed or entirely open at the rear and whether or not it has a platform raised higher than the street floor level.

1910.399

Title: Definitions applicable to this subpart.

Subpart: S

SubPart Title: Electrical - Definitions

Signaling circuit. Any electric circuit that energizes signaling equipment.

Storable swimming or wading pool. A pool that is constructed on or above the ground and is capable of holding water to a maximum depth of 1.07 m (42 in.), or a pool with nonmetallic, molded polymeric walls or inflatable fabric walls regardless of dimension.

Switchboard. A large single panel, frame, or assembly of panels on which are mounted, on the face or back, or both, switches, overcurrent and other protective devices, buses, and (usually) instruments. Switchboards are generally accessible from the rear as well as from the front and are not intended to be installed in cabinets. (See Panelboard.)

Switch.

- **(1) General-use switch.** A switch intended for use in general distribution and branch circuits. It is rated in amperes, and it is capable of interrupting its rated current at its rated voltage.
- **(2) General-use snap switch.** A form of general-use switch so constructed that it can be installed in flush device boxes or on outlet box covers, or otherwise used in conjunction with wiring systems recognized by this subpart.
- **(3) Isolating switch.** A switch intended for isolating an electric circuit from the source of power. It has no interrupting rating, and it is intended to be operated only after the circuit has been opened by some other means.
- **(4) Motor-circuit switch.** A switch, rated in horsepower, capable of interrupting the maximum operating overload current of a motor of the same horsepower rating as the switch at the rated voltage.

Switching devices. (Over 600 volts, nominal.) Devices designed to close and open one or more electric circuits. Included in this category are circuit breakers, cutouts, disconnecting (or isolating) switches, disconnecting means, interrupter switches, and oil (filled) cutouts.

Transportable X-ray. X-ray equipment installed in a vehicle or that may readily be disassembled for transport in a vehicle.

Utilization equipment. Equipment that utilizes electric energy for electronic, electromechanical, chemical, heating, lighting, or similar purposes.

Ventilated. Provided with a means to permit circulation of air sufficient to remove an excess of heat, fumes, or vapors.

Volatile flammable liquid. A flammable liquid having a flash point below 38 °C (100 °F), or a flammable liquid whose temperature is above its flash point, or a Class II combustible liquid having a vapor pressure not exceeding 276 kPa (40 psia) at 38 °C (100 °F) and whose temperature is above its flash point.

Voltage (of a circuit). The greatest root-mean-square (rms) (effective) difference of potential between any two conductors of the circuit concerned.

Voltage, nominal. A nominal value assigned to a circuit or system for the purpose of conveniently designating its voltage class (as 120/240 volts, 480Y/277 volts, 600 volts). The actual voltage at which a circuit operates can vary from the nominal within a range that permits satisfactory operation of equipment.

Voltage to ground. For grounded circuits, the voltage between the given conductor and that point or conductor of the circuit that is grounded; for ungrounded circuits, the greatest voltage between the given conductor and any other conductor of the circuit.

Watertight. So constructed that moisture will not enter the enclosure.

Weatherproof. So constructed or protected that exposure to the weather will not interfere with successful operation. Rainproof, raintight, or watertight equipment can fulfill the requirements for weatherproof where varying weather conditions other than wetness, such as snow, ice, dust, or temperature extremes, are not a factor.

Wireways. Sheet-metal troughs with hinged or removable covers for housing and protecting electric wires and cable and in which conductors are laid in place after the wireway has been installed as a complete system.

§ 1910, Appendix A to Subpart S
Referenced Standards

[46 FR 4056, Jan. 16, 1981; 46 FR 40185, Aug. 7, 1981; 55 FR 32020, Aug. 6, 1990; 72 FR 7221, Feb. 14, 2007]

Appendix A - References for Further Information

The references contained in this appendix provide nonmandatory information that can be helpful in understanding and complying with Subpart S of this Part. However, compliance with these standards is not a substitute for compliance with Subpart S of this Part.

ANSI/API RP 500-1998 (2002) Recommended Practice for Classification of Locations for Electrical Installations at Petroleum Facilities Classified as Class I Division 1 and Division 2.

ANSI/API RP 505-1997 (2002) Recommended Practice for Classification of Locations for Electrical Installations at Petroleum Facilities Classified as Class I, Zone 0, Zone 1 and Zone 2.

ANSI/ASME A17.1-2004 Safety Code for Elevators and Escalators.

ANSI/ASME B30.2-2005 Overhead and Gantry Cranes (Top Running Bridge, Single or Multiple Girder, Top Running Trolley Hoist).

ANSI/ASME B30.3-2004 Construction Tower Cranes.

ANSI/ASME B30.4-2003 Portal, Tower, and Pedestal Cranes.

ANSI/ASME B30.5-2004 Mobile and Locomotive Cranes.

ANSI/ASME B30.6-2003 Derricks.

ANSI/ASME B30.7-2001 Base Mounted Drum Hoists.

ANSI/ASME B30.8-2004 Floating Cranes and Floating Derricks.

ANSI/ASME B30.11-2004 Monorails and Underhung Cranes.

ANSI/ASME B30.12-2001 Handling Loads Suspended from Rotorcraft.

ANSI/ASME B30.13-2003 Storage/Retrieval (S/R) Machines and Associated Equipment.

ANSI/ASME B30.16-2003 Overhead Hoists (Underhung).

ANSI/ASME B30.22-2005 Articulating Boom Cranes.

ANSI/ASSE Z244.1-2003 Control of Hazardous Energy Lockout/Tagout and Alternative Methods.

ANSI/ASSE Z490.1-2001 Criteria for Accepted Practices in Safety, Health, and Environmental Training.

ANSI/IEEE C2-2002 National Electrical Safety Code.

ANSI K61.1-1999 Safety Requirements for the Storage and Handling of Anhydrous Ammonia.

ANSI/UL 913-2003 Intrinsically Safe Apparatus and Associated Apparatus for Use in Class I, II, and III, Division 1, Hazardous (Classified) Locations.

ASTM D3176-1989 (2002) Standard Practice for Ultimate Analysis of Coal and Coke.

ASTM D3180-1989 (2002) Standard Practice for Calculating Coal and Coke Analyses from As-Determined to Different Bases.

NFPA 20-2003 Standard for the Installation of Stationary Pumps for Fire Protection.

NFPA 30-2003 Flammable and Combustible Liquids Code.

NFPA 32-2004 Standard for Drycleaning Plants.

NFPA 33-2003 Standard for Spray Application Using Flammable or Combustible Materials.

NFPA 34-2003 Standard for Dipping and Coating Processes Using Flammable or Combustible Liquids.

NFPA 35-2005 Standard for the Manufacture of Organic Coatings.

NFPA 36-2004 Standard for Solvent Extraction Plants.

NFPA 40-2001 Standard for the Storage and Handling of Cellulose Nitrate Film.

NFPA 58-2004 Liquefied Petroleum Gas Code.

NFPA 59-2004 Utility LP-Gas Plant Code.

NFPA 70-2002 National Electrical Code. (See also NFPA 70-2005.)

NFPA 70E-2000 Standard for Electrical Safety Requirements for Employee Workplaces. (See also NFPA 70E-2004.)

NFPA 77-2000 Recommended Practice on Static Electricity.

NFPA 80-1999 Standard for Fire Doors and Fire Windows.

NFPA 88A-2002 Standard for Parking Structures.

NFPA 91-2004 Standard for Exhaust Systems for Air Conveying of Vapors, Gases, Mists, and Noncombustible Particulate Solids.

NFPA 101-2006 Life Safety Code.

NFPA 496-2003 Standard for Purged and Pressurized Enclosures for Electrical Equipment.

NFPA 497-2004 Recommended Practice for the Classification of Flammable Liquids, Gases, or Vapors and of Hazardous (Classified) Locations for Electrical Installations in Chemical Process Areas.

NFPA 505-2006 Fire Safety Standard for Powered Industrial Trucks Including Type Designations, Areas of Use, Conversions, Maintenance, and Operations.

NFPA 820-2003 Standard for Fire Protection in Wastewater Treatment and Collection Facilities.

NMAB 353-1-1979 Matrix of Combustion-Relevant Properties and Classification of Gases, Vapors, and Selected Solids.

NMAB 353-2-1979 Test Equipment for Use in Determining Classifications of Combustible Dusts.

NMAB 353-3-1980 Classification of Combustible Dust in Accordance with the National Electrical Code.

Appendix B
Subpart S Old to New Cross-Reference

Old Section	New Section	Changes and Rationale
Sec. 1910.303 General	**Sec. 1910.303 General**	
1910.303(a)	1910.303(a)	No substantive change. A reference to Sec. 1910.399 definition of "approved" is added for clarification.
1910.303(b)(1), introductory text	1910.303(b)(1), introductory text	No substantive change.
1910.303(b)(1)(i)	1910.303(b)(1)(i)	No substantive change.
1910.303(b)(1)(ii)	1910.303(b)(1)(ii) 1910.303(b)(1)(iii)	*1910.303(b)(1)(ii) – No substantive change* **Adds wire-bending and connection space to the explicit list of things to consider when judging equipment.
1910.303(b)(1)(iii)	1910.303(b)(1)(iv)	No substantive change.
1910.303(b)(1)(iv)	1910.303(b)(1)(v)	No substantive change.
1910.303(b)(1)(v)	1910.303(b)(1)(vi)	No substantive change.
1910.303(b)(1)(vi)	1910.303(b)(1)(vii)	No substantive change.
1910.303(b)(1)(vii)	1910.303(b)(1)(viii)	No substantive change.
1910.303(b)(2)	1910.303(b)(2)	No substantive change.
	1910.303(b)(3)	**Adds a requirement for completed wiring to be free from short circuits and grounds other than those required in the standard.
	1910.303(b)(4)	**Adds requirements for equipment intended to interrupt current to have adequate interrupting ratings.
	1910.303(b)(5)	**Adds requirements for the coordination of overcurrent protection for circuits and equipment.
	1910.303(b)(6)	**Adds a requirement for conductors and equipment to be identified for the purpose when installed in an environment containing deteriorating agents.
	1910.303(b)(7)	**Adds requirements for installing electric equipment in a neat and workmanlike manner.
	1910.303(b)(8)	**Adds requirements for equipment to be mounted securely and to allow for proper cooling.
	1910.303(c)(1)	**Adds requirements to ensure that electrical connections are secure and electrically safe.
	1910.303(c)(2)	**Adds requirements for connections at terminals and for the identification of terminals intended for connection to more than one conductor or to aluminum.
1910.303(c)	1910.303(c)(3)(i)	No substantive change.
	1910.303(c)(3)(ii)	**Adds a requirement that wire connector or splicing means installed on directly buried conductors be listed for such use.

Old Section	New Section	Changes and Rationale
1910.303(d)	1910.303(d)	No substantive change.
1910.303(e)	1910.303(e)	No substantive change. (Individual requirements are placed in separate paragraphs).
1910.303(f)	1910.303(f)(1), (f)(2), and (f)(3)	No substantive change. (Individual requirements are placed in separate paragraphs).
	1910.303(f)(4)	Adds a requirement for disconnecting means required by Subpart S to be capable of accepting a lock. This provision is added to make the Subpart S requirements on disconnecting means consistent with § 1910.147(c)(2)(iii), which requires energy isolating devices (a generic term, which includes electrical disconnecting means) to be designed to accept a lockout device.
	1910.303(f)(5)	**Adds marking requirements for series combination ratings of circuit breakers or fuses.
1910.303(g)(1), introductory text	1910.303(g)(1), introductory text	No substantive change.
1910.303(g)(1)(i)	1910.303(g)(1)(i) Table S-1, Note 3	**The final rule revises the language to clarify how wide and high the clear space must be. (See detailed explanation later in the preamble).
1910.303(g)(1)(ii)	1910.303(g)(1)(ii)	No substantive change.
1910.303(g)(1)(iii)	1910.303(g)(1)(iii)	No substantive change.
	1910.303(g)(1)(iv)	**Adds a requirement for a second entrance on equipment rated 1200 amperes under certain conditions.
1910.303(g)(1)(iv)	1910.303(g)(1)(i)(B)	**Reduces the minimum width of the clear space to 762 mm.
1910.303(g)(1)(v)	1910.303(g)(1)(v)	**Adds a prohibition against controlling illumination for working spaces by automatic means only.
1910.303(g)(1)(vi)	1910.303(g)(1)(vi)	**Increased the minimum height of the working space from 1.91 m (6.25 ft) to 1.98 m (6.5 ft) for new installations.
	1910.303(g)(1)(vii)	**Adds requirements for switchboards, panelboards, and distribution boards installed for the control of light and power circuits, and motor control centers to be installed in dedicated space and to be protected against damage.
1910.303(g)(2)	1910.303(g)(2)	No substantive change.
1910.303(h)(1)	1910.303(h)(1)	No substantive change.
1910.303(h)(2), introductory text	1910.303(h)(2)(i) and (h)(2)(ii)	**The minimum height of fences restricting access to electrical installations over 600 V is reduced from 2.44 m to 2.13 m (7.0 ft).
1910.303(h)(2)(i) and (h)(2)(ii)	1910.303(h)(2)(iii), (h)(2)(iv), (h)(2)(v), and (h)(5)(iii)	**1. The final rule organizes these requirements based on whether the installations are indoors or outdoors. (The existing standard organizes them based on whether or not the installations are accessible to unqualified employees). 2. Adds requirements intended to prevent tampering by the general public. 3. Removes requirement to lock underground

Old Section	New Section	Changes and Rationale
		box covers weighing more than 45.4kg.
1910.303(h)(3), introductory text	1910.303(h)(3)	No substantive change.
1910.303(h)(3)(i)	1910.303(h)(5)(i), Table S-2, Note 3	**The distances in Table S-2 for the depth of working space in front of electric equipment are increased for new installations to match the distances in NFPA 70E-2000.
1910.303(h)(3)(ii)	1910.303(h)(5)(iv)	No substantive change.
1910.303(h)(3)(iii)	1910.303(h)(5)(v)	**The distances in Table S-3 for the elevations of unguarded live parts are increased for new installations to match the distances in NFPA 70E-2000.
1910.303(h)(4)(i)	1910.303(h)(4)(i)	**The existing standard requires a second entrance to give access to the working space about switchboards and control panels over 600 V if the equipment exceeds 1.22 mm (24 in.) in width if it is practical to install a second entrance. The final rule requires an entrance on each end of switchboards and panelboards exceeding 1.83 m (6 ft.) unless the working space permits a continuous and unobstructed way of travel or the working space is doubled. In addition, the final rule requires the lone entrance permitted under either of these exceptions to be at least the distance specified in Table S -2 from exposed live parts.
1910.303(h)(4)(ii)	1910.303(h)(4)(ii)	No substantive change.
	1910.303(h)(5)(ii)	**Adds requirements for equipment operating at 600 V or less installed in rooms or enclosures containing exposed live parts or exposed wiring operating at more than 600 V.
	1910.303(h)(5)(vi)	**Adds requirements limiting the installation of pipes or ducts that are foreign to electrical installation operating at more than 600 V.
Sec. 1910.304 Wiring design and protection.	**Sec. 1910.304 Wiring design and protection.**	
1910.304(a)(1)	1910.304(a)(1)	No substantive change. (Individual requirements are placed in separate paragraphs).
1910.304(a)(2)	1910.304(a)(2)	No substantive change.
1910.304(a)(3)	1910.304(a)(3)	No substantive change.
	1910.304(b)(1)	**Adds requirements for the identification of multi-wire branch circuits.
	1910.304(b)(2)(i)	**Adds requirements that receptacles installed on 15 – and 20 – ampere circuits be of the grounding type and that grounding-type receptacle be installed in circuits within their rating.
	1910.304(b)(2)(ii)	**Adds a requirement for grounding contacts on receptacles to be effectively grounded.
	1910.304(b)(2)(iii)	**Adds requirements on the methods used to ground receptacles and cord connectors.
	1910.304(b)(2)(iv)	**Adds requirements on the replacement of receptacles.

Old Section	New Section	Changes and Rationale
	1910.304(b)(2)(v)	**Adds a requirement that receptacles installed on branch circuits having different voltages, frequencies, or types of current be non-interchangeable.
	1910.304(b)(3)	**Adds requirements for ground fault circuit interrupter protection. (See the discussion of these requirements later in this section of the preamble).
1910.304(b)(2)	1910.304(b)(4), introductory text	No significant change.
	1910.304(b)(4)(i)	**Adds requirements for ratings of lampholders.
	1910.304(b)(4)(ii)	**Adds requirements for ratings of receptacles.
	1910.304(b)(5)	**Adds requirements for receptacles to be installed wherever cords with attachment plugs are used.
1910.304(c), introductory text	1910.304(c), introductory text	No significant change (The requirements in existing in existing paragraph (c)(5) are placed in a separate paragraph (d).
1910.304(c)(1)	1910.304(c)(1)	**Adds a requirement for the separation of conductors on poles.
1910.304(c)(2)	1910.304(c)(2)	**Increases the minimum clearances for new installation of open conductors and service drops to match those in NFPA 70E – 2000.
1910.304(c)(3)	1910.304(c)(3)(i)	No substantive change. (The final rule clarifies that paragraph (c)(2) applies to platforms, projections, or surfaces from which runs of open conductors can be reached).
	1910.304(c)(3)(ii)	**Adds restrictions for installing overhead service conductors near building openings through which materials may be moved.
1910.304(c)(4)	1910.304(c)(4)	**Adds an exception to the minimum clearance requirement for conductors attached to the side of a building. (The final rule also clarifies that paragraph (c)(2) applies to roof surfaces that are subject to pedestrian or vehicular traffic).
1910.304(c)(5)	1910.304(d)	No substantive change.
1910.304(d)(1)(i)	1910.304(e)(1)(i)	No substantive change.
1910.304(d)(1)(ii)	1910.304(e)(1)(ii)	No substantive change.
	1910.304(e)(1)(iii)	**Adds a requirement for service disconnecting means to be suitable for the prevailing conditions.
1910.304(d)(2)	1910.304(e)(2)	No substantive change.
1910.304(e)(1), introductory text	1910.304(f)(1), introductory text	No substantive change.
1910.304(e)(1)(i)	1910.304(f)(1)(i)	No substantive change.
1910.304(e)(1)(ii)	1910.304(f)(1)(ii)	No substantive change.
1910.304(e)(1)(iii)	1910.304(f)(1)(iii)	**The types of circuits that are allowed to have a single switch disconnect for multiple fuses are now specified in the standard.
1910.304(e)(1)(iv)	1910.304(f)(1)(iv)	No substantive change.
1910.304(e)(1)(v)	1910.304(f)(1)(v)	**Adds a requirement to clarify that handles of circuit breakers and similar moving parts also

Old Section	New Section	Changes and Rationale
		need to be guarded so that they do not injure employees.
1910.304(e)(1)(vi)(A)	1910.304(f)(1)(vi)	No substantive change.
1910.304(e)(1)(vi)(B)	1910.304(f)(1)(vii)	No substantive change.
1910.304(e)(1)(vi)(C)	1910.304(f)(1)(viii)	**Adds circuit breakers used on 277 – volt fluorescent lighting circuits to the types of breakers required to be marked "SWD."
	1910.304(f)(1)(ix)	**Adds a requirement to clarify ratings of circuit breakers.
1910.304(e)(2)	1910.304(f)(2)	**Adds specific requirements on how to protect feeders and branch circuits energized at more than 600 volts.
1910.304(f), introductory text	1910.304(g), introductory text	No substantive change.
1910.304(f)(1), introductory text	1910.304(g)(1), introductory text	No substantive change.
1910.304(f)(1)(i)	1910.304(g)(1)(i)	No substantive change.
1910.304(f)(1)(ii)	1910.304(g)(1)(ii)	No substantive change
1910.304(f)(1)(iii)	1910.304(g)(1)(iii)	No substantive change
1910.304(f)(1)(iv)	1910.304(g)(1)(iv)	No substantive change (The specific voltage ratings in existing paragraphs (g)(1)(iv)(B) and (g)(1)(iv)(C) are being removed. However, this is not a substantive change as those are the voltages used in the described systems).
1910.304(f)(1)(v)	1910.304(g)(1)(v)	**Adds an exception to the requirement to ground systems for high-impedance grounded systems or 480 V to 1000 V under certain conditions.
1910.304(f)(2)	1910.304(g)(2)	**No substantive change. (The standard adds descriptions of which conductor is to be grounded for the different systems.
	1910.304(g)(3)	**Changes requirements for grounding portable and vehicle mounted generators so that the requirements are equivalent to those in OSHA's Construction Standards (Sec 1926.404(f)(3)). The sentence in the construction standard reading "No other [nonneutral] conductor need be bonded to the generator frame" has been dropped from the general industry version. The sentence is not regulatory in nature, and its omission has no effect on the requirement.
1910.304(f)(3)	1910.304(g)(4)	**No longer allows employers to use a cold water pipe as a source of ground for installations made or modified after the effective date.
1910.304(f)(4)	1910.304(g)(5)	**Adds a requirement that the path to ground be effective.
1910.304(f)(5)(i)	1910.304(g)(6)(i)	No substantive change.
1910.304(f)(5)(ii)	1910.304(g)(6)(ii)	No substantive change.
1910.304(f)(5)(iii)	1910.304(g)(6)(iii)	No substantive change
1910.304(f)(5)(iv)	1910.304(g)(6)(iv) and (g)(6)(v)	**The exceptions for grounding fixed equipment operating at more than 150 V are extended to all fixed electric equipment regardless of voltage.

Old Section	New Section	Changes and Rationale
		Also, the final rule includes a new exception for double-insulated equipment.
1910.304(f)(5)(v)	1910.304(g)(6)(vi) and (g)(6)(vii)	**Adds the following equipment to the list of cord – and plug – connected equipment required to be grounded: stationary and fixed motor-operated tools and light industrial motor-operated tools.
1910.304(f)(5)(vi)	1910.304(g)(7)	**Adds frames and tracks electrically operated hoists to the list of non-electrical equipment required to be grounded.
1910.304(f)(6)	1910.304(g)(8)	No substantive change.
1910.304(f)(7)(i)	1910.304(g)(9), introductory text	No substantive change.
1910.304(f)(7)(ii)	1910.304(g)(9)(i)	No substantive change.
1910.304(f)(7)(iii)	1910.304(g)(9)(ii)	No substantive change.
Sec. 1910.305 Wiring methods, components, and equipment for general use.	**Sec. 1910.305 Wiring methods, components, and equipment for general use.**	
1910.305(a), introductory text	1910.305(a), introductory text	No substantive change.
1910.305(a)(1)(i)	1910.305(a)(1)(i)	**Adds a requirement that equipment be bonded so as to provide adequate fault-current-carrying capability. Also clarifies that nonconductive coatings need to be removed unless the fittings make this unnecessary.
	1910.305(a)(1)(ii)	**Adds an exception to the bonding requirement for the reduction of electrical noise.
1910.305(a)(1)(ii)	1910.305(a)(1)(iii)	No substantive change.
1910.305(a)(2), introductory text	1910.305(a)(2), introductory text	No substantive change. Removes the provision allowing temporary wiring to be of a class less than permanent wiring per the 2002 NEC. The change has no substantive effect because: (1) The term "a class less than" is not defined, and (2) temporary wiring is required to meet the same requirements regardless of the deleted language. (Both the final rule and the existing standard contain the following requirement: "Except as specifically modified in this paragraph, all other requirements of this subpart for permanent wiring shall apply to temporary wiring installations").
1910.305(a)(2)(i), introductory text	1910.305(a)(2)(i), introductory text	No substantive change.
1910.305(a)(2)(i)(A)	1910.305(a)(2)(i)(A)	Removes demolition from the list of activities for which temporary wiring is permitted. Demolition is a form of construction work, which is not covered by the Subpart S installation requirements.
1910.305(a)(2)(i)(B)	1910.305(a)(2)(i)(C)	**Adds emergencies to the list of activities for which temporary wiring is permitted.
1910.305(a)(2)(i)(C)	1910.305(a)(2)(i)(B)	No substantive change.
	1910.305(a)(2)(ii)	**Clarifies that temporary wiring must be removed when the project or purpose for which it was used has been completed.

Old Section	New Section	Changes and Rationale
1910.305(a)(2)(ii)	1910.305(a)(2)(iii)	**Adds "construction-like activities" to the list of permitted uses for temporary electrical installations over 600 volts.
1910.305(a)(2)(iii)(A)	1910.305(a)(2)(iv)	**Feeders may now only be run as single insulated conductors when accessible to qualified employees only and used for experiments, development work, or emergencies. (Individual requirements are placed in separate paragraphs).
1910.305(a)(2)(iii)(B)	1910.305(a)(2)(v)	No substantive change. (Individual requirements are placed in separate paragraphs).
1910.305(a)(2)(iii)(C)	1910.305(a)(2)(vi)	No substantive change.
1910.305(a)(2)(iii)(D)	1910.305(a)(2)(vii)	No substantive change.
1910.305(a)(2)(iii)(E)	1910.305(a)(2)(viii)	**Adds a requirement that disconnecting means for a multiwire circuit simultaneously disconnect all ungrounded conductors of the circuit.
1910.305(a)(2)(iii)(F)	1910.305(a)(2)(ix)	**This provision no longer allows installing fixtures or lampholders more than 2.1 meters above the working surface as a means of guarding. Also, the final rule adds a requirement for grounding metal-case sockets.
1910.305(a)(2)(iii)(G)	1910.305(a)(2)(x)	No substantive change.
	1910.305(a)(2)(xi)	**Adds requirements for cable assemblies and flexible cords and cables to be adequately supported.
1910.305(a)(3)(i)(a)	1910.305(a)(3)(i)	No substantive change. (Some raceway and cable types that were included in generic terms have been explicitly added to the list of wiring methods acceptable in cable trays).
1910.305(a)(3)(i)(b)	1910.305(a)(3)(ii)	**Adds several types of cables and single insulated conductors to the list of types permitted in industrial establishments.
	1910.305(a)(3)(iii)	**Adds a requirement limiting the use of metallic cable trays as an equipment grounding conductor.
1910.305(a)(3)(i)(c)	1910.305(a)(3)(iv)	No substantive change.
1910.305(a)(3)(ii)	1910.305(a)(3)(v)	No substantive change.
1910.305(a)(4)(i)	1910.305(a)(4)(i)	No substantive change.
1910.305(a)(4)(ii)	1910.305(a)(4)(ii)	**Adds specific support requirements and limits the application of these requirements to conductors smaller than No. 8.
1910.305(a)(4)(iii)	1910.305(a)(4)(iii)	No substantive change.
1910.305(a)(4)(iv)	1910.305(a)(4)(iv)	No substantive change.
1910.305(a)(4)(v)	1910.305(a)(4)(v)	No substantive change.
1910.305(b)(1)	1910.305(b)(1)(i) and (b)(1)(ii)	No substantive change. (Individual requirements are placed in separate paragraphs).
	1910.305(b)(1)(iii)	**Adds requirements for supporting cables entering cabinets, cutout boxes, and meter sockets.
1910.305(b)(2)	1910.305(b)(2)(i)	No substantive change.
	1910.305(b)(2)(ii)	**Adds a requirement for any exposed edge of a combustible ceiling finish at a fixture canopy or

Old Section	New Section	Changes and Rationale
		pan to be covered with noncombustible material.
1910.305(b)(3)	1910.305(b)(3)	No substantive change. (Individual requirements are placed in separate paragraphs).
1910.305(c)(1)	1910.305(c)(1), (c)(2), and (c)(3)(i)	No substantive change. (Individual requirements are placed in separate paragraphs).
	1910.305(c)(3)(ii)	**Adds a requirement for load terminals on switches to be deenergized when the switches are open except under limited circumstances.
	1910.305(c)(4)	**Adds a specific requirement for flush-mounted switches to have faceplates that completely cover the opening and that seat against the finished surface.
1910.305(c)(2)	1910.305(c)(5)	**Adds a requirement to ground faceplates for snap switches.
1910.305(d)	1910.305(d)	No substantive change. (Individual requirements are placed in separate paragraphs).
1910.305(e)(1)	1910.305(e)(1)	**Adds a requirement for metallic cabinets, cutout boxes, fittings, boxes, and panelboard enclosures installed in damp or wet location to have an air space between the enclosure and the mounting surface.
1910.305(e)(2)	1910.305)(e)(2)	No substantive change.
1910.305(f)	1910.305(f)	No substantive change. (Individual requirements are placed in separate paragraphs).
1910.305(g)(1)(i)	1910.305(g)(1)(i) and (g)(1)(ii)	**Adds the following to the types of connections permitted for flexible cords and cables: Portable and mobile signs and connection of moving parts. The final rule also clarifies that flexible cords and cables may be used for temporary wiring as permitted in final Sec. 1910.305(a)(2).
1910.305(g)(1)(ii)	1910.305(g)(1)(iii)	No substantive change.
1910.305(g)(1)(iii)	1910.305(g)(1)(iv)	No substantive change. (Clarifies that flexible cords and cables may not be installed inside raceways).
1910.305(g)(1)(iv)	1910.305(g)(1)(v)	**Permits additional cords to be used in show windows and show cases.
1910.305(g)(2)(i)	1910.305(g)(2)(i)	**Adds new types of cords to the list of those that must be marked with their type designation.
1910.305(g)(2)(ii)	1910.305(g)(2)(ii)	**Changes the minimum size of hard service and junior hard service cords that may be spliced from No. 12 to 14.
1910.305(g)(2)(iii)	1910.305(g)(2)(iii)	No substantive change.
1910.305(h)	1910.305(h), introductory text, (h)(1), (h)(2), (h)(3), (h)(6), (h)(7), and (h)(8)	**Permits the minimum size of the insulated ground-check conductor G-GC cables to be No. 10 rather than No. 8. (Individual requirements are placed in separate paragraphs).
	1910.305(h)(4)	**Adds a requirement for shields to be grounded.
	1910.305(h)(5)	**Adds minimum bending radii requirements for portable cables.
1910.305(i)(1)	1910.305(i)(1)	No substantive change.
1910.305(i)(2)	1910.305(i)(2)	No substantive change.

Old Section	New Section	Changes and Rationale
1910.305(i)(3)	1910.305(i)(3)	**Also permits fixture wire to be used in fire alarm circuits.
1910.305(j)(1)(i)	1910.305(j)(1)(i)	No substantive change.
1910.305(j)(1)(ii)	1910.305(j)(1)(ii)	No substantive change. (Clarifies that metal-shell paper-lined lampholders may not be used for handlamps.
1910.305(j)(1)(iii)	1910.305(j)(1)(iii)	**Adds a requirement that the grounded circuit conductor where present, be connected to the screw shell.
1910.305(j)(1)(iv)	1910.305(j)(1)(iv)	No substantive change.
	1910.305(j)(2)(i)	**Adds requirements to ensure that attachment plugs and connectors have no exposed live parts.
1910.305(j)(2)(i)	1910.305(j)(2)(ii)	No substantive change.
	1910.305(j)(2)(iii)	**Clarifies that nongrounding-type receptacles may not be used with grounding-type attachment plugs.
1910.305(j)(2)(ii)	1910.305(j)(2)(iv)	No substantive change.
	1910.305(j)(2)(v), (j)(2)(vi), and (j)(2)(vii)	**Adds requirements for receptacles outdoors to be installed in weatherproof enclosures appropriate for the use of the receptacle and for the location.
1910.305(j)(3)(i)	1910.305(j)(3)(i)	No substantive change.
1910.305(j)(3)(ii)	1910.305(j)(3)(ii)	**Adds a requirement to group and identify disconnecting means for appliances supplied by more than one source.
1910.305(j)(3)(iii)	1910.305(j)(3)(iii)	**Adds requirements for marking frequency and required external overload protection for appliances.
	1910.305(j)(3)(iv)	**Clarifies that markings must be visible or easily accessible after installation.
1910.305(j)(4), introductory text	1910.305(j)(4), introductory text	No substantive change.
1910.305(j)(4)(i)	1910.305(j)(4)(i)	No substantive change.
1910.305(j)(4)(ii)(A)	1910.305(j)(4)(ii)	No substantive change.
1910.305(j)(4)(ii)(B)	1910.305(j)(4)(iii)	No substantive change.
1910.305(j)(4)(ii)(C)		Removed. All disconnecting means must be capable of being locked in the open position by Sec. 1910.302(c), and 1910.303(f)(4).
1910.305(j)(4)(ii)(D)	1910.305(j)(4)(iv)	No substantive change.
1910.305(j)(4)(ii)(E)	1910.305(j)(4)(v)	No substantive change.
1910.305(j)(4)(ii)(F)	1910.305(j)(4)(vi)	No substantive change.
1910.305(j)(4)(iii)	1910.305(j)(4)(vii)	No substantive change.
1910.305(j)(4)(iv)(A)		Removed. Covered by Sec. 1910.303(g)(2), (h)(2), and(h)(4)(iii).
1910.305(j)(4)(iv)(B)	1910.305(j)(4)(viii)	No substantive change.
1910.305(j)(5)(i)	1910.305(j)(5)(i)	No substantive change.
1910.305(j)(5)(ii)	1910.305(j)(5)(ii)	No substantive change.
1910.305(j)(5)(iii)	1910.305(j)(5)(iii)	No substantive change.
1910.305(j)(5)(iv)	1910.305(j)(5)(iv)	No substantive change. (Oil-insulated transformers installed indoors are presumed to present a hazard to employees since a transformer

Old Section	New Section	Changes and Rationale
		failure will lead to a fire within the building unless the transformer is installed in a vault).
1910.305(j)(5)(v)	1910.305(j)(5)(v)	No substantive change.
1910.305(j)(5)(vi)	1910.305(j)(5)(vi)	No substantive change.
1910.305(j)(5)(vii)	1910.305(j)(5)(vii)	No substantive change.
1910.305(j)(5)(viii)	1910.305(j)(5)(viii)	No substantive change.
1910.305(j)(6)(i)	1910.305(j)(6)(i)	No substantive change.
1910.305(j)(6)(ii), introductory text	1910.305(j)(6)(ii), introductory text	No substantive change.
	1910.305(j)(6)(ii)(A) and (j)(6)(ii)(B)	**Adds requirements to provide disconnecting means of adequate capacity for capacitors operating at more than 600 V.
1910.305(j)(6)(ii)(A)	1910.305(j)(6)(ii)(C)	No substantive change.
1910.305(j)(6)(ii)(B)	1910.305(j)(6)(ii)(D)	No substantive change.
1910.305(j)(7)	1910.305(j)(7)	No substantive change.
Sec. 1910.306 Specific purpose equipment and installations.	**Sec. 1910.306 Specific purpose equipment and installations.**	
1910.306(a)(1)	1910.306(a)(1)(i), (a)(2)(i), and (a)(2)(ii)	**Reorganized and clarified the requirements for disconnecting means for signs. The final rule does not apply these requirements to exit signs.
	1910.306(a)(1)(ii)	**Adds a requirement for the disconnects for signs located within fountains to be at least 1.52m (5.0 ft) from the fountain wall.
1910.306(a)(2)	1910.306(a)(2)(iii)	No substantive change.
1910.306(b), introductory text	1910.306(b), introductory text	No substantive change.
1910.306(b)(1)(i)	1910.306(b)(1)	**Adds specific requirements for the type and location of disconnecting means for runway conductors.
1910.306(b)(1)(ii)	1910.306(b)(2)	No substantive change. (The final rule reorganizes these requirements).
1910.306(b)(2)	1910.306(b)(3)	No substantive change.
1910.306(b)(3)	1910.306(b)(4)	No substantive change.
1910.306(c)	1910.306(c), introductory text	**This paragraph now covers wheelchair lifts, and stairway chair lifts.
1910.306(c)(1)	1910.306(c)(1)	No substantive change.
1910.306(c)(2)	1910.306(c)(8)	No substantive change.
1910.306(c)(3)	1910.306(c)(2)	No substantive change.
	1910.306(c)(3)	**Adds requirements for the type of disconnecting means.
	1910.306(c)(4)	**Adds requirements for the operation of disconnecting means.
	1910.306(c)(5)	**Adds requirements for the location of disconnecting means.
	1910.306(c)(6)	**Adds requirements for the identification of disconnecting means.
	1910.306(c)(7)	**Adds requirements for disconnecting means for single car and multicar installations supplied by more than one source.
	1910.306(c)(9)	**Adds requirements for warning signs for interconnected multicar controllers.
	1910.306(c)(10)	**Adds exceptions related to the location of mo-

Old Section	New Section	Changes and Rationale
		tor controllers.
1910.306(d)(1)	1910.306(d)(1)	**Adds requirements for the type and rating of the disconnecting means.
1910.306(d)(2)	1910.306(d)(2)	Clarifies that a supply circuit switch may be used as a disconnecting means if the circuit supplies only one welder.
1910.306(e)	1910.306(e)	**Adds a requirement to group the disconnecting means for the HVAC systems serving information technology rooms with the disconnecting means for the information technology equipment. The final rule exempts integrated electrical systems covered by Sec.1910.308(g). (The existing standard refers to this equipment as data processing equipment).
1910.306(f), introductory text	1910.306(f), introductory text	**Adds coverage of X-rays for dental or medical use.
1910.306(f)(1)(i)	1910.306(f)(1)(i)	No substantive change.
1910.306(f)(1)(ii)	1910.306(f)(1)(ii)	No substantive change.
1910.306(f)(2)(i)	1910.306(f)(2)(i)	No substantive change.
1910.306(f)(2)(ii)	1910.306(f)(2)(ii)	No substantive change.
1910.306(g)(1)	1910.306(g), introductory text	No substantive change.
1910.306(g)(2)(i)	1910.306(g)(1)(i)	No substantive change.
1910.306(g)(2)(ii)	1910.306(g)(1)(ii)	No substantive change.
1910.306(g)(2)(iii)	1910.306(g)(1)(iii)	**Adds a requirement for the installation of doors or detachable panels to provide access to internal parts. Adds a requirement that detachable panels not be readily removable.
1910.306(g)(2)(iv)	1910.306(g)(1)(iv)	No substantive change.
1910.306(g)(2)(v)	1910.306(g)(1)(v)	No substantive change. (Individual requirements are place in separate paragraphs).
1910.306(g)(2)(vi)	1910.306(g)(1)(vi)	**Adds a requirement to ensure adequate rating of disconnecting means. The final rule also clarifies when the supply circuit disconnecting means may be used as the disconnecting means for induction and dielectric heating equipment.
1910.306(g)(3)	1910.306(g)(2)	No substantive change. (Individual requirements are places in separate paragraphs).
1910.306(h)(1)	1910.306(h), introductory text	No substantive change.
1910.306(h)(2)	1910.399	No substantive change.
1910.306(h)(3)	1910.306(h)(1)	No substantive change.
1910.306(h)(4)(i) and (h)(4)(ii)	1910.306(h)(2)	No substantive change. (The two provisions are combined into one paragraph.
1910.306(h)(5)(i)	1910.306(h)(3)(i)	No substantive change.
1910.306(h)(5)(ii)	1910.306(h)(3)(ii)	No substantive change.
1910.306(h)(6)(i)	1910.306(h)(4)(i)	**Adds requirements limiting primary and secondary voltage on isolating transformers supplying receptacles for ungrounded cord – and plug – connected equipment. Also, adds requirement for overcurrent protection for circuits supplied by these transformers.
1910.306(h)(6)(ii)	1910.306(h)(4)(ii)	No substantive change.

Old Section	New Section	Changes and Rationale
1910.306(h)(6)(iii)	1910.306(h)(4)(iii)	No substantive change. (Individual requirements are placed in separate paragraphs).
1910.306(h)(7)(i) and (h)(7)(ii)	1910.306(h)(5)(i)	No substantive change.
1910.306(h)(7)(iii)	1910.306(h)(5)(ii)	No substantive change.
1910.306(h)(7)(iv)	1910.306(h)(5)(iii)	No substantive change.
1910.306(h)(8)	1910.306(h)(6)	No substantive change.
1910.306(h)(9)	1910.306(h)(7)	No substantive change.
1910.306(i)(1)	1910.306(i)(1)	No substantive change.
1910.306(i)(2)	1910.306(i)(2)	**Allows the disconnecting means for a center pivot irrigation machine to be located not more than 15.m (50 ft) from the machine if the disconnecting means is visible from the machine. (Individual requirements are placed in separate paragraphs).
1910.306(j)(1)	1910.306(j), introductory text	**Clarifies that hydro-massage bathtubs are covered by this paragraph.
1910.306(j)(2)(i)	1910.306(j)(1)(i)	No substantive change.
	1910.306(j)(1)(ii)	**Extends the boundary within which receptacles require ground-fault circuit interrupter protection from 4.57 m (15 ft) to 6.08, (20 ft) for new installations.
	1910.306(j)(1)(iii)	**Adds requirements for the installation of at least one receptacle near permanently installed pools at dwelling units.
1910.306(j)(2)(ii)(A)	1910.306(j)(2)(i)	**Clarifies that ceiling suspended (paddle) fans are covered by this requirement.
1910.306(j)(2)(ii)(B)	1910.306(j)(2)(ii)	No substantive change.
1910.306(j)(3)	1910.306(j)(3)	No substantive change.
1910.306(j)(4)(i)	1910.306(j)(4)(i)	No substantive change.
1910.306(j)(4)(ii)	1910.306(j)(4)(ii)	No substantive change.
	1910.306(j)(4)(iii)	**Adds a requirement to guard lighting fixtures facing upward.
1910.306(j)(5)	1910.306(j)(5)	No substantive change.
	1910.306(k)	**Adds requirements for carnivals, circuses, fairs, and similar events.
Sec. 1910.307 Hazardous (classified locations)	**Sec. 1910.307 Hazardous (classified locations)**	
1910.307(a)	1910.307(a)	**Adds the Zone classification system for Class I locations. (See detailed discussion later in this section of the preamble).
	1910.307(b)	**Adds documentation requirements for hazardous locations classified using either the division or zone classification system. (See detailed discussion later in this section of the preamble).
1910.307(b), introductory text	1910.307(c), introductory text	No substantive change.
1910.307(b)(1)	1910.307(c)(1)	No substantive change.
1910.307(b)(2)(i)	1910.307(c)(2)(i)	No substantive change.
1910.307(b)(2)(ii), introductory text	1910.307(c)(2)(ii), introductory text	No substantive change.
1910.307(b)(2)(ii)(A)	1910.307(c)(2)(ii)(A)	No substantive change.
1910.307(b)(2)(ii)(B)	1910.307(c)(2)(ii)(B)	**Also permits fixtures approved for Class II, Division 2 locations to omit the group marking.

Old Section	New Section	Changes and Rationale
1910.307(b)(2)(ii)(C)	1910.307(c)(2)(ii)(C)	No substantive change.
1910.307(b)(2)(ii)(D)	1910.307(c)(2)(ii)(D)	No substantive change.
	1910.307(c)(2)(ii)(E)	**Adds a requirement that electric equipment suitable for an ambient temperature exceeding 40 [deg] C (104[deg]F) be marked with the maximum ambient temperature.
1910.307(b)(3)	1910.307(c)(3)	No substantive change.
1910.307(b)(3), Note	1910.307(c)(3), Note	The last sentence of the note is removed to make it clear that the OSHA standard does not incorporate the National Electrical Code by reference. The NEC continues to be a guideline that employers may reference in determining the type and design of equipment and installations that will meet the OSHA standard.
1910.307(c)	1910.307(d)	No substantive change.
1910.307(d)	1910.307(e)	No substantive change.
	1910.307(f)	**The final rule adds a list of specific protective techniques for electrical installations in hazardous locations classified under the division classification system.
	1910.307(g)	**Adds the zone classification system as an alternative method of installing electric equipment in hazardous locations. This paragraph sets the protective techniques and other requirements necessary for safe installation of electric equipment in hazardous locations classified under the zone classification system. (See detailed discussion later in this section of the preamble).
1910.308 Special systems.	**1910.308 Special systems.**	
1910.308(a), introductory text	1910.308(a), introductory text	No substantive change.
1910.308(a)(1)(i)	1910.308(a)(1)(i), and (a)(3)(ii)	**Adds the following wiring methods to those acceptable for installation operating at more than 600 V: Electric metallic tubing, rigid nonmetallic conduit, busways, and cable bus. The proposal also removes the specific requirement to support cables having a bare lead sheath or braided outer covering in a manner to prevent damage to the braid or sheath. This hazard is covered by Sec. 1910.303(b)(1) and (b)(8)(i) and new Sec. 1910.308(a)(4).
1910.308(a)(1)(ii)	1910.308(a)(1)(ii)	No substantive change.
	1910.308(a)(2) and (a)(3)(i)	** Adds requirements to ensure that high-voltage cables can adequately handle the voltage stresses placed upon them and to ensure that any coverings are flame retardant.
	1910.308(a)(4)	**Adds requirements for the protection of high-voltage cables against moisture and physical damage where the cable conductors emerge from a metal sheath.
1910.308(a)(2)(i)	1910.308(a)(5)(i)	No substantive change.
	1910.308(a)(5)(ii)	**Adds requirements for fuses to protect each ungrounded conductor, for adequate ratings of

Old Section	New Section	Changes and Rationale
		fuses installed in parallel, and for the protection of employees from power fuses of the vented type.
1910.308(a)(2)(ii)	1910.308(a)(5)(iii)	**Clarifies that distribution cutouts are not suitable for installation in buildings or transformer vaults.
	1910.308(a)(5(iv)	**Adds requirements for fused cutouts to either be capable of interrupting load current or be supplemented by a means of interrupting load current. In addition, a warning sign would be required for fused cutouts that cannot interrupt load current.
	1910.308(a)(5)(v)	**Adds a requirement for guarding nonshielded cables and energized parts of oil-filled cutouts.
	1910.308(a)(5)(vi)	**Adds requirements to ensure that load interrupting switches will be protected against interrupting fault current and to provide for warning signs for backfed switches.
1910.308(a)(2)(iii)	1910.308(a)(5)(vii)	No substantive change.
1910.308(a)(3)	1910.308(a)(6)	No substantive change.
1910.308(a)(4)(i)	1910.308(a)(7), introductory text.	No substantive change.
1910.308(a)(4)(ii)	1910.308(a)(7)(i), (a)(7)(iii)	No substantive change. (Individual requirements are placed in separate paragraphs).
	1910.308(a)(7)(ii)	**Clarifies that multiconductor portable cable may supply mobile equipment.
1910.308(a)(4)(iii)	1910.308(a)(7)(iv), and (a)(7)(v)	No substantive change. (Individual requirements are placed in separate paragraphs).
	1910.308(a)(7)(vi)	**Limits the conditions under which switch or contactor enclosures may be used as junction boxes or raceways.
1910.308(a)(4)(iv)	1910.308(a)(7)(vii)	No substantive change.
1910.308(a)(4)(v)	1910.308(a)(7)(viii)	No substantive change.
1910.308(b)(1)	1910.308(b), introductory text	No substantive change.
1910.308(b)(2)	1910.308(b)(1)	No substantive change.
1910.308(b)(3)	1910.308(b)(2)	**Clarifies that emergency illumination includes all required means of egress lighting, illuminated exit signs, and all other lights necessary to provide required illumination.
	1910.308(b)(3)	**Adds requirements to provide signs indicating the presence and location of on-site emergency power sources under certain conditions.
1910.308(c)(1), introductory text	1910.308(c)(1), introductory text	No substantive change.
1910.308(c)(1)(i), (c)(1)(ii), and (c)(1)(iii)	1910.308(c)(1)(i), (c)(1)(ii), and(c)(1)(iii)	**Clarifies the power limitations of Class1, 2, and 3 remote control, signaling, and power-limited circuits based on equipment listing.
1910.308(c)(2)	1910.308(c)(2)	No substantive change.
	1910.308(c)(3)	**Adds requirements for the separation of cables and conductors of Class 2 and Class 3 circuits from cables and conductors of other types of

Old Section	New Section	Changes and Rationale
		circuits.
1910.308(d)(1)	1910.308(d)(1)	No substantive change.
1910.308(d)(2), introductory text	1910.308(d)(2), introductory text	No substantive change.
1910.308(d)(2)(i)	1910.308(d)(2)(i)	No substantive change.
1910.308(d)(2)(ii)	1910.308(d)(2)(ii)	**Adds a requirement for power-limited fire alarm circuit power sources to be listed and marked as such.
1910.308(d)(3)	1910.308(d)(3)(i)	No substantive change.
1910.308(d)(4)	1910.308(d)(3)(ii), (d)(3)(iii), and (d)(3)(iv)	**Clarifies the requirements for installing power-limited fire-protective signaling circuits with other types of circuits. (Individual requirements are placed in separate paragraphs).
1910.308(d)(5)	1910.308(d)(4)	No substantive change.
1910.308(e)(1)	1910.308(e), introductory text	No substantive change.
1910.308(e)(2)	1910.308(e)(1)	**Clarifies the requirement for listed primary protectors to make it clear that circuits confined within a block do not need protectors.
1910.308(e)(3)(i)	1910.308(e)(2)(i), (e)(2)(ii)	No substantive change.
1910.308(e)(3)(ii)	1910.308(e)(2)(iii)	No substantive change.
1910.308(e)(3)(iii)	1910.308(e)(2)(iv)	No substantive change.
1910.308(e)(4)	1910.308(e)(3)	No substantive change.
1910.308(e)(5)	1910.308(e)(4)	No substantive change.
	1910.308(f)	**Adds requirements to separate conductors of solar photovoltaic systems from conductors of other systems to provide a disconnecting means for solar photovoltaic systems.
	1910.308(g)	**Adds an exception to the provisions on the location of overcurrent protective devices for integrated electrical systems.

Note to table:
**These new and revised provisions are included in the 2000 and 2004 editions of NFPA 70E standard. The NFPA 70E Committee believes that these provisions, which were taken from the 1999 and 2002 NEC®, respectively, are essential to employee safety. OSHA agrees with the consensus of NFPA's expert opinion that these requirements are reasonably necessary to protect employees and has included them in the final rule. On occasion, OSHA has rewritten the provision to lend greater clarity to its requirements. However, these editorial changes to the language of NFPA 70E do not represent substantive differences. NFPA's handling of these provisions and the rationale underpinning them is a matter of public record for the NEC and NFPA 70E and is part of the record for this rulemaking (Exs. 2-9 through 2-18). OSHA agrees with the rationale in this record as it pertains to the new and revised provisions the Agency is adopting.

Appendix C
NEMA Plug Configurations

ANSI / NEMA
WD 6–1996
Page 141

NEMA CONFIGURATIONS FOR SPECIFIC PURPOSE PLUGS AND RECEPTACLES

	DESCRIPTION	NEMA NUMBER	15 AMPERE		30 AMPERE		50 AMPERE	
			RECEPTACLE	PLUG	RECEPTACLE	PLUG	RECEPTACLE	PLUG
MIDGET LOCKING	125V, 2 POLE, 2 WIRE	ML1	ML1-15R	ML1-15P				
	125V, 2 POLE, 3 WIRE GROUNDING	ML2	ML2-15R	ML2-15P				
	125/250V, 3 POLE, 3 WIRE	ML3	ML3-15R	ML3-15P				
FSL CONFIGURATIONS	28V DC, 2 POLE, 3 WIRE GROUNDING	FSL1			FSL1-30R	FSL1-30P		
	120V, 400HZ, 2 POLE, 3 WIRE GROUNDING	FSL2			FSL2-30R	FSL2-30P		
	120V, 400 HZ, 3-PHASE 3 POLE, 4 WIRE GROUNDING	FSL3			FSL3-30R	FSL3-30P		
	120/208V, 3 Ø Y, 400 HZ 4 POLE, 5 WIRE GROUNDING	FSL4			FSL4-30R	FSL4-30P		
MARINE SHIP-TO-SHORE	125V, 2 POLE, 3 WIRE GROUNDING	SS1					SS1-50R	SS1-50P
	125/250V, 3 POLE, 4 WIRE GROUNDING	SS2					SS2-50R	SS2-50P
TRAVEL TRAILER	120V AC, 2 POLE, 3 WIRE GROUNDING	TT			TT-30R	TT-30P		

NOTE: BLANK SPACES RESERVED FOR FUTURE CONFIGURATIONS

ANSI / NEMA
WD 6-1996
Page 142

NEMA CONFIGURATIONS FOR STRAIGHT BLADE PLUGS AND RECEPTACLES

DESCRIPTION		NEMA NUMBER	15 AMPERE RECEPTACLE	15 AMPERE PLUG	20 AMPERE RECEPTACLE	20 AMPERE PLUG	30 AMPERE RECEPTACLE	30 AMPERE PLUG	50 AMPERE RECEPTACLE	50 AMPERE PLUG	60 AMPERE RECEPTACLE	60 AMPERE PLUG
2-POLE 2-WIRE	125V	1	1-15R	1-15P POLARIZED / NON POLAR		1-20P MATES WITH 5-20R		1-30P MATES WITH 5-30R				
2-POLE 2-WIRE	250V	2		2-15P MATES WITH 6-15R	2-20R	2-20P	2-30R	2-30P				
2-POLE 2-WIRE	277V AC	3										
2-POLE 2-WIRE	600V	4										
2-POLE 3-WIRE GROUNDING	125V	5	5-15R	5-15P	5-20R	5-20P	5-30R	5-30P	5-50R	5-50P		
2-POLE 3-WIRE GROUNDING	125V	5ALT			5ALT-20R							
2-POLE 3-WIRE GROUNDING	250V	6	6-15R	6-15P	6-20R	6-20P	6-30R	6-30P	6-50R	6-50P		
2-POLE 3-WIRE GROUNDING	250V	6ALT			6ALT-20R							
2-POLE 3-WIRE GROUNDING	277V AC	7	7-15R	7-15P	7-20R	7-20P	7-30R	7-30P	7-50R	7-50P		
2-POLE 3-WIRE GROUNDING	347V AC	24	24-15R	24-15P	24-20R	24-20P	24-30R	24-30P	24-50R	24-50P		
2-POLE 3-WIRE GROUNDING	480V AC	8										
2-POLE 3-WIRE GROUNDING	600V AC	9										
3-POLE 3-WIRE	125 / 250V	10			10-20R	10-20P	10-30R	10-30P	10-50R	10-50P		
3-POLE 3-WIRE	3 Ø 250V	11	11-15R	11-15P	11-20R	11-20P	11-30R	11-30P	11-50R	11-50P		
3-POLE 3-WIRE	3 Ø 480V	12										
3-POLE 3-WIRE	3 Ø 600V	13										
3-POLE 4-WIRE GROUNDING	125 / 250V	14	14-15R	14-15P	14-20R	14-20P	14-30R	14-30P	14-50R	14-50P	14-60R	14-60P
3-POLE 4-WIRE GROUNDING	3 Ø 250V	15	15-15R	15-15P	15-20R	15-20P	15-30R	15-30P	15-50R	15-50P	15-60R	15-60P
3-POLE 4-WIRE GROUNDING	3 Ø 480V	16										
3-POLE 4-WIRE GROUNDING	3 Ø 600V	17										
4-POLE 4-WIRE	3 Ø Y 120 / 208V	18	18-15R	18-15P	18-20R	18-20P	18-30R	18-30P	18-50R	18-50P	18-60R	18-60P
4-POLE 4-WIRE	3 Ø Y 277 / 480V	19										
4-POLE 4-WIRE	3 Ø Y 347 / 600V	20										
4-POLE 5-WIRE GROUNDING	3 Ø Y 120 / 208V	21										
4-POLE 5-WIRE GROUNDING	3 Ø Y 277 / 480V	22										
4-POLE 5-WIRE GROUNDING	3 Ø Y 347 / 600V	23										

NOTE: BLANK SPACES RESERVED FOR FUTURE CONFIGURATIONS

ANSI / NEMA
WD 6-1996
Page 143

NEMA CONFIGURATIONS FOR LOCKING PLUGS AND RECEPTACLES

DESCRIPTION		NEMA NUMBER	15 AMPERE		20 AMPERE		30 AMPERE		50 AMPERE		60 AMPERE	
			RECEPTACLE	PLUG	RECEPTACLE	PLUG	RECEPTACLE	PLUG	RECEPTACLE	PLUG	RECEPTACLE	PLUG
2-POLE 2-WIRE	125V	1	L 1-15R	L 1-15P								
	250V	2			L 2-20R	L 2-20P						
	277V AC	3										
	600V	4										
2-POLE 3-WIRE GROUNDING	125V	5	L 5-15R	L 5-15P	L 5-20R	L 5-20P	L 5-30R	L 5-30P	L 5-50R	L 5-50P	L 5-60R	L 5-60P
	250V	6	L 6-15R	L 6-15P	L 6-20R	L 6-20P	L 6-30R	L 6-30P	L 6-50R	L 6-50P	L 6-60R	L 6-60P
	277V AC	7	L 7-15R	L 7-15P	L 7-20R	L 7-20P	L 7-30R	L 7-30P	L 7-50R	L 7-50P	L 7-60R	L 7-60P
	347V AC	24			L24-20R	L24-20P						
	480V AC	8			L 8-20R	L 8-20P	L 8-30R	L 8-30P	L 8-50R	L 8-50P	L 8-60R	L 8-60P
	600V AC	9			L 9-20R	L 9-20P	L 9-30R	L 9-30P	L 9-50R	L 9-50P	L 9-60R	L 9-60P
3-POLE 3-WIRE	125 / 250V	10			L10-20R	L10-20P	L10-30R	L10-30P				
	3 Ø 250V	11	L11-15R	L11-15P	L11-20R	L11-20P	L11-30R	L11-30P				
	3 Ø 480V	12			L12-20R	L12-20P	L12-30R	L12-30P				
	3 Ø 600V	13					L13-30R	L13-30P				
3-POLE 4-WIRE GROUNDING	125 / 250V	14			L14-20R	L14-20P	L14-30R	L14-30P	L14-50R	L14-50P	L14-60R	L14-60P
	3 Ø 250V	15			L15-20R	L15-20P	L15-30R	L15-30P	L15-50R	L15-50P	L15-60R	L15-60P
	3 Ø 480V	16			L16-20R	L16-20P	L16-30R	L16-30P	L16-50R	L16-50P	L16-60R	L16-60P
	3 Ø 600V	17					L17-30R	L17-30P	L17-50R	L17-50P	L17-60R	L17-60P
4-POLE 4-WIRE	3 Ø Y 120 / 208V	18			L18-20R	L18-20P	L18-30R	L18-30P				
	3 Ø Y 277 / 480V	19			L19-20R	L19-20P	L19-30R	L19-30P				
	3 Ø Y 347 / 600V	20			L20-20R	L20-20P	L20-30R	L20-30P				
4-POLE 5-WIRE GROUNDING	3 Ø Y 120 / 208V	21			L21-20R	L21-20P	L21-30R	L21-30P	L21-50R	L21-50P	L21-60R	L21-60P
	3 Ø Y 277 / 480V	22			L22-20R	L22-20P	L22-30R	L22-30P	L22-50R	L22-50P	L22-60R	L22-60P
	3 Ø Y 347 / 600V	23			L23-20R	L23-20P	L23-20R	L23-20P	L23-50R	L23-50P	L23-60R	L23-60P

NOTE: BLANK SPACES RESERVED FOR FUTURE CONFIGURATIONS

Appendix D
Commentary Submissions on Grounding

Below are the factual arguments for the proposed removal of this reference that the editors made to OSHA during the comment period of the standard rulemaking process, taken directly from the Subpart S preamble; You, the reader, can judge the preponderance of the evidentiary arguments. These comments were submitted to OSHA by the Editors when the proposed electrical rule was published in the Federal Register. Unfortunately in the Editors' view, OSHA decided to retain the antiquated grounding references in it's final rule. It is recommended that the reader revisit the detailed explanation on grounding previously described in this book for the rationale of why water pipes provide inadequate equipment grounds.

Objection # 1:
Strongly recommend that this permission be entirely deleted.

Page 17820 – 1910.304(g)(3) "Grounding connections.

(iii) On extensions of existing branch circuits that do not have an equipment grounding conductor, grounding type receptacles may be grounded to a grounded cold water pipe near the equipment."

The comments below relate specifically to equipment grounding. The lack of effective equipment grounding leads to electrocutions.

According to the background on page 17776 of the proposed rule, OSHA states "This rulemaking would update those requirements to make them consistent with the latest edition of NFPA 70E."

The rule proposed in 1910.304(g)(3)(iii) Grounding path, suggests that a cold water pipe may be used as the equipment grounding means on branch circuits where an equipment grounding conductor is not present. Water piping used as an equipment grounding conductor is not only recognized as an improper electrical engineering practice, but is a violation the current NFPA 70 (National Electrical Code® (2002)) and NFPA 70E (2004).

The permission to use a cold water pipe near the equipment as proposed by OSHA was removed from the National Electrical Code® nearly a quarter of a century ago, and last appeared as proposed in the 1981 edition. The wording surrounding permission to use a water pipe for equipment grounding remained somewhat convoluted for the 1984, 1987 and 1990 edition of the Code, but was completely removed from the NEC® for the 1993 Edition.

Further proof that a water pipe cannot serve as a proper equipment grounding conductor can be found in the current NEC® handbook, 2002 edition, published by NFPA, which explains in an editorial note under Article 250.130 that "because of the requirements of 250.52(A)(1) an interior metal water pipe more than 5 ft from the point of entrance of the water pipe into the building is no longer allowed to serve as a connection to the grounding electrode conductor."

The 1990 NEC® Handbook published by NFPA provided additional clarification on the issue of permitting "the grounding conductor of a grounding type receptacle outlet to be grounded to a water pipe which is bonded in accordance with Section 250-80(a)." The editorial notes under 250-80(a) as written by NFPA engineers states "The intent of the exception is to recognize the fact that increased use of nonmetallic water piping systems underground makes it virtually impossible to comply with the main rule of section 250-80 in the circumstances described."

Although the use of a cold water pipe was permitted under NFPA 70E in the 2000 edition it has been removed from the 2004 Edition. The removal was based upon a comment which was published in the Report on Proposals (ROP) 1/31/02, Log #24 –

RECOMMENDATION:

Delete Part II, 2-6.2.3.

SUBSTANTIATION:

Using an isolated equipment grounding conductor such as the cold water pipe risks increased "reluctance" (substitute reactance) along the equipment grounding conductor where AC fault is involved. This violates the NEC® 300.3(B) and also NFPA 70E Part II, 2-6.5 that requires all circuit conductors to be grouped together so magnetic fields are offset and reluctance (substitute reactance) is minimized. We use the same rationale in Part II, 2-6.5(a) by prohibiting the metal frame of a building to serve as an equipment grounding conductor because of reluctance (reactance). IAEI studies have shown isolating grounding paths have increased resistance and overcurrent protective devices will not clear the ground faults.

It is unclear as to why the NFPA Technical Correlating Committee did not refer this deletion to the NFPA 70E committee for deletion when it was completely removed from the NEC® in 1993. However, as stated above, it has been removed from the 2004 Edition and is no longer part of any NFPA electrical code. OSHA would actually be stepping backwards to reinstate a standard requirement that is contrary to NFPA codes as well as a recognized hazard.

The equipment grounding conductor has been a requirement in the NEC® for over half of a century. Branch circuits in this day and age lacking the safety control of equipment grounding conductors should be updated electrically and should not substitute other ineffective systems such as water piping.

As mentioned in the proposal referenced above, NEC® 300.3(B) and also NFPA 70E Part II, 2-6.5 require all circuit conductors "and the equipment grounding conductors and bonding conductors shall be contained within the same raceway, auxiliary gutter, cable tray, cable bus, assembly, trench, cable or cord." This requirement is to keep conductors grouped close together so magnetic fields generated by the flow of ac electricity, which reacts with the circuit conductors (called reactance), may cancel each other out, thereby minimizing the total circuit impedance for safety reasons. This is most specifically related to equipment grounding, which is to prevent electrocution. When we defeat the equipment grounding and have an equipment failure electrically, portions of the equipment, which are not supposed to be energized, can become energized. Typically this would be electrical equipment with metallic parts such as electrical tool enclosures and metal pipes electrically connected to these enclosures. Energized metallic surfaces obviously expose persons to serious electrocution hazards.

Furthermore, the use of water pipes to serve as an equipment grounding conductor is a violation of OSHA's own regulation found at 1910.303(a) Approval. Water pipes are never permitted to serve as conduits or raceways for electrical conductors, as they are not listed by any OSHA Nationally Recognized Testing Laboratories (NRTL's) for such use. None of the associated fittings and/or couplings for water pipes are listed for electrical use or as electrical conduit. Water pipes are not tested to any NRTL (i.e. Underwriters Laboratories (UL)) electrical standards and are not listed to safely carry electrical fault currents likely to be imposed upon them in the event of an electrical fault or short circuit in the equipment utilizing the branch circuit. Therefore, using water pipes as part of an electrical system would seem contrary to the OSHA's own rule.

Additionally, spaces defined in NEC® Article 110, which extend from the floor of electrical equipment to a height of 6' above the equipment, are considered dedicated electrical spaces and any foreign piping systems such as plumbing, heating, ventilation, and air-conditioning ducts are prohibited from even encroaching the dedicated electrical space surrounding the electrical equipment. In other words, if the National Electrical Code® does not even want such systems in the proximity of electrical equipment, why would a direct electrical connection be permitted?

Water pipes cannot be counted upon to serve the same function as an equipment grounding conductor, which is to prevent electrocution due to malfunctioning equipment on the branch circuit by allowing large amounts of current to flow and trip the overcurrent device. The use of water pipes as equipment grounding conductors is actually ***more*** *likely to cause an electrocution in the event that a plumber, pipe-fitter or similar professional working on the water piping system would break a pipe connection involved in a fault, thereby exposing themselves to the full lethal circuit voltage and providing a path for current to flow. Unlike electrical workers working on branch circuits, there are no specific requirements for plumbers, pipe fitters or similar professionals to deenergize and lock out electrical circuits in order to work on plumbing systems, nor should there be one.*

The advent of current technology and practice of using nonmetallic pipe in all or part of a plumbing system would cause metallic parts of equipment or sections of the water piping to become energized if a tool or equipment were to malfunction and expose anyone (plumber, pipe-fitter, general plant employee) to an electrocution hazard from simple contact with the piping system.

Besides fault currents introduced by faults and short circuits, these same branch circuits can, and often do, have leakage currents on them due to the age and condition of equipment plugged in, which can provide similar electrocution and/or shock hazards. Old refrigerators provide one very common example.

The solution to preventing new and additional hazards in the workplace (thus insuring safety), providing proper electrical system installations, and remaining up-to-date with the long-established NFPA requirements is to remove this section.